T0331012

# Relativistic Quantum Mechanics

Written by two of the most prominent leaders in particle physics, Relativistic Quantum Mechanics: An Introduction to Relativistic Quantum Fields provides a classroom-tested introduction to the formal and conceptual foundations of quantum field theory. Designed for advanced undergraduate- and graduate-level physics students, the text only requires previous courses in classical mechanics, relativity and quantum mechanics.

The introductory chapters of the book summarise the theory of special relativity and its application to the classical description of the motion of a free particle and a field. The authors then explain the quantum formulation of field theory through the simple example of a scalar field described by the Klein–Gordon equation as well as its extension to the case of spin ½ particles described by the Dirac equation. They also present the elements necessary for constructing the foundational theories of the standard model of electroweak interactions, namely quantum electrodynamics and the Fermi theory of neutron beta decay. Many applications to quantum electrodynamics and weak interaction processes are thoroughly analysed. The book also explores the timely topic of neutrino oscillations.

Logically progressing from the fundamentals to recent discoveries, this textbook provides students with the essential foundation to study more advanced theoretical physics and elementary particle physics. It will help them understand the theory of electroweak interactions and gauge theories.

View the second and third books in this collection: Electroweak Interactions and An Introduction to Gauge Theories.

**Key Features of the New Edition:**

Besides a general revision of text and formulae, three new chapters have been added.

- Chapter 17 introduces and discusses double beta decay processes with and without neutrino emission, the latter being the only process able to determine the Dirac or Majorana nature of the neutrino (discussed in Chapter 13). A discussion of the limits to the Majorana neutrino mass obtained recently in several underground laboratories is included.
- Chapter 18 illustrates the calculation of the mass spectrum of "quarkonia" (mesons composed by a pair of heavy, charm or beauty quarks), in analogy with the positronium spectrum discussed in Chapter 12. This calculation has put into evidence the existence of "unexpected" states and has led to the new field of "exotic hadrons", presently under active theoretical and experimental scrutiny.
- Chapter 19 illustrates the Born-Oppenheimer approximation, extensively used in the computation of simple molecules, and its application to the physics of exotic hadrons containing a pair of heavy quarks, with application to the recently observed doubly charmed baryons.

# Relativistic Quantum Mechanics

## An Introduction to Relativistic Quantum Fields

Second Edition

Luciano Maiani and Omar Benhar

Second edition published 2024
by CRC Press
2385 NW Executive Center Drive, Suite 320, Boca Raton FL 33431

and by CRC Press
4 Park Square, Milton Park, Abingdon, Oxon, OX14 4RN

*CRC Press is an imprint of Taylor & Francis Group, LLC*



First edition published by CRC Press 2015



ISBN: 978-1-032-56594-1 (hbk)
ISBN: 978-1-032-55940-7 (pbk)
ISBN: 978-1-003-43626-3 (ebk)

DOI: 10.1201/9781003436263

Typeset in CMR10
by KnowledgeWorks Global Ltd.

*Publisher's note:* This book has been prepared from camera-ready copy provided by the authors.

# Contents

Preface xi

The authors xiii

CHAPTER 1 ▪ The Symmetries of Space-Time 1

1.1 THE PRINCIPLE OF RELATIVITY 1
1.2 PHYSICAL LORENTZ TRANSFORMATIONS 5
1.3 CAUSAL STRUCTURE OF SPACE-TIME 6
1.4 CONTRAVARIANT AND COVARIANT VECTORS 7
1.5 PROBLEMS FOR CHAPTER 1 8

CHAPTER 2 ▪ The Classical Free Particle 10

2.1 SPACE–TIME MOTION 10
2.2 PARTICLE OF ZERO MASS 12
2.3 ACTION PRINCIPLE FOR THE FREE PARTICLE 13
2.4 THE MASS–ENERGY RELATION 15
2.5 PROBLEMS FOR CHAPTER 2 17

CHAPTER 3 ▪ The Lagrangian Theory of Fields 18

3.1 THE ACTION PRINCIPLE 18
3.2 HAMILTONIAN AND CANONICAL FORMALISM 20
3.3 TRANSFORMATION OF FIELDS 24
3.4 CONTINUOUS SYMMETRIES 28
3.5 NOETHER'S THEOREM 29
3.6 ENERGY–MOMENTUM TENSOR 31
3.7 PROBLEMS FOR CHAPTER 3 36

CHAPTER 4 ■ Klein–Gordon Field Quantisation 39

4.1 THE REAL SCALAR FIELD 39
4.2 GREEN'S FUNCTIONS OF THE SCALAR FIELD 42
4.3 QUANTISATION OF THE SCALAR FIELD 46
4.4 PROBLEMS FOR CHAPTER 4 50

CHAPTER 5 ■ Electromagnetic-Field Quantisation 52

5.1 MAXWELL'S EQUATIONS IN COVARIANT FORM 52
5.2 GREEN'S FUNCTIONS OF THE ELECTROMAGNETIC FIELD 54
5.3 THE MAXWELL–LORENTZ EQUATIONS 57
5.4 HAMILTON FORMALISM AND MINIMAL SUBSTITUTION 63
5.5 QUANTISATION OF THE ELECTROMAGNETIC FIELD IN VACUUM 67
5.6 THE SPIN OF THE PHOTON 73
5.7 PROBLEMS FOR CHAPTER 5 75

CHAPTER 6 ■ The Dirac Equation 78

6.1 FORM AND PROPERTIES OF THE DIRAC EQUATION 79
6.1.1 Spin 82
6.1.2 Relativistic Invariance 83
6.1.3 Boost 89
6.1.4 Solutions of the Dirac Equation for a Free Particle 91
6.1.5 The Magnetic Moment of the Electron 95
6.2 THE RELATIVISTIC HYDROGEN ATOM 98
6.2.1 Factorisation of the Dirac Equation in Polar Coordinates 98
6.2.2 Separation of Variables 99
6.2.3 Eigenvalues of the Hamiltonian 103
6.3 TRACES OF THE $\gamma$ MATRICES 106
6.4 PROBLEMS FOR CHAPTER 6 107

CHAPTER 7 ▪ Quantisation of the Dirac Field 110

7.1 PARTICLES AND ANTIPARTICLES 110
7.2 SECOND QUANTISATION: HOW IT WORKS 113
7.3 CANONICAL QUANTISATION OF THE DIRAC FIELD 115
7.4 THE REPRESENTATION OF THE LORENTZ GROUP 120
7.5 MICROCAUSALITY 122
7.6 THE RELATION BETWEEN SPIN AND STATISTICS 124
7.7 PROBLEMS FOR CHAPTER 7 126

CHAPTER 8 ▪ Free Field Propagators 128

8.1 THE TIME-ORDERED PRODUCT 128
8.2 PROPAGATORS OF THE SCALAR FIELD 129
8.3 PROPAGATORS OF THE DIRAC FIELD 131
8.4 THE PHOTON PROPAGATOR 133
8.5 PROBLEMS FOR CHAPTER 8 135

CHAPTER 9 ▪ Interactions 136

9.1 QUANTUM ELECTRODYNAMICS 137
9.2 THE FERMI INTERACTION FOR $\beta$ DECAY 142
9.3 STRONG INTERACTIONS 143
9.4 HADRONS, LEPTONS AND FIELDS OF FORCE 144
9.5 PROBLEM FOR CHAPTER 9 146

CHAPTER 10 ▪ Time Evolution of Quantum Systems 147

10.1 THE SCHRÖDINGER REPRESENTATION 147
10.2 THE HEISENBERG REPRESENTATION 149
10.3 THE INTERACTION REPRESENTATION 150
10.3.1 Theory of Time-dependent Perturbations 152
10.3.2 Time-ordered Products 152
10.4 SYMMETRIES AND CONSTANTS OF THE MOTION 153

CHAPTER 11 ▪ Relativistic Perturbation Theory 157

11.1 THE DYSON FORMULA 159
11.2 CONSERVATION LAWS 160
11.3 COLLISION CROSS SECTION AND LIFETIME 161
11.4 PROBLEMS FOR CHAPTER 11 166

CHAPTER 12 ▪ The Discrete Symmetries: P, C, T 168

12.1 PARITY 168
12.2 CHARGE CONJUGATION 170
12.3 TIME REVERSAL 172
12.4 TRANSFORMATION OF THE STATES 176
12.5 SOME APPLICATIONS 180
12.5.1 Furry's Theorem 180
12.5.2 Symmetries of Positronium 180
12.6 THE CPT THEOREM 183
12.6.1 Equality of Particle and Antiparticle Masses 187
12.7 PROBLEMS FOR CHAPTER 12 189

CHAPTER 13 ▪ Weyl and Majorana Neutrinos 191

13.1 THE WEYL NEUTRINO 191
13.2 THE MAJORANA NEUTRINO 195
13.3 RELATIONSHIPS AMONG WEYL, MAJORANA AND DIRAC NEUTRINOS 197
13.4 PROBLEM FOR CHAPTER 13 200

CHAPTER 14 ▪ Applications: QED 201

14.1 SCATTERING IN A CLASSICAL COULOMB FIELD 201
14.2 ELECTROMAGNETIC FORM FACTORS 206
14.3 THE ROSENBLUTH FORMULA 208
14.4 COMPTON SCATTERING 214
14.5 INVERSE COMPTON SCATTERING 221
14.6 THE PROCESSES $\gamma\gamma \to e^+e^-$ AND $e^+e^- \to \gamma\gamma$ 224
14.7 $e^+e^- \to \mu^+\mu^-$ ANNIHILATION 227

14.8 PROBLEMS FOR CHAPTER 14 232

CHAPTER 15 ■ Applications: Weak Interactions 233

15.1 NEUTRON DECAY 233
15.2 MUON DECAY 240
15.3 UNIVERSALITY, CURRENT × CURRENT THEORY 244
15.4 TOWARDS A FUNDAMENTAL THEORY 247
15.5 PROBLEMS FOR CHAPTER 15 249

CHAPTER 16 ■ Neutrino Oscillations 250

16.1 OSCILLATIONS IN VACUUM 252
16.2 NATURAL AND ARTIFICIAL NEUTRINOS 255
16.3 INTERACTION WITH MATTER: THE MSW EFFECT 259
16.4 ANALYSIS OF THE EXPERIMENTS 263
16.5 OPEN PROBLEMS 270
16.6 PROBLEM FOR CHAPTER 16 271

CHAPTER 17 ■ Neutrinoless Double-Beta Decay 272

17.1 DOUBLE BETA DECAY 272
17.1.1 Two-neutrino Double Beta Decay 274
17.1.2 Neutrinoless Double Beta Decay 277
17.2 EXPERIMENTAL STUDIES OF DOUBLE BETA DECAY 281

CHAPTER 18 ■ A Leap Forward: Charmonium 284

18.1 A PRIMER: BARYONS, MESONS, QUARKS AND QCD 284
18.1.1 Conserved Quantum Numbers 285
18.1.2 Quarks and QCD 285
18.1.3 Infrared Confinement and Asymptotic Freedom 288
18.2 CHARMONIA 291
18.2.1 The Cornell Potential and Its Relativistic Corrections 292
18.2.2 Strategy and Numerical Results 295
18.3 CHARMONIA END EXOTICS 296
18.4 PROBLEMS FOR CHAPTER 18 299

CHAPTER 19 ▪ The Born-Oppenheimer Approximation for the Doubly 300

19.1 BORN-OPPENHEIMER APPROXIMATION IN BRIEF 301
19.2 COLOUR GYMNASTIC FOR QUARK-QUARK POTENTIALS 302
19.3 THE DOUBLY CHARMED BARYON 304
19.3.1 The BO Approximation for $\Xi_{cc}^{++}$ 304
19.3.2 Numerical Results 308
19.3.3 About the BO Approximation Error in QCD 309
19.4 PROBLEMS FOR CHAPTER 19 310

APPENDIX A ▪ Basic Elements of Quantum Mechanics 311

A.1 THE PRINCIPLE OF SUPERPOSITION 311
A.2 LINEAR OPERATORS 313
A.3 OBSERVABLES AND HERMITIAN OPERATORS 315
A.4 THE NON-RELATIVISTIC SPIN 0 PARTICLE 316
A.4.1 Translations and Rotations 318
A.4.2 Spin 321

APPENDIX B ▪ The Non-Relativistic Hydrogen Atom 323

B.1 FACTORISATION OF THE LAPLACIAN 323
B.2 SEPARATION OF VARIABLES 324
B.3 EIGENVALUES OF THE HAMILTONIAN 326
B.4 EIGENFUNCTIONS 328

Bibliography 331

Index 339

# Preface

This book is based on the relativistic quantum mechanics lecture course given since 2006 to first year master's degree (*Laurea Magistrale*) students in the physics department of the University of Rome, "La Sapienza".

The course, which is mandatory for all students irrespective of their chosen field of study, is intended to provide an introduction to the formal and conceptual foundations of quantum field theory.

The requirement to be relevant to students with different interests, combined with the fact that the course takes place in the first session of their formative year, has had an important role in the choice of the subjects covered, as well as how they are taught.

The only prerequisite for the understanding of the contents of this volume is to have taken courses in classical mechanics, relativity and quantum mechanics, as given in the first three years of the bachelor's degree (*Laurea*) course.

The introductory chapters are devoted to a summary of the theory of special relativity and its application to the classical description of the motion of a free particle and a field. Within the Lagrangian formulation of classical field theory, particular emphasis is placed on the relationships between symmetry and conservation laws, whose fundamental role is demonstrated in many examples in the next volume of this series.

The quantum formulation of field theory is introduced through the simple example of a scalar field described by the Klein–Gordon equation, showing that generalisation of the Schrödinger equation to the case of a relativistic particle renders a probabilistic interpretation of its solution impossible. The alternative interpretation as a quantum field emerges in a natural way from the necessity to describe processes characterised by particle creation and destruction, made possible by the equivalence between mass and energy established by the theory of relativity.

The elements necessary for the construction of the theories which are the foundation of the standard model of electroweak interactions, namely quantum electrodynamics and the Fermi theory of neutron beta decay, are discussed in the central chapters. Problems connected with the quantisation of the electromagnetic field are analysed and spinor fields described by the Dirac equation are introduced. The structure of the interaction term between charged particles and the electromagnetic field is discussed showing how the form which is obtained from classical theory through the so-called minimal substitution is in reality, prescribed by the symmetry of the Lagrangian.

An important part of this volume is dedicated to the derivation of relativistic perturbation theory and its application to the calculation of observable quantities like the collision cross-sections and decay lifetimes. The steps necessary to arrive at the final results are derived in a detailed way, demonstrating the mathematical tools that by the end of the course the students should have learned to use.

The relativistic quantum mechanics course is one of a series of theoretical physics courses offered by the physics department of "La Sapienza". It should provide students who plan to study theoretical physics and elementary particle physics with the essential foundations to be able to follow subsequent modules, having as an objective the theory of electroweak interactions and gauge theories. It is in this spirit that a chapter dedicated to neutrino oscillations has been included, as one of the most topical and stimulating subjects in elementary particle physics, and one which is not generally treated in introductory field theory texts.

Finally, we would like to acknowledge how much we owe to the collaboration between the authors and Nicola Cabibbo over more than twenty years, as well as our gratitude for comments we have received from our students.

# The Authors

**Luciano Maiani**, born in 1941, is emeritus professor of theoretical physics at the University of Rome, "La Sapienza", and author of more than two hundred scientific publications on the theoretical physics of elementary particles. He, together with S. Glashow and J. Iliopoulos, made the prediction of a new family of particles, those with "charm", which form an essential part of the unified theory of the weak and electromagnetic forces. He has been president of the Italian Institute for Nuclear Physics (INFN), Director-General of CERN in Geneva and president of the Italian National Council for Research (CNR). He promoted the development of the Virgo Observatory for gravitational wave detection, the neutrino beam from CERN to Gran Sasso and at CERN directed the crucial phases of the construction of the Large Hadron Collider. He has taught and worked in numerous foreign institutes. He was head of the theoretical physics department at the University of Rome, "La Sapienza", from 1976 to 1984 and held the chair of theoretical physics from 1984 to 2011. He is a member of the Italian Lincean Academy and a Fellow of the American Physical Society.

**Omar Benhar**, born in 1953, is an INFN research director and teaches gauge theories at the University of Rome, "La Sapienza". He has worked extensively in the USA as a visiting professor, at the University of Illinois and the Old Dominion University, as well as an associate scientist at the Thomas Jefferson National Accelerator Facility. Since 2013, he has served as an adjunct professor at the Centre for Neutrino Physics of Virginia Polytechnic Institute and State University. He is the author of more than a hundred scientific papers on the theory of many-particle systems, the structure of compact stars and electroweak interactions of nuclei.

CHAPTER 1

# THE SYMMETRIES OF SPACE-TIME

## 1.1 THE PRINCIPLE OF RELATIVITY

The starting point of the theory of special relativity is the possibility to identify *inertial frames* of reference (IF), defined as those systems in which Newton's first law of motion (below) is valid:

*(1) A body not subject to a force in an IF is in a state of rest or performs uniform straight-line motion.*

On a closer look, we are in the presence of a circular argument: the absence of forces can be ascertained only by observing uniform straight-line motion, which, however, requires the *a priori* definition of a reference frame. Physicists resolve the problem in a pragmatic manner, starting from those systems which are clearly not inertial (the system made by my car on a rough road is certainly not inertial; the motion of the objects in the car is strongly influenced by *apparent forces*) and identifying systems in nature which are gradually better approximations to an ideal IF:

- our house (over intervals which are short compared to the Earth's period of rotation),
- the Earth (for durations short compared to the solar year),
- the Sun (if we ignore the orbital motion around the centre of the galaxy),
- the galaxy...

Once an IF has been identified, it is possible to construct an infinite number of them, in fact, $\infty^6$, which differ in the position of the origin (three coordinates) and by a constant relative velocity (three components). The principle

DOI: 10.1201/9781003436263-1

of special relativity formulated by Galileo states that:

*(2) The laws of physics are invariant under a change of IF.*

In a given IF, physical phenomena can be analysed in terms of *events*; occurrences identified with a point $\boldsymbol{x}$ and a certain time $t$. An event is therefore characterised by a 4-vector which provides the coordinates of the event in a given IF:

$$\text{coordinates} = (ct, \boldsymbol{x}) = x^\mu \quad (\mu = 0, .., 3) .$$

In the time coordinate, we have inserted a factor $c$ (the velocity of light in free space) so that $x^0 = ct$ has the same physical dimensions, of a length, as $\boldsymbol{x}$. In this way, time can be measured in terms of length (the distance travelled by light in the given interval). This definition of time forms part of the so-called natural units, which we introduce later. Astronomers follow the opposite convention, measuring distances in terms of the time needed to traverse them (for example, light years).

To give meaning to the principle of special relativity, we must establish the rules governing the transformation of the coordinates of a given event in one IF, $O$, to another, $O'$.

Let us assume, for simplicity, that the origins of $O$ and of $O'$ coincide at time $t = 0$. The requirement that uniform straight line motion in $O$ should look similar in $O'$ constrains the transformation to be linear and homogeneous:

$$(x')^\mu = \Lambda^\mu_\nu x^\nu, \tag{1.1}$$

where repeated indices (from 0 to 3) indicate a summation and $\Lambda$ is independent of $x^\mu$.

To principles *(1)* and *(2)*, Einstein added:

*(3) The speed of light in free space c is a universal constant, independent of the reference frame.*

Let us consider two events which differ by $\Delta\boldsymbol{x}$ and $\Delta t$, which we suppose to be infinitesimal. We define the *squared invariant length* of the interval $(\Delta t, \Delta\boldsymbol{x})$ by the quantity[1]:

$$\Delta s = (c\Delta t)^2 - (\Delta\boldsymbol{x})^2 = (\Delta x^0)^2 - (\Delta\boldsymbol{x})^2.$$

If $|\Delta\boldsymbol{x}| = c|\Delta t|$, the two events are connected through the propagation of a light ray which leaves from the first event and arrives in exact coincidence with the second. Clearly this coincidence must occur in all reference frames and, given the invariance of the speed of light, this implies that the difference should

[1] $\Delta s$ is often referred to as the *line element*.

also correspond to a zero invariant length in the transformed coordinates. As a formula:

$$\Delta s = (c\Delta t)^2 - (\Delta \boldsymbol{x})^2 = 0 \rightarrow \Delta s' = (c\Delta t')^2 - (\Delta \boldsymbol{x}')^2 = 0.$$

This condition can hold only if the transformation rule is such that:

$$\Delta s' = \lambda \Delta s,$$

where $\lambda$ must be independent of the coordinates and can only depend on the magnitude of the velocity. Now we take into account that the relationship between $O$ and $O'$ is completely symmetric; seen from $O$, $O'$ moves at the same speed at which $O'$ sees $O$ travel. Therefore, by exchanging the roles of the two systems, it must also be true that:

$$\Delta s = \lambda \Delta s', \tag{1.2}$$

so that $\lambda = \pm 1$. The case $-1$ is excluded by the fact that the coordinate transformations are continuously connected by the identity transformation, which clearly has $\lambda = +1$; therefore, we conclude that the transformations (1.1) must *conserve the invariant length* of the interval between two events (hence the name given to $\Delta s$). Given the linearity of the transformations, the condition can immediately be extended to finite intervals.

To formalise the condition (1.2), we introduce the metric tensor $g_{\mu\nu}$, which allows us to rewrite $\Delta s$ as:

$$\Delta s = g_{\mu\nu} \Delta x^\mu \Delta x^\nu \tag{1.3}$$

$$g_{\mu\nu} = \text{diag}(+1, -1, -1, -1). \tag{1.4}$$

Condition (1.2) is rewritten as follows:

$$s' = g_{\mu\nu} x'^\mu x'^\nu = g_{\mu\nu} \Lambda^\mu_\rho \Lambda^\nu_\sigma x^\rho x^\sigma = (\Lambda^\mu_\rho g_{\mu\nu} \Lambda^\nu_\sigma) x^\rho x^\sigma = s \tag{1.5}$$

which then implies:

$$(\Lambda^\mu_\rho g_{\mu\nu} \Lambda^\nu_\sigma) = g_{\rho\sigma}. \tag{1.6}$$

Expressed as a matrix product of $\Lambda$ and $g$, the equation can be rewritten:

$$\Lambda^T g \Lambda = g. \tag{1.7}$$

This equation defines a *group* of matrices which, in mathematical terms, represent the *group of Lorentz transformations.* The coordinate transformations are only one example of these transformations, as we will soon see. First, let us see how these considerations apply in the concrete example of *special Lorentz transformations.*

The system $O'$ moves in the direction of the positive $x$ axis of $O$ with speed $v$. The explicit form of (1.1) is:

$$\begin{aligned} \Delta x' &= \alpha \Delta x - \delta c \Delta t \\ c\Delta t' &= -\epsilon \Delta x + \zeta c \Delta t \\ \Delta y' &= \Delta y; \;\; \Delta z' = \Delta z, \end{aligned} \tag{1.8}$$

with $\alpha, \delta, \epsilon, \zeta$ to be determined. Moreover,

$$s' = (\zeta^2 - \delta^2)(c\Delta t)^2 - (\alpha^2 - \epsilon^2)(\Delta x)^2 - 2(\epsilon\zeta - \alpha\delta)(\Delta x)(c\Delta t) - (\Delta y')^2 - (\Delta z')^2.$$

Setting $\lambda = 1$ in (1.2), we must require $s' = s$ so that:

$$\begin{aligned} \zeta^2 - \delta^2 &= \alpha^2 - \epsilon^2 = 1 \\ \epsilon\zeta - \alpha\delta &= 0. \end{aligned} \tag{1.9}$$

Equation (1.8) can be solved by substituting:

$$\begin{aligned} \alpha &= \zeta = \cosh(\theta) \\ \delta &= \epsilon = \sinh(\theta), \end{aligned} \tag{1.10}$$

where $\theta$ is a real parameter (the *rapidity*) connected to the relative velocity between $O$ and $O'$. If we set $\Delta x' = 0$ in (1.8), the second equation of (1.8) should define the motion of a stationary object in $O'$ seen from $O$, so $\Delta x = v \cdot \Delta t$.

Comparing, we obtain:

$$\tanh(\theta) = \frac{v}{c}, \tag{1.11}$$

and therefore, from (1.8), we obtain the well known special Lorentz transformations:

$$\begin{aligned} \Delta x' &= \gamma(\Delta x - \beta c\Delta t) \\ c\Delta t' &= \gamma(-\beta\Delta x + c\Delta t) \\ \beta &= \frac{v}{c}, \quad \gamma = \frac{1}{\sqrt{1 - \frac{v^2}{c^2}}}. \end{aligned} \tag{1.12}$$

Comments.

- From (1.11) we see that the speed of a physical system cannot exceed $c$.
- Newton hypothesised a universal time, which flows equally in every frame of reference. If, instead of using *principle (3)*, we set $t' = t$, the special Lorentz transformations reduce to *Galilean transformations*:

$$\begin{aligned} x' &= x - v \cdot t \\ t' &= t \end{aligned}$$

  which are obtained from (1.12) in the limit $c \to \infty$ (*non-relativistic limit*).
- Rotations in a Euclidean space leave invariant the squared Pythagorean length: $s_E = x^2 + y^2 + \ldots$. Correspondingly, they satisfy the orthogonality relation $R \cdot R^T = 1$, analogous to (1.7) except for the replacement of the matrix $g$ with $\mathbb{1}$.

## 1.2 PHYSICAL LORENTZ TRANSFORMATIONS

From (1.7), taking the determinant of both sides, we obtain the condition:

$$\det(\Lambda) \cdot \det(g) \cdot \det(\Lambda^T) = \det(\Lambda)^2 \det(g) = \det(g). \tag{1.13}$$

Therefore it must be the case that $\det(\Lambda) = \pm 1$; the Lorentz group comprises at least two disconnected components. Only those elements with determinant equal to +1 (the *proper* Lorentz transformations) are able to be continuously connected to the identity transformation. The transformations with determinant −1 are known as *improper*. An important example of an improper transformation is the *parity operation*, denoted by $P$:

$$P : \boldsymbol{x}' = -\boldsymbol{x}; \; t' = t. \tag{1.14}$$

The proper transformations are, for their part, constituted of two disconnected components, which are distinguished according to whether or not they invert the direction of time. Events with coordinates $(t, \mathbf{0})$ in $O$ describe, as a function of t, the history of a stationary clock at the origin of $O$. If we apply the relation (1.1), we find:

$$t' = \Lambda^0_0 \cdot t,$$

therefore, the sign of $\Lambda^0_0$ determines whether the clocks of $O'$ run in the same direction as in $O$. The transformations characterised by:

$$\Lambda^0_0 > 0 \tag{1.15}$$

are known as *orthochronous.*

Clearly, if an orthochronous transformation takes a system $O'$ to another frame of reference $O''$, the overall transformation: $O \to O' \to O''$ is also orthochronous. Hence the transformations characterised by (1.15) form a subgroup of the Lorentz group. An important example of an improper non-orthochronous transformation is *time inversion*, denoted by $T$:

$$T : \boldsymbol{x}' = \boldsymbol{x}; \;\; t' = -t. \tag{1.16}$$

Combining both $P$ and $T$ transformations, *total inversion*, $I$, is obtained, which is a non-orthochronous but also proper transformation:

$$I : x'^{\mu} = -x^{\mu}. \tag{1.17}$$

The changes between physically realisable IFs which are continuously connected to the identity transformation are orthochronous. The principle of relativity must apply strictly only to these transformations.

*The laws of physics are invariant under proper and orthochronous Lorentz transformations.* The corresponding group is usually denoted by $L^{\uparrow}_{+}$.

Comment. In classical physics this principle is implicitly extended to $P$ and $T$ transformations. However, some nuclear and subnuclear processes are *not* invariant under $P$ ($\beta$ decays) or under $T$ (decays of neutral K mesons); Nature does not respect them. Total inversion, $I$, is worth further discussion. In quantum mechanics, invariance of the laws of physics under the action of inversion, $I$, combined with the operation exchanging particle with antiparticle, $C$, can be *demonstrated* (CPT theorem).

## 1.3 CAUSAL STRUCTURE OF SPACE-TIME

The invariance of the length of the interval between two events, $s$, under the Lorentz transformation, gives rise to a factorisation of space-time into regions with a different causal connection to a given event. For convenience, let us set an event $A$ at the origin of the spatial coordinates at time $t = 0$ in a given IF. The space-time of the events $(t, \boldsymbol{x})$ factorises into four regions *in a way which is independent of the chosen reference frame.*

- Events with $s = 0$. They are found on the surfaces of two cones (*light cones*), which represent the trajectories of light rays emerging from $A$ (*future cone*) or converging on A (*past cone*). They are known as *lightlike.*
- Events with $s > 0, t > 0$. They fill the interior of the future light cone; they are events which can be influenced by event $A$, since they can be reached by physically realisable signals originating from $A$ which travel at speeds less than $c$.
- Events with $s > 0, t < 0$. They fill the interior of the past light cone; they are events which can influence event $A$, by means of physically realisable signals which travel at speeds less than c. These events and those of the preceding paragraph are known as *timelike* (see Problem 1).
- Events with $s < 0$. These are outside the light cones; they are events which cannot have any causal relationship with $A$, because signals with speeds exceeding $c$ would be required. The events in this region form the *absolute present* of $A$. They are known as *spacelike* events (see Problem 2).

In quantum mechanics the absence of the possibility to observe simultaneously two physical quantities are connected to the causal influence which the measurement of one of them can exercise on the measurement of the other (for example the measurement of $p$ and $x$). Let us consider two observable quantities whose measuring apparatus are localised in finite regions of space-time (local observables). When the relevant region of one of the two observables is located entirely in the present of the other, the principle of causality requires that the two observables *should be simultaneously measurable* and the corresponding operators should commute with each other. This principle is known as *microcausality.*

## 1.4 CONTRAVARIANT AND COVARIANT VECTORS

Let $x^\mu$ and $y^\mu$ be two 4-vectors which transform according to (1.1). In the literature, these objects are known as *contravariant vectors*, indicated by upper indices. The scalar product formed with the metric tensor is invariant:

$$\begin{aligned}(x\cdot y) &= g_{\mu\nu}x^\mu y^\nu = x^\mu y_\mu = y^\mu x_\mu \\ (x'\cdot y') &= (x\cdot y)\ .\end{aligned} \tag{1.18}$$

In equation (1.18) we introduced vectors with *lower indices*, commonly known as *covariant* 4-vectors. Clearly, the transformation rule for these vectors must be such as to provide an invariant when a covariant vector is multiplied by a contravariant vector. In fact, this rule can immediately be obtained starting from (1.18):

$$y'_\mu = g_{\mu\nu}y'^\nu = g_{\mu\nu}\Lambda^\nu_\rho y^\rho.$$

Multiplying (1.6) by $(\Lambda^{-1})^\rho_\tau$ we find:

$$g_{\tau\nu}\Lambda^\nu_\sigma = (\Lambda^{-1})^\rho_\tau g_{\rho\sigma}; \tag{1.19}$$

$$\text{or}\quad y'_\mu = g_{\mu\nu}\Lambda^\nu_\rho y^\rho = (\Lambda^{-1})^\rho_\mu g_{\rho\sigma}y^\sigma = (\Lambda^{-1})^\rho_\mu y_\rho.$$

Therefore *covariant vectors transform using the matrix* $\Lambda^{-1}$.

The four-vector $\partial/\partial x^\mu$ transforms according to

$$\frac{\partial}{\partial x^\mu} \to \frac{\partial}{\partial x'^\mu} = \frac{\partial x^\nu}{\partial x'^\mu}\frac{\partial}{\partial x^\nu} = (\Lambda^{-1})^\nu_\mu \frac{\partial}{\partial x^\nu}, \tag{1.20}$$

that is like a *covariant* four-vector. Therefore, we can write

$$\frac{\partial}{\partial x^\mu} \equiv \left(\frac{1}{c}\frac{\partial}{\partial t},\ \boldsymbol{\nabla}\right) = \partial_\mu, \quad \frac{\partial}{\partial x_\mu} \equiv \left(\frac{1}{c}\frac{\partial}{\partial t},\ -\boldsymbol{\nabla}\right) = \partial^\mu, \tag{1.21}$$

from which it follows that, given a four-vector $u$

$$\partial_\mu u^\mu = \frac{1}{c}\frac{\partial u^0}{\partial t} + \boldsymbol{\nabla}\cdot\boldsymbol{u} = \partial^\mu u_\mu, \tag{1.22}$$

and the d'Alambertian operator is defined as

$$\Box = \partial_\mu\partial^\mu = \frac{1}{c^2}\frac{\partial^2}{\partial t^2} - \nabla^2. \tag{1.23}$$

In formal terms, covariant vectors are the elements of the *dual space* of the vector space of contravariant vectors. In general, the dual $\widetilde{V}$ of a vector space $V$ is defined as the space of *linear functionals of vectors* $y^\mu$, of functions, that is, defined for every $y$ and $z$, such that:

$$\begin{aligned}&f = f(y), \\ &f(\alpha y + \beta z) = \alpha f(y) + \beta f(z)\end{aligned}$$

with $\alpha$ and $\beta$ numbers (in our case, real numbers). It is easy to show that every element $f$ of $\widetilde{V}$ can be written as:

$$f(y) = \sum_{\mu} f_\mu y^\mu = f_\mu y^\mu = (f \cdot y), \tag{1.24}$$

hence $\widetilde{V}$ has the same dimensions as $V$ and *there is a one-to-one mapping* between the elements of $V$ and those of $\widetilde{V}$ (theorem due to Riesz).

In our case, this mapping is realised with the metric matrix $g_{\mu\nu}$ which gives the element of $\widetilde{V}$, $y_\mu = g_{\mu\nu} y^\mu$, which corresponds to $y^\mu$. With this relation, the elements of $\widetilde{V}$ are invariant functions under Lorentz transformations:

$$f'(y') = f(y), \text{ if } y' = \Lambda y \, .$$

Using the relation (A.5) it is immediately seen that the preceding equation requires two transformation rules:

$$y'^\mu = \Lambda^\mu_\nu y^\nu, \quad f'_\mu = (\Lambda^{-1})^\rho_\mu f_\rho$$

such that:

$$f'(y') = f'_\mu y'^\mu = f_\rho (\Lambda^{-1})^\rho_\mu \Lambda^\mu_\nu y^\nu = f_\rho (\Lambda^{-1} \cdot \Lambda)^\nu_\rho y^\nu = f_\rho \delta^\rho_\nu y^\nu = f_\rho y^\rho = f(y)$$

where $\delta^\rho_\nu$ is the Kronecker delta.

## 1.5 PROBLEMS FOR CHAPTER 1

### Sect. 1.1

1. From the definition of velocity, (1.11) prove that the speed of a physical system cannot exceed $c$.

2. Newton hypothesised a universal time, which flows equally in every frame of reference. Work out the transformations obtained by setting set $t' = t$, thereby obtaining the *Galilean transformations*:

$$x' = x - v \cdot t$$
$$t' = t \, .$$

3. Prove that the Galilean transformations of Problem **1** are obtained from the Lorentz transformations (1.12) in the limit $c \to \infty$ (i.e. the *non-relativistic limit*).

4. Show that the rotation matrices in a Euclidean space satisfy the orthogonality relation

$$R \cdot R^T = 1$$

as a consequence of them leaving invariant the squared Pythagorean length: $s_E = x^2 + y^2 + ....$ Note that the above relation is analogous to (1.7) except for the replacement of the matrix $\mathbb{1}$ with $g$.

5. Show that, if $\Lambda_1$ and $\Lambda_2$ are Lorentz matrices satisfying the Lorentz condition (1.7), $\Lambda^T g \Lambda = g$, their matrix product $\Lambda_1 \cdot \Lambda_2$ satisfies the same condition.

6. Show that, by carrying out two special Lorentz transformations with rapidities $\theta_1$ and $\theta_2$, a Lorentz transformation with rapidity $\theta_1 + \theta_2$ is obtained. From this, derive the rule for the combination of velocities:

$$v_1 \oplus v_2 = \frac{v_1 + v_2}{1 + \frac{v_1 \cdot v_2}{c^2}} .$$

Sect. 1.2

1. Show that if an event has $s > 0$ it is possible to find an inertial frame in which $x^\mu = (t, \mathbf{0})$. (Consequently, these events are called *timelike*).

2. Show that if an event has $s < 0$ it is possible to find an inertial frame in which $x^\mu = (0, \mathbf{x})$. (Consequently, these events are called *spacelike*).

3. Show that if an event has $s = 0$ it is possible to find an inertial frame in which $x^\mu = (|\mathbf{x}|, \mathbf{x})$ and $\mathbf{x}$ aligned along the $z$ axis. (These events are called *lightlike*).

CHAPTER 2

# THE CLASSICAL FREE PARTICLE

## 2.1 SPACE–TIME MOTION

The motion in space-time of a classical particle (that is to say, non-quantum mechanical) is described by four functions of time: $x^\mu = x^\mu(t)$. We consider the interval which separates two positions very close together in time and its invariant length:

$$\Delta x^\mu = (c\Delta t, \boldsymbol{v} \cdot \Delta t) \tag{2.1}$$
$$\Delta s = (\Delta x \cdot \Delta x) = (1 - \frac{v^2}{c^2})(c\Delta t)^2.$$

Obviously, $\Delta x^\mu$ is a timelike interval, because the particle travels at a speed less than that of light, hence an IF exists in which $\Delta x^\mu$ has only a temporal component. In this IF the particle is stationary (in the time interval under consideration) and $\sqrt{\Delta s}/c$ gives the time interval measured by a clock at rest with respect to it at that moment. The new variable is called the *proper time* of the particle, denoted by $\tau$. By definition, an interval of proper time is a relativistic invariant and the relation between $\tau$ and the time $t$ in the chosen IF is given by the relation:

$$d\tau = \sqrt{1 - \beta^2}\, dt = \frac{dt}{\gamma(t)}. \tag{2.2}$$

If we use the proper time to characterise the motion of the particle, we can define a velocity four-vector, $u^\mu$:

$$u^\mu = \frac{d\, x^\mu(\tau)}{d\tau} = \gamma \frac{dx^\mu(t)}{dt}, \tag{2.3}$$


DOI: 10.1201/9781003436263-2

in which $u^\mu$ is obviously a covariant 4-vector, since it transforms like $x^\mu$. We note the value of its components and of its invariant magnitude:

$$u^\mu = \gamma(c, \boldsymbol{v}); \tag{2.4}$$

$$g_{\mu\nu}u^\mu u^\nu = u^\mu u_\mu = c^2. \tag{2.5}$$

In non-relativistic mechanics, the momentum is defined as $\boldsymbol{p} = m\boldsymbol{v}$, where $m$ is the inertial mass. We define the 4-vector momentum analogously as:

$$p^\mu = mu^\mu \tag{2.6}$$

where $m$ is a constant which characterises the particle and which obviously coincides with the inertial mass at low velocity. For this reason, $m$ is called the *rest mass* of the particle. The rest mass is a relativistic invariant, as is seen from the relation:

$$p^2 = p^\mu p_\mu = (mc)^2. \tag{2.7}$$

The time component of $p^\mu$ is related to the energy of the particle. We calculate it explicitly starting from the relation:

$$p^2 = (mc)^2 = (p^0)^2 - (\boldsymbol{p})^2 \tag{2.8}$$

from which:

$$p^0 = \sqrt{(mc)^2 + (\boldsymbol{p})^2} \simeq \frac{1}{c}(mc^2 + \frac{\boldsymbol{p}^2}{2m}) \tag{2.9}$$

where the final step is valid for low velocity. The temporal component of $p^\mu$, in this limit, is equal to the kinetic energy divided by $c$ plus a constant related to the rest mass. In classical mechanics, the energy is always defined to within a constant, so we can identify $p^0$ with the energy of the particle divided by $c$:

$$p^\mu = (\frac{\epsilon}{c}, \boldsymbol{p}). \tag{2.10}$$

Thus we arrive at the important conclusion that in a relativistic formulation *momentum and energy are parts of a single physical entity*, the energy-momentum 4-vector (or 4-momentum) $p^\mu$.

The rules determining the components of $p^\mu$ after transformation of the coordinate system follow immediately from the nature of the 4-vector $p^\mu$. In the specific case of a special Lorentz transformation along the $x$-axis, we have (the quantities with accents refer to the IF $O'$, those without an accent to $O$):

$$\begin{aligned} p'^1 &= \gamma(p^1 - \beta\frac{\epsilon}{c}) \\ \frac{\epsilon'}{c} &= \gamma(-\beta p^1 + \frac{\epsilon}{c}) \\ p'^2 &= p^2; \;\; p'^3 = p^3. \end{aligned} \tag{2.11}$$

If the particle is at rest in $O$, we find, in particular:

$$\epsilon' = \gamma mc^2;\ p' = \gamma\beta mc. \tag{2.12}$$

The velocity of the particle, in any IF, is given by:

$$\beta = \frac{cp'}{\epsilon'}. \tag{2.13}$$

Comment. In deriving the value of $p^0$ in (2.9) starting from (2.8), we have chosen the positive root. This seems an innocuous step, but we can examine it more closely. The negative root is separated from the positive one by a gap of at least $2mc^2$. In classical mechanics the energy varies continuously; thus, if we assume that the particle has $E > mc^2$ at the start, it can never finish in a state with $E < -mc^2$. In quantum mechanics, however, the energy can change discontinuously and we cannot exclude transitions from states with $E > mc^2$ to others with $E < -mc^2$. The resolution of this difficulty takes us directly to the concept of *antiparticles*.

## 2.2 PARTICLE OF ZERO MASS

The condition (2.7) allows us to consider the limit $m \to 0$. In this case, in any IF:

$$\begin{aligned} |\boldsymbol{p}| &= p^0; \\ \beta &= 1, \end{aligned} \tag{2.14}$$

the particle travels at the speed of light in any IF, for any state of motion. Obviously photons, the quanta of light, have this property; *the photon is a particle with zero rest mass.*

Despite the fact that (2.7) varies smoothly as $m \to 0$, particles with mass $m > 0$ are intrinsically different from those of zero mass, however, small the value of $m$; *the limit $m \to 0$ is intrinsically discontinuous.*

The simplest way to obtain this result is by verifying that the symmetry group for the momentum (the *little group* introduced by Wigner) is different in the two cases.

- If $p^\mu$ is the momentum of a particle of non-zero mass we can find an IF in which $p^\mu$ has only a temporal component (its *rest system*). In the rest system, $p^\mu = (mc^2, \mathbf{0})$ and the group of transformations which leave the momentum invariant is the entire group of three-dimensional spatial rotations, $O(3)$, the group of orthogonal matrices in three dimensions.

- If $p^\mu$ is the momentum of a zero-mass particle, we can write it in the form: $p^\mu = (|\boldsymbol{p}|, \boldsymbol{p})$. The invariance group is now that of rotations in the plane orthogonal to the direction of $\boldsymbol{p}$, the orthogonal matrices in two dimensions, $O(2)$, a commutative group much smaller than $O(3)$.

One sees therefore that when letting $m \to 0$ the little group changes discontinuously from $O(3)$ to $O(2)$. In quantum mechanics, the little group determines the value of the *spin* of the particle. For a particle with mass, the spin coincides with the angular momentum at rest and the states therefore form a representation of the group of 3-dimensional rotations; a particle of spin $S$ possesses $2S+1$ spin states.

Conversely, a particle of zero mass can never be at rest, and its intrinsic angular momentum is defined by rotations around $\boldsymbol{p}$, the elements of $O(2)$, which allow one-dimensional representations corresponding to $S_z = \frac{1}{2}n$ for a given integer $n$. If we include parity, $P$, as a good symmetry, two states should exist with $S_z = \pm\frac{1}{2}n$. For example, the photon has two spin states, corresponding to $S_z = \pm 1$ and not three, as we would have expected for a particle of spin 1.

## 2.3 ACTION PRINCIPLE FOR THE FREE PARTICLE

The results of the previous section are of such importance as to merit an independent derivation. At the same time, this permits the introduction of the formulation of relativistic dynamics based on the *action principle*, which has a fundamental importance in quantum mechanics.

We consider a trajectory in space-time, $x^\mu(t)$, which starts and ends in two fixed events:

$$\begin{aligned} x^\mu(t_1) &= x_1^\mu = (ct_1, \boldsymbol{x}_1) \\ x^\mu(t_2) &= x_2^\mu = (ct_2, \boldsymbol{x}_2) \end{aligned} \tag{2.15}$$

Given $x^\mu(t)$, we can define the *action*, $S(x_1^\mu, x_2^\mu)$:

$$S = \int_{t_1}^{t_2} L(\boldsymbol{x}(t), \boldsymbol{v}(t))\, dt \tag{2.16}$$

The trajectory actually travelled is determined by the action principle:

- *The particle trajectory corresponds to the minimum value of the action.*

From the Lagrangian function, the principle of least action derives the Lagrange equations, which in classical mechanics completely replace Newton's equations, $\boldsymbol{F} = m\boldsymbol{a}$:

$$\frac{d}{dt}\frac{\partial L}{\partial \boldsymbol{v}} = \frac{\partial L}{\partial \boldsymbol{x}} \tag{2.17}$$

The requirement that the laws of motion should be invariant following a change of IF becomes the simple statement:

- *The action must be invariant under Lorentz transformations.*

Applied in our case, this means:

$$L(\boldsymbol{x}(t), \boldsymbol{v}(t))\, dt = \textit{invariant}.$$

For a free particle, the only invariant non-trivially constant is the proper time, $d\tau$ and therefore it must be true that:

$$L(\boldsymbol{x}(t), \boldsymbol{v}(t))\, dt = -\alpha\, d\tau = -\alpha\sqrt{1 - \frac{v^2(t)}{c^2}}\, dt \tag{2.18}$$

with $\alpha$ a constant. The value of $\alpha$ is determined by the non-relativistic limit, in which $L$ should tend to the kinetic energy of the particle, $\frac{1}{2}mv^2$, to within an irrelevant additive constant. For small values of $v(t)$ we find:

$$\begin{aligned} L &\to -\alpha + \frac{\alpha v^2}{2c^2}; \\ &\to \alpha = mc^2 \end{aligned} \tag{2.19}$$

and therefore the Lagrangian of a free particle is:

$$L = -mc^2\sqrt{1 - \frac{v(t)^2}{c^2}}. \tag{2.20}$$

The kinematic momentum is given by the conjugate momentum with respect to $\boldsymbol{x}$, while the energy is given by the Hamiltonian. Therefore we find:

$$\begin{aligned} \boldsymbol{p} &= \frac{\partial L}{\partial \boldsymbol{v}} = m\gamma\boldsymbol{v} \\ H &= \boldsymbol{p}\cdot\boldsymbol{v} - L = \gamma mc^2. \end{aligned} \tag{2.21}$$

The results in (2.21) confirm the definition of the 4-momentum given in (2.6).

The Lagrange equations which follow from (2.20) and (2.17) are:

$$\frac{d}{dt}p^i = 0; \quad (i = 1, 2, 3) \tag{2.22}$$

Furthermore, from condition (2.7) we find:

$$p^0 dp^0 - \boldsymbol{p}\cdot d\boldsymbol{p} = 0 \tag{2.23}$$

and therefore:

$$\frac{d}{dt}p^0 = 0. \tag{2.24}$$

Finally, we have obtained the covariant equations of motion:

$$\frac{d}{dt}p^\mu = 0 = \frac{d}{d\tau}p^\mu \tag{2.25}$$

which express the conservation of momentum and energy.

## 2.4 THE MASS–ENERGY RELATION

To illustrate the meaning of the rest energy, we consider a system composed of many particles (in short, a gas). The total 4-momentum is given, naturally, by:

$$P^\mu = \sum_i p_i^\mu; \tag{2.26}$$

$$\text{i.e.} \quad \boldsymbol{P} = \sum_i \boldsymbol{p}_i; \; cP^0 = E = \sum \epsilon_i.$$

For a system of this type an IF exists in which $\boldsymbol{P}' = 0$ (the *centre of mass* system). This can be seen from (2.11): we orient $\boldsymbol{P}$ along the x-axis and require that $P'^1 = 0$. We find:

$$\beta = \frac{cP}{E} \tag{2.27}$$

which always has a modulus less than unity, given that:

$$c|\boldsymbol{P}| < c\sum_i |\boldsymbol{p}_i| < \sum \epsilon_i = E. \tag{2.28}$$

The total energy, $E_0$, in this frame of reference defines the rest mass of the system

$$E_0 = M_0 c^2 \tag{2.29}$$

while the energy in another IF, $E_0'$, is given by an analogous formula to (2.12):

$$E_0' = \gamma M_0 c^2 \tag{2.30}$$

We now look more closely at $M_0$. In the limit in which the gas particles are non-relativistic, we have:

$$M_0 = \frac{1}{c^2}\sum_i \epsilon_i \simeq \sum m_i + \frac{T}{c^2} \tag{2.31}$$

where $T$ is the kinetic energy contained in the gas. If we increase or reduce this energy, by heating or cooling the gas, the rest mass varies accordingly, as:

$$\Delta M_0 = \frac{\Delta E}{c^2}. \tag{2.32}$$

If we introduce an interaction between the gas particles, this adds to the second term of (2.31) a quantity $V$ which can be positive or negative.

The conclusion is that the rest mass of a composite system differs from the sum of the rest masses of its constituents and the difference can be released (or must be provided) in the form of an amount of energy again given by relation (2.32):

$$\Delta E = Q = (M_0 - \sum_i m_i)c^2. \tag{2.33}$$

Equations (2.32) and (2.33) provide Einstein's mass–energy relationship, with innumerable applications in physics theory and practice.

From now on, the masses of atomic and subatomic particles will always be given in terms of energy, according to (2.32), using the units 1 MeV = 1000 keV = $10^6$ eV.

As a reminder:

- electron: $M_e = 0.511$ MeV;
- proton: $M_p = 938.27$ MeV;
- neutron: $M_N = 939.57$ MeV;
- deuteron: $M_D = 1875.61$ MeV.

*An important example.* The energy emitted by the Sun is produced by the fusion of four protons into one helium nucleus. The fusion takes place via a series of reactions, the so-called *p–p* sequence (this is the principal sequence; there are several secondary series studied originally by Bethe in the 1930s; recall that *proton* = $^1$H, *deuteron* = $^2$H):

$$p + p \to {}^2\mathrm{H} + e^+ + \nu \tag{2.34}$$
$${}^2\mathrm{H} + p \to {}^3\mathrm{He} + \gamma \tag{2.35}$$
$${}^3\mathrm{He} + {}^3\mathrm{He} \to {}^4\mathrm{He} + p + p. \tag{2.36}$$

The overall reaction can be written as:

$$4p + 2e^- \to {}^4\mathrm{He} + 2e^+ + 2e^- + 2\nu. \tag{2.37}$$

The positrons annihilate with the electrons of the medium releasing energy. Therefore the total thermal energy released by each 4-proton reaction is:

$$E_{thermal} = Q - 2 < E_\nu > \tag{2.38}$$
$$Q = 4M_p + 2M_e - M_{\mathrm{He}} = 26.7 \text{ MeV}.$$

$< E_\nu >$ is the average energy carried away by neutrinos, which leave the Sun undisturbed. The neutrino energy is continuously distributed between zero (assuming negligible neutrino mass) and the $Q$ value in the formation of deuterium:

$$Q({}^1\mathrm{H} + {}^1\mathrm{H}) \to {}^2\mathrm{H} = 2M_p - M_D - M_e = 0.42 \text{ MeV}. \tag{2.39}$$

Comment. According to relativistic quantum mechanics, reactions exist in which particles can be created or destroyed; for example, proton-antiproton annihilation. In these cases the variation of energy in equation (2.32) should include the entire rest mass of the particles involved.

## 2.5 PROBLEMS FOR CHAPTER 2

### Sect. 2.1

1. The $\mu$-particle, or *muon*, has a rest mass $m_\mu \sim 106$ MeV and a lifetime at rest $\tau_\mu \sim 2.197$ s. For a muon produced with energy $E = 1$ GeV, calculate: (i) the values of $\beta$ and $\gamma$ (ii) the average length (in km) covered before decay. Conversely, what is the minimum energy needed for a muon produced at the top of the atmosphere (20 km above sea level) to arrive at sea level before decay?

### Sect. 2.4

1. Knowing that: (i) the Sun contains around $N = 10^{56}$ protons in the core, i.e. available for fusion; (ii) the solar constant (flux of solar energy onto the Earth) is $K_0 = 3.3 \cdot 10^{-2}\text{cal cm}^{-2}\ \text{s}^{-1}$; (iii) the Earth–Sun distance is 8 light-minutes; (iv) the energy carried by neutrinos is negligible, estimate the lifetime of the Sun from (2.37) and (2.38).

CHAPTER 3

# THE LAGRANGIAN THEORY OF FIELDS

In classical mechanics two types of physical systems are distinguished: material points (particles) which have spatial coordinates $\boldsymbol{x}(t)$ as dynamic variables, or fields (waves), which are dynamic systems described by one or more continuous functions of the coordinates and of time.

$$\phi = \phi(\boldsymbol{x}, t) = \phi(x). \tag{3.1}$$

The most important example of this second kind of system is the *electromagnetic field* described at every point of space by two vectors corresponding to the values of the electric field $\boldsymbol{E}(\boldsymbol{x}, t)$ and the magnetic field $\boldsymbol{B}(\boldsymbol{x}, t)$.

## 3.1 THE ACTION PRINCIPLE

In analogy with the mechanics, of systems with a finite number of degrees of freedom, it is natural to derive the field equations from an action principle. The action is defined as the time integral of the Lagrangian, between two fixed instants, $t_1 < t_2$:

$$S = \int_{t_1}^{t_2} \boldsymbol{L}\; dt. \tag{3.2}$$

The Lagrangian of a system of particles is the sum of the many different degrees of freedom. In the case of a field, the degrees of freedom are localised at every point of space, therefore:

$$\boldsymbol{L} = \int d^3x\; \mathcal{L}(\phi, \phi_\mu, x), \tag{3.3}$$

where we have denoted the derivatives of the field with respect to the coordinates as $\phi_\mu$:

$$\phi_\mu(x) = \frac{\partial \phi}{\partial x^\mu}.$$


DOI: 10.1201/9781003436263-3

The function $\mathcal{L}$ is given the name *Lagrangian density* or, more simply for brevity, the Lagrangian and depends on the fields (the dynamic variables) and their derivatives. The time derivatives are the generalisation of velocity, while the dependence of the Lagrangian on spatial derivatives allows to couple the degrees of freedom among nearby points in space.

We have considered a possible explicit dependence of the Lagrangian on the space-time coordinates, to allow for the effect of possible agents external to the system of fields. For an isolated system, this dependence cannot be present and the Lagrangian depends on the coordinates only through the fields and their derivatives.

In terms of $\mathcal{L}$:

$$S = \int_{V_4} d^4x \, \mathcal{L}(\phi, \phi_\mu, x) \tag{3.4}$$

where $V_4$ is the region of space-time limited by the hypersurfaces $\Gamma_1 : t = t_1$ and $\Gamma_2 : t = t_2$ .

We assume the values of the fields are fixed on $\Gamma_{1,2}$. The principle of the least action states that:

- the evolution of the field between these values is given by the function $\phi(x) = \bar{\phi}(\boldsymbol{x}, t)$ which minimises $S$, with boundary conditions fixed on $\Gamma_{1,2}$.

We note that the Lagrangian density is not unique. Because the principle of least action stipulates that the fields have defined values on the boundaries of $V_4$, we can add to the Lagrangian the divergence of any 4-vector without changing the minimum of the action or, hence, the equations of motion.

To derive the differential equations which determine the evolution of the field, we set:

$$\phi(x) = \bar{\phi}(x) + \delta\phi(x); \quad \delta\phi(\boldsymbol{x}, t_1) = \delta\phi(\boldsymbol{x}, t_2) = 0. \tag{3.5}$$

The condition of least action becomes the equation:

$$\begin{aligned} \delta S = 0 = & \int d^4x \, [\frac{\partial \mathcal{L}}{\partial \phi}\delta\phi + \frac{\partial \mathcal{L}}{\partial\partial_\mu \phi}\delta(\partial_\mu)\phi] \\ = & \int d^4x \, [\frac{\partial \mathcal{L}}{\partial \phi}\delta\phi + \frac{\partial \mathcal{L}}{\partial\partial_\mu \phi}\partial_\mu\delta\phi] \\ = & \int d^4x \, [\frac{\partial \mathcal{L}}{\partial \phi} - \partial_\mu(\frac{\partial \mathcal{L}}{\partial\partial_\mu \phi})]\delta\phi + \int d^4x \, \partial_\mu(\frac{\partial \mathcal{L}}{\partial\partial_\mu \phi}\delta\phi) \end{aligned} \tag{3.6}$$

because $\delta(\partial_\mu)\phi = \partial_\mu\delta\phi$.

Given that the field variations vanish at the edges of the region of integration, the final term in (3.6) is zero. Moreover, (3.6) must apply for arbitrary variations $\delta\phi$, so the function in square brackets must cancel identically in $x$. Thus we find the *Euler–Lagrange equations*:

$$\partial_\mu(\frac{\partial \mathcal{L}}{\partial\partial_\mu \phi}) = \frac{\partial \mathcal{L}}{\partial \phi} \tag{3.7}$$

a system of partial differential equations. Naturally, if we have several fields $\phi^i$, $i = 1, ..., N$, we have an equation for each component.

In Newtonian mechanics, the motion of a particle is determined if we define its position and velocity at a fixed instant in time. The extension of this principle gives rise to the requirement that $\mathcal{L}$ should be at most quadratic in the derivatives of the fields and we will follow this principle. In that case, $\partial\mathcal{L}/\partial\partial_\mu\phi$ is linear in $\partial_\mu\phi$, the Euler–Lagrange equation is of second order in the derivatives and the solution is determined *once the field and its time derivative are assigned values* on the hypersurface $t = t_1$.

Relativistic Invariance. The relativistic invariance of the theory is embodied in the simple requirement that the action should be relativistically invariant. The size of space-time is itself invariant, given that, for a transformation $\Lambda$:

$$d^4x' = det(\Lambda)d^4x = d^4x. \tag{3.8}$$

Therefore

- *the relativistic invariance of the action requires that the Lagrangian density should itself be invariant.*

## 3.2 HAMILTONIAN AND CANONICAL FORMALISM

The route to the canonical formalism begins with the definition of the momentum conjugate to each dynamic variable. In the case of a field theory, we define the *conjugate momentum density*

$$\pi(\boldsymbol{x}, t) = \frac{\partial\mathcal{L}}{\partial\partial_t\phi} \tag{3.9}$$

and stipulate that equation (3.9) should be used to express $\partial_t\phi$ as a function of $\phi, \boldsymbol{\nabla}\phi, \pi$. Subsequently one can define the *Hamiltonian density.*

$$\mathcal{H}(\pi, \phi) = \pi\partial_t\phi - \mathcal{L}; \;\; \boldsymbol{H} = \int d^3x\mathcal{H}. \tag{3.10}$$

It should be noted that the Hamiltonian density is an auxiliary quantity; only the Hamiltonian is physically relevant while the Hamiltonian density is defined to within the 3-divergence of a vector, which integrates to zero when the fields vanish at infinity.

The equations of motion are obtained simply by differentiating equation (3.10) and using the Euler–Lagrange equations. We put:

$$\phi(x) = \bar{\phi}(x) + \delta\phi(x); \;\; \pi(x) = \bar{\pi}(x) + \delta\pi(x); \;\; t = \bar{t} + \delta t. \tag{3.11}$$

We then find:

$$\delta \boldsymbol{H} = \int d^3x \left( \partial_t\phi\ \delta\pi + \pi\delta(\partial_t\phi) - \frac{\partial \mathcal{L}}{\partial\phi}\delta\phi - \frac{\partial \mathcal{L}}{\partial \boldsymbol{\nabla}\phi}\cdot\delta(\boldsymbol{\nabla}\phi) - \frac{\partial \mathcal{L}}{\partial\partial_t\phi}\delta(\partial_t\phi) \right)$$
$$- \int d^3x\ \frac{\partial \mathcal{L}}{\partial t}\delta t =$$
$$= \int d^3x \left( \partial_t\phi\ \delta\pi - \partial_t\pi\ \delta\phi - \boldsymbol{\nabla}\cdot(\frac{\partial \mathcal{L}}{\partial \boldsymbol{\nabla}\phi}\delta\phi) \right) - \int d^3x\ \frac{\partial \mathcal{L}}{\partial t}\delta t. \tag{3.12}$$

where we have used the Euler–Lagrange equations in the form:

$$\frac{\partial \mathcal{L}}{\partial\phi} - \boldsymbol{\nabla}\frac{\partial \mathcal{L}}{\partial\boldsymbol{\nabla}\phi} = \partial_t\frac{\partial \mathcal{L}}{\partial\partial_t\phi} = \partial_t\pi \tag{3.13}$$

We can discard the 3-divergence from the differential of the Hamiltonian, so we obtain:

$$\delta \boldsymbol{H} = \int d^3x\,(\partial_t\phi\ \delta\pi - \partial_t\pi\ \delta\phi) - \int d^3x\ \frac{\partial \mathcal{L}}{\partial t}\delta t. \tag{3.14}$$

On the other hand, from the fact that the Hamiltonian density depends on $\phi$, $\boldsymbol{\nabla}\phi$ and $\pi$, we obtain (continuing to discard the 3-divergences):

$$\delta \boldsymbol{H} = \int d^3x \left[ \frac{\partial \mathcal{H}}{\partial\phi}\delta\pi + \left( \frac{\partial \mathcal{H}}{\partial\phi} - \boldsymbol{\nabla}\cdot\frac{\partial \mathcal{H}}{\partial\boldsymbol{\nabla}\phi} \right)\delta\phi \right] + \int d^3x\ \frac{\partial \mathcal{H}}{\partial t}\delta t. \tag{3.15}$$

Comparing the coefficients of the differentials in (3.14) and (3.15), we find Hamilton's equations:

$$\begin{aligned} \partial_t\phi &= \frac{\partial \mathcal{H}}{\partial\pi}; \\ \partial_t\pi &= -(\frac{\partial \mathcal{H}}{\partial\phi} - \boldsymbol{\nabla}\cdot\frac{\partial \mathcal{H}}{\partial\boldsymbol{\nabla}\phi}); \\ \frac{\partial \mathcal{H}}{\partial t} &= -\frac{\partial \mathcal{L}}{\partial t}. \end{aligned} \tag{3.16}$$

In the second equation the derivative with respect to $\boldsymbol{\nabla}\phi$ appears, owing to the fact that in $\mathcal{L}$, and therefore in $\mathcal{H}$, we treated the dependence on $\phi$ and $\boldsymbol{\nabla}\phi$ separately. In fact all the terms in brackets correspond to the quantity $(\partial H/\partial q)$ in the case of one degree of freedom. From the third equation we find:

$$\frac{d\boldsymbol{H}}{dt} = \int d^3x\ \frac{\partial \mathcal{H}}{\partial t} = -\int d^3x\ \frac{\partial \mathcal{L}}{\partial t} \tag{3.17}$$

For an isolated system, as we have seen, the Lagrangian density can depend on the space-time location only through the fields. Consequently, for an isolated system, we recover the law of conservation of energy: *the Hamiltonian of an isolated system is a constant of the motion.* The same thing holds under the weaker condition that the system should be independent of time.

Functionals and Functional Derivatives. From a mathematical point of view, the Hamiltonian in (3.10) is a *functional* of the Hamiltonian density: a rule which associates a number to every given function, $\mathcal{H}$. The functional is therefore a function defined on the space of functions, rather than on the space of real or complex numbers. Analogously to what is done with functions, we can introduce the concept of the derivative of a functional in the following way:

Let $\boldsymbol{H}[f]$ be a functional of the function $f(\boldsymbol{x})$[1]. We define the derivative of the functional starting from the variation:

$$\delta \boldsymbol{H} = \boldsymbol{H}[f(\boldsymbol{x}) + \delta^{(3)}(\boldsymbol{x} - \boldsymbol{y})\epsilon] - \boldsymbol{H}[f(\boldsymbol{x})] \tag{3.18}$$

setting:

$$\frac{\delta \boldsymbol{H}}{\delta f(\boldsymbol{y})} = lim_{\epsilon \to 0} \frac{\boldsymbol{H}[f(\boldsymbol{x}) + \delta^{(3)}(\boldsymbol{x} - \boldsymbol{y})\epsilon] - \boldsymbol{H}[f(\boldsymbol{x})]}{\epsilon}. \tag{3.19}$$

In the case of the Hamiltonian, we have

$$\boldsymbol{H} = \int d^3x \, \mathcal{H}(\phi, \pi). \tag{3.20}$$

from which, for example:

$$\frac{\delta \boldsymbol{H}}{\delta \pi} = \frac{\partial \mathcal{H}}{\partial \pi}. \tag{3.21}$$

A normal function can also be considered as a particular example of a functional, according to the formula:

$$f(y) = \int dx \, \delta(x - y) f(x). \tag{3.22}$$

In this case, the symbol $y$ which appears in the first term is simply a dummy variable, and $f(x)$ is the argument of the functional. For the functional derivative, it is easily shown that:

$$\frac{\delta f(y)}{\delta f(x)} = \delta(x - y). \tag{3.23}$$

If we introduce the concept of the functional derivative, the second of Hamilton's equations takes an aspect more similar to that of the case of a finite number of degrees of freedom. Applying the definition (3.19), Hamilton's equations (3.17) can be put in the form[2]:

$$\begin{aligned} \partial_t \phi(x) &= \frac{\delta \boldsymbol{H}}{\delta \pi(x)} \\ \partial_t \pi(x) &= -\frac{\delta \boldsymbol{H}}{\delta \phi(x)}. \end{aligned} \tag{3.24}$$

[1]As in the example given in this section, we consider the functions of a three-dimensional variable. The generalisation to more dimensions is obvious.

[2]We note, on the basis of (3.19), that the physical dimensions of the functional derivative which appear in the preceding equations are equal to dim[$\boldsymbol{H}$] -dim[$\phi$] (or $\pi$) - [length$^3$], or, equivalently, to dim[$\mathcal{H}$] -dim[$\phi$] (or $\pi$).

Poisson Brackets. The observables in the canonical formalism are, in general, functionals of $\pi(x)$ and $\phi(x)$. Given two observables, $A$ and $B$, we can introduce the *Poisson bracket*, in analogy to the finite dimensional case, defined as:

$$\{A, B\} = \int d^3x \left( \frac{\delta A}{\delta\phi(x)} \frac{\delta B}{\delta\pi(x)} - \frac{\delta B}{\delta\phi(x)} \frac{\delta A}{\delta\pi(x)} \right) = -\{B, A\}. \tag{3.25}$$

It is easily shown that:

$$\begin{aligned} \frac{\delta\phi(x)}{\delta\phi(y)} &= \frac{\delta\pi(x)}{\delta\pi(y)} = \delta(x-y) \\ \frac{\delta\phi(x)}{\delta\pi(y)} &= \frac{\delta\pi(x)}{\delta\phi(y)} = 0. \end{aligned} \tag{3.26}$$

Using Poisson brackets we can give yet another form to Hamilton's equations:

$$\begin{aligned} \partial_t\phi &= \{\phi, \boldsymbol{H}\} \\ \partial_t\pi &= \{\pi, \boldsymbol{H}\}. \end{aligned} \tag{3.27}$$

The equations (3.26) and (3.27) are the starting point for the canonical quantisation of a field theory.

Poisson brackets are antisymmetric under the exchange of the two terms, equation (3.25), like the *commutator of two matrices*:

$$[\boldsymbol{A}, \boldsymbol{B}] = \mathbf{AB} - \boldsymbol{BA} \tag{3.28}$$

and, similarly to commutators, Poisson brackets satisfy a Jacobi identity:

$$\{A, \{B, C\}\} + \{B, \{C, A\}\} + \{C, \{A, B\}\} = 0. \tag{3.29}$$

The Jacobi identity and equations (3.26) and (3.27) are the starting point of the canonical quantisation of a field theory (see the problems for Section 3.2), to be discussed in Section 4.3.

Comment. The quantum canonical field variables, $\boldsymbol{\phi}(x)$ and $\boldsymbol{\pi}(x)$, are operators (i.e. infinite dimensional matrices) depending upon the space-time coordinates $x = (\boldsymbol{x}, x^0)$. In the limit $\hbar \to 0$, these operators have to go into the classical field variables, $\phi(x)$ and $\pi(x)$, commuting functions of $x$. The limit (at equal time):

$$lim_{\hbar\to 0} \frac{[\phi(\boldsymbol{x}, 0), \pi(\boldsymbol{y}, 0)]}{i\hbar} \tag{3.30}$$

is classically a function which obeys the algebraic properties of the commutator. In view of the antisymmetry of the Poisson bracket and of the Jacobi identity (3.29), one is naturally led to assume that:

$$lim_{\hbar\to 0} \frac{[\phi(\boldsymbol{x}, 0), \pi(\boldsymbol{y}, 0)]}{i\hbar} = \{\phi(\boldsymbol{x}, 0), \pi(\boldsymbol{y}, 0)\} = \delta^{(3)}(\boldsymbol{x} - \boldsymbol{y}). \tag{3.31}$$

Thus, the Heisenberg equal-time relations for the canonical variables $\boldsymbol{q}$ and $\boldsymbol{p}$, translated into field theory:

$$[\phi(\boldsymbol{x},0),\pi(\boldsymbol{y},0)] = i\hbar\delta^{(3)}(\boldsymbol{x}-\boldsymbol{y}) \tag{3.32}$$

guarantee that quantum field theory goes into classical field theory for vanishing Planck's constant.

## 3.3 TRANSFORMATION OF FIELDS

To construct relativistically invariant Lagrangians, we must begin from the rules of field transformations, the relations which connect the fields observed in IF $O$, $\phi(x)$, to the values observed in $O'$, $\phi'(x')$, which correspond to the same event in space-time, described by the coordinates $x^\mu$ and $x'^\mu$ in $O$ and in $O'$:

$$x'^\mu = \Lambda^\mu_\nu\, x^\nu. \tag{3.33}$$

The simplest case is that of a *scalar field* in which the values are the same:

$$\phi'(x') = \phi(x). \tag{3.34}$$

The derivatives of $\phi$ transform like *covariant* vectors, Section 1.4:

$$\frac{\partial\phi'(x')}{\partial x'^\mu} = \frac{\partial\phi(x)}{\partial x'^\mu} = \frac{\partial x^\lambda}{\partial x'^\mu}\frac{\partial\phi(x)}{\partial x^\lambda} = (\Lambda^{-1})^\lambda_\mu\,\frac{\partial\phi}{\partial x^\lambda}. \tag{3.35}$$

It is immediately verified that, instead, the derivatives with respect to $x_\mu$ transform like *contravariant* vectors:

$$\frac{\partial\phi'(x')}{\partial x'_\mu} = \Lambda^\mu_\lambda\,\frac{\partial\phi}{\partial x_\lambda}. \tag{3.36}$$

Consequently, we denote the derivatives as:

$$\frac{\partial\phi}{\partial x^\mu} = \phi_\mu = \partial_\mu\phi;\quad \frac{\partial\phi}{\partial x_\mu} = \phi^\mu = \partial^\mu\phi. \tag{3.37}$$

With further derivatives, multi-index tensors can be constructed, with the corresponding transformation properties:

$$\phi^{\nu_1,\nu_2,\dots}_{\mu_1,\mu_2,\dots}(x) = \frac{\partial}{\partial x^{\mu_1}}\frac{\partial}{\partial x^{\mu_2}}\cdots\frac{\partial}{\partial x_{\nu_1}}\frac{\partial}{\partial x_{\nu_2}}\dots\phi(x); \tag{3.38}$$

$$(\phi')^{\nu_1,\nu_2,\dots}_{\mu_1,\mu_2,\dots}(x') = (\Lambda^{-1})^{\lambda_1}_{\mu_1}(\Lambda^{-1})^{\lambda_2}_{\mu_2}\dots\Lambda^{\nu_1}_{\rho_1}\Lambda^{\nu_2}_{\rho_2}\dots\phi^{\rho_1,\rho_2,\dots}_{\lambda_1,\lambda_2,\dots}(x).$$

Extending the case of the scalar field, we can define *tensor* fields, functions of $x^\mu$ provided with a certain number of upper (covariant) indices, $n_s$, and lower (contravariant) $n_g$, $F^{\nu_1,\nu_2,\dots}_{\mu_1,\mu_2,\dots}(x)$ whose transformation, rules are the same as (3.38):

$$(F')^{\nu_1,\nu_2,\dots}_{\mu_1,\mu_2,\dots}(x') = (\Lambda^{-1})^{\lambda_1}_{\mu_1}(\Lambda^{-1})^{\lambda_2}_{\mu_2}\dots\Lambda^{\nu_1}_{\rho_1}\Lambda^{\nu_2}_{\rho_2}\dots F^{\rho_1,\rho_2,\dots}_{\lambda_1,\lambda_2,\dots}(x). \tag{3.39}$$

Important examples are the antisymmetric Maxwell tensor, $F^{\mu\nu}(x) = -F^{\nu\mu}(x)$, which describes the electromagnetic field, and the vector field, $A^{\mu}(x)$, which describes the vector potential.

The rank of a tensor (the number of covariant and contravariant indices) can be reduced by contracting the indices with invariant tensors (i.e. tensors such that $T' = T$). For the Lorentz transformations there are three types of invariant operation:

- contraction of a covariant and a contravariant index with the Kronecker delta: $\delta^{\mu}_{\nu}$,
- contraction of two covariant indices with the tensor $g_{\mu\nu}$ (or two contravariant indices with $g^{\mu\nu}$),
- contraction with the completely antisymmetric Levi–Civita tensor $\epsilon_{\mu\nu\rho\sigma}$ defined as:

$$\begin{aligned} \epsilon_{0123} &= +1 \\ \epsilon_{\mu_1\mu_2\mu_3\mu_4} &= 0 \quad \text{(two equal indices)} \\ \epsilon_{\mu_1\mu_2\mu_3\mu_4} &= \pm 1 : \ \mu_1, \mu_2, \mu_3, \mu_4 = \ \text{even/odd permutations of 0123.} \end{aligned} \tag{3.40}$$

To show that the Levi–Civita tensor is invariant, we consider:

$$X_{0123} = \epsilon_{\mu_1\mu_2\mu_3\mu_4} \Lambda_0^{\mu_1} \Lambda_1^{\mu_2} \Lambda_2^{\mu_3} \Lambda_3^{\mu_4}. \tag{3.41}$$

$X$ is the sum of products of the matrix elements of $\Lambda$ from rows and columns all different to each other, multiplied by the sign of the permutation which transforms $(0, 1, 2, 3)$ into $(\mu_1, \mu_2, \mu_3, \mu_4)$. Therefore $X = det\Lambda = 1$. Furthermore, $X_{\mu\nu\rho\sigma} = 0$ if the two indices are equal, $X_{\mu\nu\rho\sigma} = \pm 1$ according to whether the permutation is even or odd. It follows that:

$$\epsilon_{\mu_1\mu_2\mu_3\mu_4} \Lambda^{\mu_1}_{\rho_1} \Lambda^{\mu_2}_{\rho_2} \Lambda^{\mu_3}_{\rho_3} \Lambda^{\mu_4}_{\rho_4} = \epsilon_{\rho_1\rho_2\rho_3\rho_4} \tag{3.42}$$

and thus the required invariance.

The tensors of a given rank $n_s, n_g$ describe a linear manifold. A tensor is said to be *reducible* if this manifold contains subspaces invariant under the transformations (3.39) which should be non-trivial (that is, different from 0 and from the same manifold). Otherwise, the tensor is irreducible.

Possible invariant subspaces can be obtained by projecting the general tensor using the invariant operations described earlier. For an irreducible tensor, on the other hand, the invariant operations give zero or project onto all the starting space.

By completely contracting the indices of products of tensor fields and their derivatives, one obtains tensors of rank zero (with no free indices) which are invariant (they transform like the scalar field in (3.34)). These invariant combinations are the building blocks with which to construct the *Lagrangian density*, which describes the field dynamics.

*Example* 1. An important case is that of tensors with two covariant antisymmetric indices. Obviously, these tensors have $4 \cdot 3/2 = 6$ independent components, which can be organised into 3-vectors in the following way:

$$F^{\mu\nu} = -F^{\nu\mu} = \begin{pmatrix} 0 & B^3 & -B^2 & E^1 \\ -B^3 & 0 & B^1 & E^2 \\ B^2 & -B^1 & 0 & E^3 \\ -E^1 & -E^2 & -E^3 & 0 \end{pmatrix}. \tag{3.43}$$

The application of the Levi–Civita tensor takes us to the definition of the *dual tensor*, $\bar{F}^{\mu\nu}$:

$$\bar{F}^{\mu\nu} = g^{\mu\mu_1} g^{\nu\nu_1} \frac{1}{2} \epsilon_{\mu_1\nu_1\lambda\rho} F^{\lambda\rho} = \frac{1}{2} \epsilon^{\mu\nu}{}_{\lambda\rho} F^{\lambda\rho}. \tag{3.44}$$

It is not difficult to see that $\bar{F}$ is obtained from $F$ with the substitutions $\boldsymbol{E} \to -\boldsymbol{B}$; $\boldsymbol{B} \to \boldsymbol{E}$:

$$\bar{F}^{\mu\nu} = \begin{pmatrix} 0 & E^3 & -E^2 & -B^1 \\ -E^3 & 0 & E^1 & -B^2 \\ E^2 & -E^1 & 0 & -B^3 \\ B^1 & B^2 & B^3 & 0 \end{pmatrix}. \tag{3.45}$$

The application of the Levi–Civita tensor transforms the space of these tensors into itself:

$$\frac{1}{2} \epsilon^{\mu\nu}{}_{\lambda\rho} F^{\lambda\rho} = \bar{F}^{\mu\nu}; \qquad \frac{1}{2} \epsilon^{\mu\nu}{}_{\lambda\rho} \bar{F}^{\lambda\rho} = -F^{\mu\nu} \tag{3.46}$$

(where we have used the relation $\epsilon^{0123} = -1$). Starting from these equations, we can define two irreducible components which correspond to the eigenvalues $\pm i$ of the duality transformation (3.46):

$$\begin{aligned} (X^{\pm})^{\mu\nu} &= F^{\mu\nu} \pm i\bar{F}^{\mu\nu} \\ \frac{1}{2} \epsilon^{\mu\nu}{}_{\lambda\rho} (X^{\pm})^{\lambda\rho} &= \mp i (X^{\pm})^{\mu\nu}. \end{aligned} \tag{3.47}$$

Because we have used only invariant operations, the factorisation (3.47) is invariant under Lorentz transformations and the space of antisymmetric tensors with two indices factorise into two invariant subspaces, each of three dimensions.

These two subspaces form a unique complex if we add the parity operation (cf. Chapter 1). Under parity, the vectors $\boldsymbol{E}$ and $\boldsymbol{B}$ behave, respectively, like a *polar* vector and an *axial* vector:

$$P : \boldsymbol{E}(\boldsymbol{x}, t) \to -\boldsymbol{E}(-x, t); \qquad \boldsymbol{B}(\boldsymbol{x}, t) \to +\boldsymbol{B}(-x, t) \tag{3.48}$$

and therefore:

$$P : X^{\pm}(\boldsymbol{x}, t) = f(\boldsymbol{E} \mp i\boldsymbol{B}) \to f(-\boldsymbol{E} \mp i\boldsymbol{B}) = -f(\boldsymbol{E} \pm i\boldsymbol{B}) = -X^{\mp}(-\boldsymbol{x}, t). \tag{3.49}$$

The same conclusion is reached by considering the two *quadratic invariants* which can be constructed starting from $F$ and $\bar{F}$:

$$\begin{aligned} \mathcal{L}_1 &= \frac{1}{2}\, F^{\mu\nu} F_{\mu\nu} = (\boldsymbol{E})^2 - (\boldsymbol{B})^2 \\ \mathcal{L}_2 &= \frac{1}{2}\, \bar{F}^{\mu\nu} F_{\mu\nu} = (\boldsymbol{E} \cdot \boldsymbol{B}). \end{aligned} \tag{3.50}$$

The form of (3.50) suggests to consider the complex vectors:

$$\begin{aligned} (\boldsymbol{Z})^{\pm} &= \boldsymbol{E} \mp i\boldsymbol{B}; \\ (\boldsymbol{Z})^{\pm} \cdot (\boldsymbol{Z})^{\pm} &= \mathcal{L}_1 \mp 2\mathcal{L}_2. \end{aligned} \tag{3.51}$$

The squares of the two 3-vectors are separately conserved by Lorentz transformations, which therefore transforms their components among each other. The parity transformation obviously exchanges $Z^+$ with $-Z^-$.

*Example* 2. Tensors with two symmetric indices, both covariant (or contravariant): $T^{\mu\nu}$ (or $T_{\mu\nu}$). In this case, the projection with $g_{\mu\nu}$ gives an invariant and the space factorises into the space of tensors *symmetric and with zero trace*, $g_{\mu\nu}\, T^{\mu\nu} = 0$, of nine dimensions, and into a unidimensional space of tensors of the form $g^{\mu\nu}\, T$.

Comment. The classification of irreducible tensors is discussed in [11]. From an algebraic point of view, it is found that the Lorentz group is equivalent to the product of two groups of rotations: $L_+^{\uparrow} = SU(2) \otimes SU(2)$. Therefore the irreducible tensors are characterised by *two angular momenta:* $j_1$ *and* $j_2$. The tensors which belong to the irreducible representation $(j_1, j_2)$ have dimensions $d = (2j_1 + 1)(2j_2 + 1)$.

The (contravariant or covariant) 4-vectors correspond to the representation $(1/2, 1/2)$ of $SU(2) \otimes SU(2)$. The multi-index tensors described earlier are obtained from tensor products of this representation. For example, the tensors with two contravariant indices $T_{\mu,\nu}$ are generated by the product:

$$(1/2, 1/2) \otimes (1/2, 1/2) = (1{+}0, 1{+}0) = [(1, 1) \oplus (0, 0)] \oplus [(1, 0) \oplus (0, 1)]. \tag{3.52}$$

We have placed the brackets to separate the symmetric tensors (those without trace plus the trace) from the asymmetric ones. We note the simplicity with which the rule for the combination of angular momenta reproduces the factorisation into irreducible components, in particular their dimensionality.

## 3.4 CONTINUOUS SYMMETRIES

In the previous section, we made reference to two observers who study the same system of events (for example a certain physical field configuration) viewed from two different inertial frames.

The principle of relativity requires that the two IFs should be completely equivalent, so that each observer is independently able to describe the field dynamics with an action and a Lagrangian density which have the same functional dependence on the fields. Relativistic invariance then requires:

$$S(\phi') = \int d^4x' \ \mathcal{L}[\phi'(x'), \phi'_\mu(x'), x'] = S(\phi) = \int d^4x \ \mathcal{L}[\phi(x), \phi_\mu(x), x]. \quad (3.53)$$

if $\phi(x)$, $x$ and $\phi'(x')$, $x'$ are the components of the field and the coordinates associated with an event in the two IFs, equations (3.33) and (3.39).

We can consider equation (3.53) from another viewpoint, as the invariance (or symmetry) of the action for the transformation which substitutes $\phi(x)$, $x$ with $\phi'(x')$, $x'$ in the same frame of reference:

$$\begin{aligned} &x \to x'; \\ &\phi(x) \to \phi'(x'). \end{aligned} \quad (3.54)$$

Viewed in this way, the relativity principle simply expresses the symmetry of the action under the symmetry group of (proper and orthochronous) Lorentz transformations and we can study the consequences of this symmetry at the same time as other possible symmetries of the action which can take the general form (3.54), with different realisations of the transformation rules.

Regarding coordinates, we are restricted to the addition of *translations from the origin of space-time.* These transformations, together with the proper and orthochronous Lorentz transformations, form the *Poincaré group* which is the group of natural space-time symmetries in special relativity.

Also included within (3.54) are transformations which change the fields but not the coordinates, $x' = x$, which are referred to as *internal symmetries*, for example phase transformations of the complex fields which we will consider extensively in what follows.

In this section, we limit the discussion to transformations which belong to *continuous groups*. In this case, infinitesimal transformations close to the identity transformation can be defined. By means of products of infinitesimal transformations we can arrive at all the transformations of the group (at least as far as the component connected to the identity is concerned). Therefore we can explore the consequences of symmetry under a continuous group restricting ourselves to infinitesimal transformations.

Thus we consider the transformation defined by infinitesimal variations:

$$\begin{aligned} &x'^\mu = x^\mu + \delta x^\mu; \quad (3.55) \\ &\phi'(x') = \phi(x) + \delta_T\phi. \end{aligned}$$

For simplicity, we have omitted in (3.55) the indices connected to the field, $\phi$, which are implied. We denote $\delta_T\phi$ as the *total variation of the field.* We can decompose $\delta_T\phi$ in the following way:

$$\begin{aligned}\delta_T\phi &= \phi'(x') \; - \; \phi(x) = \phi'(x') \; - \; \phi(x') \; + \; \phi(x') \; - \; \phi(x) \\ &= \delta\phi(x) \; + \; \partial_\mu\phi(x)\delta x^\mu.\end{aligned} \tag{3.56}$$

The total variation is the sum of the *functional variation*, $\delta\phi$, and of a translation by $\delta x^\mu$.

The Poincaré group transformations and, a fortiori, the transformations associated with internal symmetries leave the size of $d^4x$ invariant. This requires that:

$$d^4x' = ||\frac{\partial x'}{\partial x}||d^4x = \det(\delta^\lambda_\mu + \frac{\partial\delta x^\lambda}{\partial x^\mu})d^4x = d^4x. \tag{3.57}$$

If we use the identity:

$$\det(1+\epsilon) = 1 + \text{Trace}(\epsilon), \tag{3.58}$$

valid at least to terms of higher order in the infinitesimal matrix $\epsilon$, the invariance condition on the volume of integration takes the form:

$$\begin{aligned}&1 = 1 + \partial_\mu\delta x^\mu; \qquad \text{or} \\ &\partial_\mu\delta x^\mu = 0\end{aligned} \tag{3.59}$$

Under these conditions, the invariance of the action, equation (3.53) simply requires the invariance of the Lagrangian density:

$$\delta_T\mathcal{L} = \mathcal{L}(\phi', \partial\phi', x') - \mathcal{L}(\phi, \partial\phi, x) = 0. \tag{3.60}$$

In the case of Lorentz transformations, the requirement is satisfied if we construct the Lagrangian as a polynomial of the fields and their derivatives, saturating the indices in an invariant way, as discussed in the preceding section.

## 3.5 NOETHER'S THEOREM

We expand (3.60) using (3.56). We find:

$$\begin{aligned}0 &= \delta\mathcal{L} + \frac{\partial\mathcal{L}}{\partial x^\mu}\delta x^\mu = \frac{\partial\mathcal{L}}{\partial\phi}\delta\phi + \frac{\partial\mathcal{L}}{\partial\,\partial_\mu\phi}\delta(\partial_\mu\phi) + \frac{\partial\mathcal{L}}{\partial x^\mu}\delta x^\mu = \\ &= [\frac{\partial\mathcal{L}}{\partial\phi} - \partial_\mu(\frac{\partial\mathcal{L}}{\partial\,\partial_\mu\phi})]\delta\phi + \partial_\mu[(\frac{\partial\mathcal{L}}{\partial\,\partial_\mu\phi})\delta\phi] + \frac{\partial\mathcal{L}}{\partial x^\mu}\delta x^\mu\end{aligned} \tag{3.61}$$

where:

$$\frac{\partial\mathcal{L}}{\partial x^\mu} = \frac{\partial}{\partial x^\mu}\mathcal{L}[\phi(x), \phi_\mu(x), x]. \tag{3.62}$$

Using the equations of motion and (3.59), we obtain the conservation equation:

$$\partial_\mu[(\frac{\partial \mathcal{L}}{\partial\, \partial_\mu \phi})\delta\phi + \mathcal{L}\delta x^\mu] = 0. \tag{3.63}$$

We can express the infinitesimal variations of the fields and coordinates as linear combinations of the infinitesimal parameters which characterise the transformation:

$$\delta x^\mu = \sum_A \epsilon_A\, (\Delta^A)^\mu(x);\;\; \delta\phi = \sum_A \epsilon_A\, (\Sigma^A)\phi \tag{3.64}$$

where $\Delta^A$ and $\Sigma^A$ are matrices which represent the generators of the infinitesimal transformations of the coordinates and the fields. Because (3.63) must be satisfied for any arbitrary values of the infinitesimal parameters, we obtain the conservation equations:

$$\partial_\mu(J^A)^\mu = \partial_\mu[(\frac{\partial \mathcal{L}}{\partial\, \partial_\mu \phi})\Sigma^A\phi + (\Delta^A)^\mu\, \mathcal{L}] \tag{3.65}$$

for the currents associated with each infinitesimal generator. The result (3.65) is Noether's theorem:

- *Each infinitesimal generator of a continuous symmetry is associated with a conserved current.*

The conserved current is determined by the Lagrangian density, according to the canonical formula:

$$(J^A)^\mu = (\frac{\partial \mathcal{L}}{\partial\, \partial_\mu \phi})\Sigma^A\phi + (\Delta^A)^\mu\, \mathcal{L}. \tag{3.66}$$

The conserved current determines an *additive constant of the motion*, represented by the integral of its time component overall space. We write:

$$(J^A)^\mu = ((J^A)^0, \frac{1}{c}\boldsymbol{J}^A) \tag{3.67}$$

and integrate the conservation equation into a fixed volume ($x^0 = ct$):

$$\int_V d^3x[\frac{\partial}{\partial t}(J^A)^0 + \boldsymbol{\nabla}\cdot \boldsymbol{J}^A] = \frac{d}{dt}Q^A + \int_\Sigma d\sigma\, \boldsymbol{n}\cdot \boldsymbol{J}^A = 0 \tag{3.68}$$

where $Q^A = \int_V d^3x\, (J^A)^0$. Equation (3.68) expresses the fact that a variation of the charge contained inside the volume $V$ is balanced by a corresponding current density flux through the surface $\Sigma$ of $V$. If we extend the integration to the whole of space, the surface integral vanishes, given the cancellation of the fields at infinity, and we obtain the law of conservation of the *total charge*:

$$\frac{d}{dt}Q^A = 0. \tag{3.69}$$

If the symmetry transformations involve space-time, the index $A$ includes one or more vector indices and (3.66) behaves like a tensor of higher rank.

For internal symmetries, in which the transformations of the symmetry group do not involve space-time, the total charge is a *Lorentz invariant.* To confirm this property, we consider two hypersurfaces: $\Gamma_0$, corresponding to $t = 0$ in our frame of reference, and $\Gamma_1$, corresponding to $t' = constant$ in a Lorentz transformed the system. We integrate the conservation equations in the four-dimensional volume bounded by these two hypersurfaces. We obtain:

$$0 = \int \partial_\mu J^\mu \, d^4x = -\int_{\Gamma_0} n_\mu J^\mu d\Gamma_0 + \int_{\Gamma_1} n'_\mu J^\mu d\Gamma_1 + I \tag{3.70}$$

where $n^\mu$ and $n'^\mu$ are the normals to the two hypersurfaces, which point in the time direction of the two reference frames, and $I$ represents the contribution of the lateral surfaces of our 4-volume. If we let these surfaces tend to infinity, $I \to 0$, and we then have:

$$\int_{\Gamma_0} n_\mu J^\mu d\Gamma_0 = \int_{\Gamma_1} n'_\mu J^\mu d\Gamma_1; \tag{3.71}$$

$$\text{i.e.} \quad \int d^3x \; J^0(\boldsymbol{x}, t) = \int d^3x' \; J^0(\boldsymbol{x}', t').$$

Noether's theorem establishes the existence of a number of conserved currents, but it does not uniquely determine their form. We can add another 4-vector to the current in (3.66) provided that it is conserved, as a result of the equations of motion or for algebraic reasons. An example of the second case is given by the 4-divergence of an antisymmetric tensor:

$$s^\mu = \partial_\lambda \, T^{\lambda\mu}. \tag{3.72}$$

If $T^{\lambda\mu} = -T^{\mu\lambda}$ the current $s^\mu$ is trivially conserved, $\partial_\mu \partial_\lambda T^{\mu\lambda} = 0$ because of the antisymmetry of $T$. In this case, the addition of $s^\mu$ modifies the current but not the conserved charge, since:

$$\int d^3x \; s^0 = \int d^3x \; \partial_i \cdot T^{i0} = 0 \tag{3.73}$$

if the fields vanish at infinity.

## 3.6 ENERGY–MOMENTUM TENSOR

We consider explicitly the Poincaré group transformations, consisting of translations in space-time and the special Lorentz transformations.

1. *Translations in space-time.* These are the coordinate transformations:

$$x'^\mu = x^\mu + a^\mu \tag{3.74}$$

with $a^\mu$ = constant. For the fields, we set:

$$\phi'(x') = \phi(x) \tag{3.75}$$

(possible tensor indices of $\phi$ are not affected by the transformation).

Invariance under translations requires:

$$\mathcal{L}(\phi', (\partial\phi)', x') = \mathcal{L}(\phi, (\partial\phi), x) \tag{3.76}$$

or, using (3.74) and (3.75):

$$\mathcal{L}(\phi, (\partial\phi), x + a) = \mathcal{L}(\phi, (\partial\phi), x). \tag{3.77}$$

Invariance under translations thus requires that $\mathcal{L}$ does not depend explicitly on $x$.

From (3.75), we derive the form of the functional variation of the fields:

$$\begin{aligned} &\delta_T\phi = 0 = \delta\phi + (\partial_\mu\phi)a^\mu; \\ &\delta\phi = -(\partial_\mu\phi)a^\mu. \end{aligned} \tag{3.78}$$

The conserved current derived from (3.66) and (3.78) is a tensor of rank two which is called the *canonical energy-momentum tensor*

$$\begin{aligned} &T^{\mu,\nu} = \frac{\partial\mathcal{L}}{\partial\partial_\mu\phi}\partial^\nu\phi \; - \; g^{\mu\nu}\mathcal{L}; \\ &\partial_\mu \, T^{\mu,\nu} = 0. \end{aligned} \tag{3.79}$$

In fact, the 00 component is just the Hamiltonian density, defined in Section 3.2.

$$T^{0,0} = \frac{\partial\mathcal{L}}{\partial\partial_t\phi}\partial_t\phi \; - \mathcal{L} = \pi \; \partial_t\phi \; - \; \mathcal{L} \tag{3.80}$$

Correspondingly, the spatial integral of $T^{0,0}$ is the Hamiltonian of the system, the first integral associated with the time independence of the system (that is, invariance with respect to time translations: $ct \to ct + a^0$):

$$\begin{aligned} &\boldsymbol{H} = \int d^3x \; T^{0,0} = E \\ &\frac{d}{dt}E = 0. \end{aligned} \tag{3.81}$$

The energy is the time component of a 4-vector, whose spatial components are the momentum. We can thus identify:

$$\int d^3x \; T^{0,\mu} = \boldsymbol{P}^\mu \tag{3.82}$$

with the overall 4-momentum of the field.

Noether's theorem guarantees that the components of $P^\mu$ are conserved for systems independent of position in space-time. Thus we find the important result according to which *conservation of energy and momentum are consequences of invariance under space-time translations*. For an isolated system, invariance under translations is associated with the homogeneity of space-time; the conservation of 4-momentum for these systems provides a concrete proof of this important physical fact.

**2**. *Lorentz transformations*. These are associated with coordinate transformations:

$$x'^\mu = \Lambda^\mu_\nu \, x^\nu. \tag{3.83}$$

For infinitesimal transformations, we can set:

$$\Lambda^\mu_\nu = \delta^\mu_\nu + \epsilon^\mu_\nu. \tag{3.84}$$

The infinitesimal parameters $\epsilon^\mu_\nu$ are not independent, because the matrices $\Lambda$ must be such that they leave the metric tensor invariant:

$$\begin{aligned} &\Lambda^T g \Lambda = g; \; i.e. \\ &\epsilon^\lambda_\mu \, g_{\lambda\nu} \; + \; g_{\mu\lambda} \, \epsilon^\lambda_\nu = \epsilon_{\mu\nu} + \epsilon_{\nu\mu} = 0, \end{aligned} \tag{3.85}$$

where we have kept the terms to first order in $\epsilon$ and we have defined a new infinitesimal tensor, $\epsilon_{\mu\nu}$, with two contravariant indices. Condition (3.85) states that this tensor should be antisymmetric in the two indices. Equation (3.83) is rewritten as:

$$\begin{aligned} x'^\mu &= x^\mu + g^{\mu\alpha} x^\beta \epsilon_{\alpha\beta}; \\ \delta x^\mu &= \frac{1}{2}\epsilon_{\alpha\beta}(g^{\mu\alpha}x^\beta - g^{\mu\beta}x^\alpha). \end{aligned} \tag{3.86}$$

Infinitesimal Lorentz transformations therefore depend on six parameters: three for spatial rotations ($\epsilon_{ij} = -\epsilon_{ji}$; $i \neq j = 1, 2, 3$) and three for special Lorentz transformations ($\epsilon_{0i} = -\epsilon_{i0}$; $i = 1, 2, 3$).

For general tensor fields, the transformations are those given in (3.39). We will use here a more compact notation which can be applied also to the more general case of spinor fields. We define an index which runs over all the components independent of the field, which we denote as $M$, $N$, etc. The infinitesimal transformations associated with the parameters $\epsilon_{\mu\nu}$ ($\mu, \nu = 0, 1, 2, 3$) are transformations of the fields $\phi_M$ given by the linear combination of $\epsilon_{\mu\nu}$ with six matrices $\Sigma^{\mu\nu}_{MN}$, antisymmetric in $\mu\nu$, associated with the generators of the Lorentz transformations of the same fields:

$$\phi'_M(x') = (\delta_{MN} + \frac{1}{2}\epsilon_{\mu\nu}\Sigma^{\mu\nu}_{MN})\phi_N(x). \tag{3.87}$$

(The factor $\frac{1}{2}$ is conventional and the sum over repeated indices is understood in all cases).

The functional variation of the fields is obtained from $\delta_T\phi$:

$$\delta_T\phi_M = \frac{1}{2}\epsilon_{\mu\nu}\Sigma^{\mu\nu}_{MN}\phi_N(x) = \delta\phi_M + \partial_\mu\phi_M\,\delta x^\mu;\ i.e. \tag{3.88}$$
$$\delta\phi_M = \frac{1}{2}[\Sigma^{\alpha\beta}_{MN}\phi_N(x) - \partial_\mu\phi_M\,(g^{\mu\alpha}x^\beta - g^{\mu\beta}x^\alpha)]\epsilon_{\alpha\beta}.$$

With these results, we can write the conserved infinitesimal current as:

$$\begin{aligned}\delta M^\mu &= \frac{\partial\mathcal{L}}{\partial\partial_\mu\phi_M}\,\delta\phi_M + \delta x^\mu\,\mathcal{L} = \\ &= \frac{1}{2}\epsilon_{\alpha\beta}[-\frac{\partial\mathcal{L}}{\partial\partial_\mu\phi_M}\partial_\lambda\phi_M(g^{\lambda\alpha}x^\beta - g^{\lambda\beta}x^\alpha) \\ &\quad + (g^{\mu\alpha}x^\beta - g^{\mu\beta}x^\alpha)\mathcal{L} + \frac{\partial\mathcal{L}}{\partial\partial_\mu\phi_M}\Sigma^{\alpha\beta}_{MN}\,\phi_N] = \\ &= \frac{1}{2}\epsilon_{\alpha\beta}[(x^\alpha\,T^{\mu,\beta} - x^\beta\,T^{\mu,\alpha}) + \frac{\partial\mathcal{L}}{\partial\partial_\mu\phi_M}\Sigma^{\alpha\beta}_{MN}\,\phi_N].\end{aligned} \tag{3.89}$$

This result takes us to the definition of the *canonical angular momentum tensor*:

$$\begin{aligned}M^{\mu[\alpha\beta]} &= (x^\alpha\,T^{\mu,\beta} - x^\beta\,T^{\mu,\alpha}) + \Sigma^{\mu[\alpha\beta]}; \\ \Sigma^{\mu[\alpha\beta]} &= \frac{\partial\mathcal{L}}{\partial\partial_\mu\phi_M}\Sigma^{\alpha\beta}_{MN}\,\phi_N; \\ \partial_\mu M^{\mu[\alpha\beta]} &= 0.\end{aligned} \tag{3.90}$$

The components associated with spatial rotations ($\alpha\beta = ij$) give as conserved charge the total angular momentum of the field, for example:

$$\boldsymbol{J}^1 = \int d^3x\,M^{0[23]} = \int d^3x[(x^2T^{0,3} - x^3T^{0,2}) + \Sigma^{0[23]}]. \tag{3.91}$$

Comment. This might suggest to identify the first term in (3.91) as due to the orbital angular momentum and the second to the intrinsic angular momentum of the field (which in quantum mechanics would correspond to the spin of the field quanta). In fact, in the case of the scalar field, the second term is absent. However, the two terms individually are ambiguous, as we will see in the following section. For example, we can redefine the energy-momentum tensor so as to eliminate completely the second term. The only quantities uniquely defined are the constants of the motion, $\boldsymbol{J}^k$.

The Symmetric Energy-Momentum Tensor. In the theory of general relativity, the Einstein equations, which connect the energy-momentum tensor to the geometry of space-time, requires that the energy-momentum tensor should be symmetric in the two indices. In general, the canonical energy-momentum tensor defined by (3.79) is not symmetric. However, making use of the ambiguity inherent in its definition, it is possible to construct a new energy-momentum tensor, $\theta^{\mu\nu}$, which is symmetric and conserved at the same time. The general construction is due to Belinfante and Rosenfeld [1].

The starting point for the construction of $\theta^{\mu\nu}$ is the conservation equation for $M^{\mu[\alpha\beta]}$. Using (3.90), we find:

$$0 = \partial_\mu M^{\mu[\alpha\beta]} = T^{\alpha,\beta} - T^{\beta,\alpha} + \partial_\mu \Sigma^{\mu[\alpha\beta]}. \tag{3.92}$$

The antisymmetric part of $T^{\mu,\nu}$ can therefore be eliminated in favour of the 4-divergence of $\Sigma$ and we can define the *symmetric part of* $T^{\mu,\nu}$ according to:

$$S^{\mu\nu} = T^{\mu,\nu} + \frac{1}{2}\partial_\lambda \Sigma^{\lambda[\mu\nu]} = S^{\nu\mu} \tag{3.93}$$

This tensor is not yet the solution to the problem, since $S^{\mu\nu}$ is not conserved:

$$\partial_\mu S^{\mu\nu} = \frac{1}{2}\partial_\lambda \partial_\mu \Sigma^{\lambda[\mu\nu]}. \tag{3.94}$$

However, this result coincides with the 4-divergence of the symmetric tensor:

$$R^{(\mu\nu)} = \frac{1}{2}[\partial_\lambda \Sigma^{\mu[\lambda\nu]} + \partial_\lambda \Sigma^{\nu[\lambda\mu]}]; \tag{3.95}$$
$$\partial_\mu R^{(\mu\nu)} = \frac{1}{2}\partial_\lambda \partial_\mu \Sigma^{\mu[\lambda\nu]}$$

Therefore we define:

$$\theta^{\mu\nu} = T^{\mu,\nu} + \frac{1}{2}\partial_\lambda [\Sigma^{\lambda[\mu\nu]} - \Sigma^{\mu[\lambda\nu]} - \Sigma^{\nu[\lambda\mu]}]. \tag{3.96}$$

The new tensor is symmetric and conserved. We can calculate the corresponding charge:

$$\int d^3x \; \theta^{0\nu} = \int d^3x [T^{0,\nu} + \frac{1}{2}\partial_\lambda (\Sigma^{\lambda[0\nu]} - \Sigma^{0[\lambda\nu]} - \Sigma^{\nu[\lambda 0]})]. \tag{3.97}$$

In the terms with derivatives, we should keep only those with $\lambda = 0$; the terms with spatial derivatives correspond to surface terms which vanish at infinity. We obtain:

$$\int d^3x \; \theta^{0\nu} = \int d^3x [T^{0,\nu} + \frac{1}{2}(\partial_0 (\Sigma^{0[0\nu]} - \Sigma^{0[0\nu]} - \Sigma^{\nu[00]})] = \int d^3x \; T^{0,\nu}. \tag{3.98}$$

Therefore the tensor $\theta^{\mu\nu}$ is a perfectly legitimate substitute for the canonical energy-momentum tensor.

An important consequence of what we have just seen is that we can construct a new momentum tensor based on $\theta^{\mu\nu}$:

$$\widetilde{M}^{\mu[\alpha\beta]} = x^\alpha \theta^{\mu\beta} - x^\beta \theta^{\mu\alpha}. \tag{3.99}$$

This new tensor is conserved, because $\theta$ is symmetric and conserved, and represents a legitimate substitute for the canonical tensor. We note that, apparently, the angular momentum is now only "orbital", proof that the separation between orbital and spin angular momenta has no physical significance in a relativistic theory.

With the new momentum tensor, we can analyse the constants of the motion associated with the special Lorentz transformations. We find:

$$K^i = \int d^3x \widetilde{M}^{0[0i]} = \int d^3x (ct\theta^{0i} - x^i \theta^{00}) = constant. \tag{3.100}$$

That $K^i$ is a constant simply expresses the fact that the barycentre of the energy, for an isolated system, moves with uniform rectilinear motion.

$$< x^i >= \frac{\int d^3x \, x^i \theta^{00}}{\int d^3x \, \theta^{00}} \tag{3.101}$$

$$< x^i >= ct \cdot \frac{\int d^3x \, \theta^{0i}}{\int d^3x \, \theta^{00}} + constant.$$

## 3.7 PROBLEMS FOR CHAPTER 3

### Sect. 3.1

**1.** Given the complex field $\phi$ and the Lagrangian density

$$\mathcal{L} = \partial_\mu \phi \partial^\mu \phi^\star - m^2(\phi\phi^\star) - \lambda(\phi\phi^\star)^2$$

determine the equations of motions for $\phi$ and $\phi^\star$.

**2.** In the same example:

- derive the canonical energy-momentum tensor $T^{\mu\nu}$, and verify that $T^{00}$ coincides with the Hamiltonian density, $\mathcal{H}$;
- obtain the Noether current associated with the invariance of the Lagrangian under global phase transformations: $\phi \to e^{i\alpha}\phi$.

**3.** Consider two real scalar fields $\phi_1$ and $\phi_2$ and the Lagrangian density

$$\mathcal{L}(\Phi, \partial_\mu \Phi) = \frac{1}{2}(\partial_\mu \Phi)^T (\partial^\mu \Phi) - \frac{1}{2} m^2 \Phi^\dagger \Phi - \frac{1}{4}\lambda(\Phi^\dagger \Phi)^2 \,, \; \Phi = \begin{pmatrix} \phi_1 \\ \phi_2 \end{pmatrix}$$

and consider the transformations $\Phi \to \Phi' = U\Phi$, induced by the matrix

$$U = \begin{pmatrix} \cos\alpha & -\sin\alpha \\ \sin\alpha & \cos\alpha \end{pmatrix}.$$

Write down the infinitesimal form of the transformation matrix; show that $\mathcal{L}$ is invariant under infinitesimal transformations; derive the corresponding Noether current.

4. Write the complex scalar field in terms of its real and imaginary components: $\phi = \phi_1 + i\phi_2$. Prove that the theory embodied by the Lagrangian of Problem **1** is identical to the one of Problem **3**.

5. Given the Lagrangian density

$$\mathcal{L} = \frac{1}{2}\partial_\mu\phi\partial^\mu\phi - \frac{1}{2}m^2\phi^2 - \frac{\lambda}{3!}\phi^3$$

   - derive the equations of motions for the field $\phi$;
   - derive the expression of the canonical energy-momentum tensor $T^{\mu\nu}$ applying Noether's theorem in the case of translation invariance;
   - obtain the Hamiltonian density $\mathcal{H}$, and verify that $\mathcal{H} = T^{00}$.

   Based on the analysis of the energy density, discuss the consistency of the theory.

6. Scale invariance. Consider the action associated with the Lagrangian density describing a *massless* real scalar field

$$S = \frac{1}{2}\int d^4x \; g^{\mu\nu} \; \partial_\mu\phi\partial_\nu\phi$$

   and the scale transformations

$$x^\mu \to x'^\mu = e^\alpha x^\mu$$

$$\phi(x) \to \phi'(x') = e^{\gamma\alpha}\phi(x) \; .$$

   acting on both $x$ and $\phi$ .

   Determine the value of $\gamma$ that leaves the action unchanged and derives the corresponding Noether current.

## Sect. 3.2

1. Using eq. (3.19), prove that:

$$\frac{\delta\phi(x)}{\delta\phi(y)} = \frac{\delta\pi(x)}{\delta\pi(y)} = \delta(x-y)$$
$$\frac{\delta\phi(x)}{\delta\pi(y)} = \frac{\delta\pi(x)}{\delta\phi(y)} = 0$$

   and that:

$$\{\phi(x), \pi(y)\} = \delta(x-y) \; .$$

2. Demonstrate the Jacobi identity for generic Poisson brackets:

$$\{A,\{B,C\}\} + \{B,\{C,A\}\} + \{C,\{A,B\}\} = 0 \; .$$

3. Demonstrate the same identity for the commutator of two matrices $[\boldsymbol{A}, \boldsymbol{B}]$, i.e.

$$[\boldsymbol{A}, [\boldsymbol{B}, \boldsymbol{C}]] + [\boldsymbol{B}, [\boldsymbol{C}, \boldsymbol{A}]] + [\boldsymbol{C}, [\boldsymbol{A}, \boldsymbol{B}]] = 0 .$$

## Sect. 3.3

1. Demonstrate that the charges corresponding to the momentum tensor in (3.99) agree with those of the canonical tensor (3.90).

CHAPTER 4

# KLEIN–GORDON FIELD QUANTISATION

## 4.1 THE REAL SCALAR FIELD

The scalar field provides the simplest example of what has been discussed so far. In general, we can choose:

$$\mathcal{L} = \frac{1}{2}(\partial^\mu \phi)(\partial_\mu \phi) - \frac{1}{2}m^2\phi^2 \; - \; V(\phi) \tag{4.1}$$

where $m$ is a constant with dimensions of $[length]^{-1}$, and $V(\phi)$ is a function of second order in $\phi$.

From (4.1) we find the Euler–Lagrange equations:

$$\Box\phi + m^2\phi + \frac{\partial V}{\partial \phi} = 0. \tag{4.2}$$

In the case $V = 0$, we find the *Klein–Gordon equation*:

$$(\Box + m^2)\phi = 0. \tag{4.3}$$

which constitutes the simplest relativistic field equation.

To construct the Hamiltonian, we calculate the conjugate momentum density (recall that $x^0 = ct$):

$$\pi = \frac{\partial L}{\partial \partial_t \phi} = \frac{1}{c^2}\partial_t \phi \tag{4.4}$$

so that:

$$H = \pi \; \partial_t\phi - L = \frac{1}{2}[c^2\pi^2 + (\boldsymbol{\nabla}\phi)^2 + m^2\phi^2 + V(\phi)]. \tag{4.5}$$

DOI: 10.1201/9781003436263-4

The stability of the field requires that the Hamiltonian should be bounded from below, for variations of $\phi$ in all possible configurations. Restricting ourselves to spatially constant configurations, we see from (4.5) that $V(\phi)$ *must be a function bounded from below.*

Hamilton's equations are:

$$\partial_t \phi = c^2 \pi; \qquad (4.6)$$
$$\partial_t \pi = -(m^2\phi + \frac{\partial V}{\partial \phi} - \nabla \cdot \nabla \phi).$$

Substituting the first into the second we find, of course, the relativistically invariant equation (4.2).

Comment. Equation (4.1) represents the most general form of a Lagrangian which is invariant and quadratic in the derivatives. A term linear in $\phi$ could be added. Even in the presence of a term of this type, the potential $V(\phi)$ in the Hamiltonian must have an absolute minimum. Therefore we can eliminate the linear term with a change of variables: $\phi' = \phi - \phi_0$, $\partial^\mu \phi' = \partial^\mu \phi$ where $\phi_0$ is the minimum. The development of $\mathcal{L}$ in $\phi'$ no longer contains the linear term.

Now we determine the general solution of the Klein–Gordon (K–G) equation, which is again:

$$(\Box + m^2)\phi = 0$$

which we can solve with *periodic boundary conditions* at the edges of a large cubic spatial volume of side $L$:

$$\phi(x, y, z, t) = \phi(x + L, y, z, ), \text{ etc.}$$

The general solution has the form of a plane wave:

$$\phi = N e^{-ik_\mu x^\mu}; \quad k^\mu = (k^0, \boldsymbol{k})$$

where $\boldsymbol{k}$ is the vector wave number. The periodicity condition implies:

$$k^1 L = 2\pi n^1, \text{etc.} \rightarrow \quad \boldsymbol{k} = \frac{2\pi}{L}(n^1, n^2, n^3) \qquad (4.7)$$

where $n^i$ are arbitrary integers. The Klein–Gordon equation, in turn, requires:

$$k_\mu k^\mu - m^2 = 0$$

from which we can deduce the two solutions:

$$k^0 = \pm\frac{\omega}{c}; \qquad \frac{\omega}{c} = \sqrt{(m)^2 + (\boldsymbol{k})^2}.. \qquad (4.8)$$

The general solution of the K–G equation is a superposition of the plane waves just found. For every $\boldsymbol{k}$ we have two plane waves, one with positive frequency $e^{-i\omega t}$ and one with negative frequency $e^{+i\omega t}$. We write:

$$\phi(x) = \Sigma_{\boldsymbol{n}} N(\omega)[a(\boldsymbol{k})\, e^{-i(\omega t - \boldsymbol{k}\cdot\boldsymbol{x})} + c(\boldsymbol{k})\, e^{+i(\omega t + \boldsymbol{k}\cdot\boldsymbol{x})}] \tag{4.9}$$

where $N$ is a normalisation factor which we will shortly define, and the sum is over all vectors with integer components. In the second term we can sum over $-\boldsymbol{n}$ and define $c(-\boldsymbol{k}) = b^*(\boldsymbol{k})$:

$$\begin{aligned}\phi(x) &= \Sigma_{\boldsymbol{n}} N(\omega)[a(\boldsymbol{k})\, e^{-i(\omega t - \boldsymbol{k}\cdot\boldsymbol{x})} + b^*(\boldsymbol{k})\, e^{+i(\omega t - \boldsymbol{k}\cdot\boldsymbol{x})}] = \\ &= \Sigma_{\boldsymbol{n}} N(\omega)[a(\boldsymbol{k})\, e^{-ik_\mu x^\mu} + b^*(\boldsymbol{k})\, e^{+ik_\mu x^\mu}]\end{aligned} \tag{4.10}$$

where, from now on, we will put $k^\mu = (\frac{\omega}{c}, \boldsymbol{k})$.

The general solution (4.10) produces a field which is, in general, complex. For a real field, we must have $b(\boldsymbol{k}) = a(\boldsymbol{k})$ and we find:

$$\phi(x) = \Sigma_{\boldsymbol{n}} N(\omega)[a(\boldsymbol{k})\, e^{-ik_\mu x^\mu} + a^*(\boldsymbol{k})\, e^{+ik_\mu x^\mu}]. \tag{4.11}$$

The $\phi$ given by (4.11) is real and depends on two real functions of $\boldsymbol{k}$, the real and imaginary parts of $a(\boldsymbol{k})$. This corresponds to the fact that, to completely determine $\phi$, it is necessary to provide two types of initial conditions: the values of $\phi(\boldsymbol{x}, 0)$ and $\partial_t\phi(\boldsymbol{x}, 0)$.

To conclude this section, we can state explicitly the relation between $a(\boldsymbol{k})$ and the initial conditions. We consider the system of functions $f_{\boldsymbol{k}}(x)$, solutions of the K–G equation for positive frequencies:

$$f_{\boldsymbol{k}}(x) = N e^{-ik_\mu x^\mu} \tag{4.12}$$

Using the orthonormality condition for exponential functions, we can calculate the two projections (recalling that $\omega(\boldsymbol{k}) = \omega(-\boldsymbol{k})$):

$$\begin{aligned}X &= \int d^3x\, [(\partial_t f_{\boldsymbol{k}}(\boldsymbol{x}, t))^*\, \phi(\boldsymbol{x}, t)]_{t=0} = i\omega(\boldsymbol{k})N^2\, V[a(\boldsymbol{k}) + a^*(-\boldsymbol{k})];\\ Y &= \int d^3x\, [(f_{\boldsymbol{k}}(\boldsymbol{x}, t))^*\, \partial_t\phi(\boldsymbol{x}, t)]_{t=0} = -i\omega(\boldsymbol{k})N^2\, V[a(\boldsymbol{k}) - a^*(-\boldsymbol{k})];\end{aligned}$$

from which we can derive $a(\boldsymbol{k})$. If we choose $N = (2\omega(\boldsymbol{k})V)^{-1/2}$, it follows that:

$$\begin{aligned}a(\boldsymbol{k}) &= i(Y - X) = i\int d^3x[f^*_{\boldsymbol{k}} \cdot (\partial_t\phi) - (\partial_t f^*_{\boldsymbol{k}}) \cdot \phi]_{t=0}\\ \phi(x) &= \Sigma_{\boldsymbol{n}} \frac{1}{\sqrt{2\omega V}}\, [a(\boldsymbol{k})\, e^{-ik_\mu x^\mu} + a^*(\boldsymbol{k})\, e^{+ik_\mu x^\mu}].\end{aligned} \tag{4.13}$$

The complex Klein–Gordon field. The extension to the case of a complex field is straightforward. The Lagrangian is written:

$$L = (\partial^\mu \phi)(\partial_\mu \phi^*) - m^2 \phi\phi^* \;-\; V(\phi\phi^*) \tag{4.14}$$

which arrives once again at the Klein–Gordon equation (4.3) for $\phi$ and $\phi^*$.

The Lagrangian (4.14) exhibits a symmetry for changes of phase, this time an internal symmetry:

$$\phi(x) \to e^{i\alpha}\phi(x); \quad \phi(x)^* \to e^{-i\alpha}\phi(x)^*. \tag{4.15}$$

In terms of the real and imaginary parts of $\phi$:

$$\phi = \frac{\phi^1 + i\phi^2}{\sqrt{2}}$$

the Lagrangian (4.14) reduces to the sum of two identical Lagrangians for the real fields $\phi^1$ and $\phi^2$. The symmetry (4.15), in this new representation, corresponds to an orthogonal rotation of the fields $\phi^{1,2}$ among themselves:

$$\phi^i = \mathcal{O}^{ij}\phi^j \quad \mathcal{O}^T\mathcal{O} = 1. \tag{4.16}$$

The complex field $\phi$ can once again be expanded in solutions of the Klein–Gordon equation according to (4.10) but now $a(\boldsymbol{k})$ and $b(\boldsymbol{k})$ are independent:

$$\phi(x) = \Sigma_{\boldsymbol{n}} \frac{1}{\sqrt{2\omega V}}[a(\boldsymbol{k})\, e^{-ik_\mu x^\mu} + b^*(\boldsymbol{k})\, e^{+ik_\mu x^\mu}]. \tag{4.17}$$

The Continuum Limit. The inclusion of the system in a cube of side $L$ is a mathematical artifice which serves to arrive at a discrete spectrum of solutions to the K–G equation. In general, at the end, it is necessary to pass to the limit $L \to \infty$. The sum over integer vectors, in this limit, tends to an integral over the oscillator density, of which we give the explicit form. From (4.7) it can be seen that the interval $\Delta n^1$ corresponds to $\frac{L}{2\pi}\Delta k^1$, etc. Therefore:

$$\Sigma_{\boldsymbol{n}} \ldots \to V \int \frac{\Delta k^1 \Delta k^2 \Delta k^3}{(2\pi)^3} \ldots = V \int \frac{d^3k}{(2\pi)^3} \ldots$$

## 4.2 GREEN'S FUNCTIONS OF THE SCALAR FIELD

We now consider the solutions of the equation of motion of a real field in the presence of a known source, $J(\mathrm{x})$:

$$(-\Box - \mu^2)\phi(x) = J(x). \tag{4.18}$$

The associated homogeneous equation is the Klein–Gordon equation, (4.3).

The solutions of (4.18) are obtained starting from the *Green's function* for the problem, the solution to the equation in the presence of a pointlike source described by a Dirac delta function localised at the origin of space-time:

$$(-\Box - \mu^2)G(x) = \delta^{(4)}(x). \tag{4.19}$$

Given the Green's function, the solution to (4.18) is simply:

$$\phi(x) = \int d^4x' \; G(x - x')J(x') \tag{4.20}$$

as is easily verified.

There are, of course, an infinite number of Green's functions, each determined by the particular boundary conditions which are assigned to (4.19). The solutions differ among themselves by a solution to the homogeneous equation, so that the *general solution* to (4.18) is written:

$$\phi(x) = \int d^4x' \; G(x - x')J(x') + \phi_0(x) \tag{4.21}$$

where $G$ is a given Green's function and $\phi_0$ is the general solution to the homogeneous K–G equation (4.3), which we described earlier, (4.13).

To solve (4.19), the method of Fourier transforms is used. Given $f(x)$, we define the Fourier transform (immediately taking the limit $V \to \infty$) as :

$$\tilde{f}(k) = \int \; d^4x \; f(x)e^{i(k \cdot x)}; \;\; f(x) = \frac{1}{(2\pi)^4} \int d^4k \tilde{f}(k)e^{-i(k \cdot x)}. \tag{4.22}$$

Equations (4.19) and (4.20) therefore become:

$$\begin{aligned} \tilde{G}(k) &= \frac{1}{k^2 - \mu^2}; \\ \tilde{\phi}(k) &= \tilde{G}(k) \cdot \tilde{J}(k). \end{aligned} \tag{4.23}$$

A particular solution to (4.18) is found formally from (4.23):

$$\phi(x) = \int d^4k e^{-i(k \cdot x)} \frac{1}{k^2 - \mu^2} \cdot \tilde{J}(k). \tag{4.24}$$

To give a precise significance to (4.24) we must take account of the fact that the denominator in the integral is singular at the points which correspond to the propagation of free waves, (4.8). To do this, it is convenient to work in the *complex plane of the variable* $k^0$. The singularity of $\tilde{G}$(k) is found on the real axis, for $k^0 = \pm\omega$, and each particular solution is found by assigning a path in the complex plane to carry out the integral in $k^0$.

To be definite we consider the integral:

$$F(x) = \int_C d^4k \; \frac{-1}{k^2 - \mu^2} \tilde{g}(k)e^{-i(kx)} \tag{4.25}$$

with $\tilde{g}(k)$ a given function, which is analytic in $k^0$ and $C$ is a path assigned in the complex plane of $k^0$. We must separately distinguish between integration along closed and open paths.

1. Closed paths. These integrals give solutions to the homogeneous equation. Actually, applying the Klein–Gordon operator, a factor $k^2 - \mu^2$ is obtained in the numerator of the integrand, which eliminates the pole; at this point we can reduce the integration path to zero, thus obtaining:

$$(\Box + \mu^2)F(x) = 0.$$

Using the residue theorem, it is easily seen that the integral is equal to zero if the path does not include either of the two singularities, or a combination of the two homogeneous solutions, represented by the residues of the integral around each singularity.

We denote by $C^+$ a path which encircles *clockwise* (only once!) the point $k^0 = +\omega(k)$ and we define:

$$\begin{aligned} i\Delta^{(+)}(x) &= \frac{1}{(2\pi)^4}\int_{C+} d^4k\; e^{-i(k\cdot x)}\frac{i}{k^2-\mu^2} = \\ &= \frac{1}{(2\pi)^4}\int d^3k \int_{C+} dk^0 \frac{i}{(k^0-\omega)(k^0+\omega)} e^{-i(k\cdot x)}; \\ i\Delta^{(+)}(x) &= \frac{1}{(2\pi)^3}\int \frac{d^3k}{2\omega} e^{-i(\omega t - \boldsymbol{k}\cdot\boldsymbol{x})} = \frac{1}{(2\pi)^3}\int \frac{d^3k}{2\omega} e^{-i(k\cdot x)}. \end{aligned} \tag{4.26}$$

Analogously, we denote by $C^-$ the path which turns in the anticlockwise direction around the singularity at $k^0 = -\omega(k)$ and we define:

$$\begin{aligned} i\Delta^{(-)}(x) &= \frac{1}{(2\pi)^4}\int_{C-} d^4k\; e^{-i(k\cdot x)}\frac{i}{k^2-\mu^2}; \\ i\Delta^{(-)}(x) &= \frac{-1}{(2\pi)^3}\int \frac{d^3k}{2\omega} e^{-i(-\omega t - \boldsymbol{k}\cdot\boldsymbol{x})} = \\ &= \frac{-1}{(2\pi)^3}\int \frac{d^3k}{2\omega} e^{+i(\omega t - \boldsymbol{k}\cdot\boldsymbol{x})} = \frac{-1}{(2\pi)^3}\int \frac{d^3k}{2\omega} e^{+i(k\cdot x)}. \end{aligned} \tag{4.27}$$

In (4.26) and (4.27), $k^\mu = [+\omega(k), \boldsymbol{k}]$. Obviously,

$$\Delta^{(-)}(x) = -\Delta^{(+)}(-x) = -[\Delta^{(+)}(x)]^*. \tag{4.28}$$

In conclusion, the integral (4.25) on a closed path, $C_0$, gives a solution to the homogeneous equation, from a linear combination of the residues at the two poles:

$$\begin{aligned} \phi_0(x) &= \frac{1}{(2\pi)^3}\int \frac{d^3k}{2\omega(k)}\tilde{g}(\boldsymbol{k})e^{-i(k\cdot x)} + \frac{-1}{(2\pi)^3}\int \frac{d^3k}{2\omega(k)}\tilde{g}(-\boldsymbol{k})e^{+i(k\cdot x)} = \\ &= \int \Delta^{(+)}(x-x')g(x') + \int \Delta^{(-)}(x-x')h(x'). \end{aligned} \tag{4.29}$$

where we have put $\tilde{g}(\boldsymbol{k}) = \tilde{g}(\omega(k), \boldsymbol{k})$, and $g$ and $h$ are two independent functions. Equation (4.29) gives a representation of the general solution of the homogeneous equation which reproduces what was given earlier, in (4.13) in the complex case (in which $\tilde{g}(\boldsymbol{k})$ and $\tilde{g}(-\boldsymbol{k})$ are independent) and in the continuum limit, in which:

$$\tilde{g}(\boldsymbol{k}) = \sqrt{2\omega(k)V}\, a(\boldsymbol{k}). \tag{4.30}$$

(Note that in the continuum limit, $V \to \infty$, the $a(\boldsymbol{k})$ must tend to zero like $1/\sqrt{V}$ if we wish for field configurations in which the total energy of the field, rather than the energy density, remains finite).

2. Open paths. In general, these paths give a solution to the inhomogeneous equation. Two different paths give the same result if we can continuously deform one into the other without encountering any singular point of $\tilde{G}$, otherwise they differ by combinations of the integrals around the singularity, i.e. by solutions of the inhomogeneous equation.

Among the particular solutions of the homogeneous equation worthy of note are those which correspond to *retarded* or *advanced* Green's functions or to the Feynman function.

- The *retarded Green's function*, $G_{ret}$, corresponds to the condition that $G(x) = 0$ for $t < 0$, which is that the result should be different from zero only *after* switching on the source at the coordinate origin (causality condition). In this case, the integration path must be completely *above the singularity.* We put $k^0 = Re\ k^0 + i\eta$ and explicitly write the exponential in (4.25):

$$e^{-i(k\cdot x)} = e^{(-iRek^0 t + \eta t + \ldots)}.$$

For $t < 0$ the integration path must be closed in the upper part of the complex plane ($\eta > 0$). To obtain a null result, the path closed in this way should not contain the singularity, which must therefore be below the path of integration. Conversely, when $t > 0$ and the integration path is closed in the lower half-plane, the path, turning clockwise, includes the two singularities. The result is therefore:

$$i\Delta_{ret}(x) = \frac{1}{(2\pi)^4}\int_{Imk^0>0} d^4k \frac{i}{k^2-\mu^2} = \theta(t)[i\Delta^{(+)}(x) + i\Delta^{(-)}(x)]. \tag{4.31}$$

- The symmetric condition, that $G$ should vanish for $t > 0$, takes us to the *advanced* Green's function:

$$i\Delta_{adv}(x) = \frac{1}{(2\pi)^4}\int_{Imk^0<0} d^4k \frac{-1}{k^2-\mu^2} = -\theta(-t)[i\Delta^{(+)}(x) + i\Delta^{(-)}(x)]. \tag{4.32}$$

- The *Feynman propagator* is obtained from the condition that it should coincide with $i\Delta^{(+)}(\mathrm{x})$ for $t > 0$ and with $i\Delta^{(-)}(\mathrm{x})$ for $t < 0$. This condition determines an integration path, $C_F$, that originates from the negative real axis from $k^0$ passing *below* the singularity in $k^0 < 0$, and *above* the one at $k^0 > 0$. In this way, for $t > 0$, when enclosed in the lower half-plane, the path turns clockwise around the point $k^0 = \omega$ and results in $i\Delta^{(+)}(\mathrm{x})$, while for $t < 0$ the path is enclosed in the upper half-plane, turning anticlockwise around the singularity at $k^0 < 0$. The same result is obtained, obviously, integrating along the real axis, after having moved the poles in the complex planes by an infinitesimal amount, $\epsilon > 0$, in the following way:

$$\begin{aligned} k^0 &= -\omega(k) \to k^0 = -\omega + i\epsilon; \\ k^0 &= +\omega(k) \to k^0 = +\omega - i\epsilon. \end{aligned}$$

As a formula, we arrive at the following definition:

$$\begin{aligned} iD_F(x) &= \int_{C_F} \frac{d^4k}{(2\pi)^4} \frac{i}{k^2 - \mu^2} e^{-i(k\cdot x)} \\ &= \int \frac{d^4k}{(2\pi)^4} \frac{i}{(k^0 - \omega + i\epsilon)(k^0 + \omega - i\epsilon)} e^{-i(k\cdot x)} \\ &= \int \frac{d^4k}{(2\pi)^4} \frac{i}{k^2 - \mu^2 + i\epsilon} e^{-i(k\cdot x)} \\ &= \theta(t) i\Delta^{(+)}(x) - \theta(-t) i\Delta^{(-)}(x). \end{aligned} \tag{4.33}$$

The propagators $\Delta^{(\pm)}(\mathrm{x})$ can be expressed in terms of known functions [2].

## 4.3 QUANTISATION OF THE SCALAR FIELD

We recall the classical Lagrangian for the complex scalar field:

$$L = \partial_\mu \phi^\dagger \partial^\mu \phi - m^2 \phi^\dagger \phi \tag{4.34}$$

which reproduces the Klein–Gordon (K–G) equation:

$$(\partial_\mu \partial^\mu + m^2)\phi = 0 \tag{4.35}$$

The Lagrangian (4.34) is invariant for translations in space-time and for Lorentz transformations, under which $\phi$ transforms like a scalar:

$$\phi'(x') = \phi(x). \tag{4.36}$$

Moreover, the Lagrangian is invariant under the phase transformations:

$$\phi'(x) = e^{i\alpha}\phi(x); \; \phi'^\dagger(x) = e^{-i\alpha}\phi^\dagger \tag{4.37}$$

with $\alpha$ constant.

It follows from Noether's theorem that:

- the energy-momentum tensor (symmetric, because there is no spin part):

$$\theta^{\mu\nu} = \partial^\mu \phi^\dagger \partial^\nu \phi - g^{\mu\nu} L \tag{4.38}$$

- the momentum tensor:

$$M^{\mu,\alpha\beta} = x^\alpha \theta^{\mu\beta} - x^\beta \theta^{\mu\alpha} \tag{4.39}$$

- the conserved current corresponding to (4.37) (the factor $1/\hbar$ is inserted for appropriate normalisation of the charge):

$$J^\mu(x) = \frac{i}{\hbar}[\phi^\dagger(\partial^\mu \phi) - (\partial^\mu \phi^\dagger)\phi]. \tag{4.40}$$

From (4.35) the conjugate momenta and the Hamiltonian are quickly found:

$$\begin{aligned}
&\pi = \frac{\partial L}{\partial \partial_t \phi} = \partial_t \phi^\dagger;\ \pi^\dagger = \partial_t \phi \\
&H = \pi^\dagger \pi + \nabla \phi^\dagger \cdot \nabla \phi + m^2 \phi^\dagger \phi;\ \boldsymbol{H} = \int d^3x\ H \\
&\boldsymbol{Q} = \int d^3x J^0(\mathbf{x}, t) = \frac{i}{\hbar} \int d^3x\ [\phi^\dagger(\partial^0 \phi) - (\partial^0 \phi^\dagger)\phi]
\end{aligned} \tag{4.41}$$

Canonical quantisation should replace classical Poisson brackets for the canonical variables (cf. Chapter 3) with the *equal time commutators* according to the rule:

$$\{A, B\} \to \frac{[A, B]}{i\hbar}. \tag{4.42}$$

We obtain the *equal time commutators* from equation (3.26) :

$$\begin{aligned}
&[\phi(\boldsymbol{x}, t), \phi(\boldsymbol{y}, t)] = [\phi(\boldsymbol{x}, t), \phi^\dagger(\boldsymbol{y}, t)] = 0 \\
&[\pi(\boldsymbol{x}, t), \pi(\boldsymbol{y}, t)] = [\pi(\boldsymbol{x}, t), \pi^\dagger(\boldsymbol{y}, t)] = 0 \\
&[\phi(\boldsymbol{x}, t), \pi^\dagger(\boldsymbol{y}, t)] = 0 \\
&[\phi(\boldsymbol{x}, t), \pi(\mathbf{y}, t)] = [\phi^\dagger(\boldsymbol{x}, t), \pi^\dagger(\boldsymbol{y}, t)] = i\hbar\delta^{(3)}(\boldsymbol{x} - \boldsymbol{y}).
\end{aligned} \tag{4.43}$$

We can also express the non-zero commutators as:

$$[\phi(\boldsymbol{x}, t), \partial_t \phi^\dagger(\boldsymbol{y}, t)] = i\hbar\delta^{(3)}(\boldsymbol{x} - \boldsymbol{y}), \tag{4.44}$$

together with the Hermitian conjugate equation.

Equations (4.44) determine the operator structure of the theory, in particular, the commutators of the dynamic variables with the Hamiltonian and therefore the equation of motion. Hamilton's equations are[1]:

$$i\hbar\frac{\partial}{\partial t}\phi = [\phi, \boldsymbol{H}] = i\hbar\phi^{\dagger};$$
$$i\hbar\frac{\partial}{\partial t}\pi = [\pi, \boldsymbol{H}] = i\hbar(\nabla\cdot\nabla - m^2)\phi^{\dagger}, \tag{4.45}$$

from which (4.35) follows. We note the charge-field commutation rule:

$$[\phi, \boldsymbol{Q}] = +\phi \tag{4.46}$$

The solutions to the K–G equation are of the form found in section 4.1. To be exact, the K–G equation, being linear in $\phi$, has the same solutions, regardless of whether $\phi$ is a complex number or an operator. These are the amplitudes of the normal oscillation modes, $a(\boldsymbol{k})$ and $b(\boldsymbol{k})$, which now become linear operators with commutation rules determined by (A.25). (We normalise $a(\boldsymbol{k})$ and $b(\boldsymbol{k})$ so as to eliminate $\hbar$ from their commutation rules.):

$$\phi(x) = \Sigma_{\boldsymbol{k}}\sqrt{\frac{\hbar}{2\omega V}}[a(\boldsymbol{k})e^{-i(kx)} + b^{\dagger}(\boldsymbol{k})e^{i(kx)}]. \tag{4.47}$$

Inverting equation (4.47) we find:

$$a(\boldsymbol{k}) = i\int d^3x[f^*_{\boldsymbol{k}}\cdot(\partial_t\phi) - (\partial_t f^*_{\boldsymbol{k}})\cdot\phi]_{t=0} =,$$
$$b(\boldsymbol{k}) = i\int d^3x[f^*_{\boldsymbol{k}}\cdot(\partial_t\phi^{\dagger}) - (\partial_t f^*_{\boldsymbol{k}})\cdot\phi^{\dagger}]_{t=0}$$
$$f_{\boldsymbol{k}} = \sqrt{\frac{1}{2\hbar\omega(\boldsymbol{k})V}}e^{-ikx}, \tag{4.48}$$

From (4.44) we therefore find:

$$[a(\boldsymbol{k}), a^{\dagger}(\boldsymbol{k}')] = [b(\boldsymbol{k}), b^{\dagger}(\boldsymbol{k}')] = \delta_{\boldsymbol{k},\boldsymbol{k}'} \tag{4.49}$$

with all other commutators equal to zero.

- The canonical commutation rules given to the operators $a$, $a^{\dagger}$ and $b$, $b^{\dagger}$ the function of *creation and destruction operators* of quantum harmonic oscillators, with two types of oscillator for every mode of vibration of the classical field.

The Hilbert space on which the field operators act is composed of tensor products of the different oscillator states. More precisely, the space of states includes:

[1]These are obtained immediately by applying (4.42) to the Poisson brackets (3.27).

- the *vacuum state* in which all the oscillators are at the lowest level. Mathematically, $|0>$ is determined by the condition of being annihilated by the application of any destruction operator:

$$a_s(\boldsymbol{p})|0>= b_r(\boldsymbol{q})|0>= 0, \text{ for any } s, r, \boldsymbol{p}, \boldsymbol{q}; \tag{4.50}$$

- states with a certain number of excitations of different oscillators, obtained by applying the creation operators $a^\dagger$ and $b^\dagger$ to the vacuum state:

$$|n_1, n_2, \ldots; m_1, m_2, \cdots >= \frac{1}{\sqrt{n_1!n_2!\ldots m_1!m_2!\ldots}} \cdot \\ \cdot [a^\dagger_{s_1}(\boldsymbol{p}_1)]^{n_1}[a^\dagger_{s_2}(\boldsymbol{p}_2)]^{n_2} \ldots [b^\dagger_{r_1}(\boldsymbol{q}_1)]^{m_1}[b^\dagger_{r_2}(\boldsymbol{q}_2)]^{m_2} \ldots |0> . \tag{4.51}$$

The physical nature of these operators is clarified by the consideration of the conserved quantities: the energy and momentum of the field, $\boldsymbol{H}$ and $\boldsymbol{P}$, and the conserved charge, $\boldsymbol{Q}$. Substituting the expansion (4.47) into (4.38) and (4.40) and using the orthogonality of the plane waves, we find (without changing the order in which the operators appear in the various expressions):

$$\begin{aligned} \boldsymbol{H} &= \int d^3x\ \theta^{00} = \Sigma_{\boldsymbol{k}}\ \hbar\omega(k)\ [a^\dagger(\boldsymbol{k})a(\boldsymbol{k}) + b(\boldsymbol{k})b^\dagger(\boldsymbol{k})] \\ \boldsymbol{P}^i &= \int d^3x\ \theta^{0i} = \Sigma_{\boldsymbol{k}}\ \hbar k^i\ [a^\dagger(\boldsymbol{k})a(\boldsymbol{k}) + b(\boldsymbol{k})b^\dagger(\boldsymbol{k})] \\ \boldsymbol{Q} &= \int d^3x\ J^0 = \Sigma_{\boldsymbol{k}}\ [a^\dagger(\boldsymbol{k})a(\boldsymbol{k}) - b(\boldsymbol{k})b^\dagger(\boldsymbol{k})]. \end{aligned} \tag{4.52}$$

We can now reorder the operators so as always to have the destruction operators on the right, finding:

$$\begin{aligned} \boldsymbol{H} &= \Sigma_{\boldsymbol{k}}\ \hbar\omega(k)\ [a^\dagger(\boldsymbol{k})a(\boldsymbol{k}) + b^\dagger(\boldsymbol{k})b(\boldsymbol{k})]\ +\ \text{constant} \\ \boldsymbol{P}^i &= \Sigma_{\boldsymbol{k}}\ \hbar k^i\ [a^\dagger(\boldsymbol{k})a(\boldsymbol{k}) + b^\dagger(\boldsymbol{k})b(\boldsymbol{k})] \\ \boldsymbol{Q} &= \Sigma_{\boldsymbol{k}}\ [a^\dagger(\boldsymbol{k})a(\boldsymbol{k}) - b^\dagger(\boldsymbol{k})b(\boldsymbol{k})]. \end{aligned} \tag{4.53}$$

The ordered operators always give zero on the vacuum state and give the occupation number of the corresponding oscillator, when applied to the states of (4.51).

The (infinite) constant in the expression for the energy represents the energy of the vacuum state, which is unobservable (while we remain within the limits of special relativity). Measuring the energy starting from the energy of the vacuum, the first equation of (4.53) shows that all the states have *positive energy*.

The values of energy and momentum corresponding to the states $a^\dagger(\boldsymbol{k})|0>$ are those of a relativistic particle with 4-momentum $p^\mu = (\hbar\omega(k), \hbar\boldsymbol{k})$ and $p_\mu p^\mu = (\hbar m)^2$. The set of states which we have found is that of a perfect quantum gas formed of identical particles of two types (the particles created by the operators $a^\dagger$ and $b^\dagger$), with equal mass and charge $\pm 1$ respectively.

From the canonical quantisation rules (4.44) we see that the particles obey the rules of Bose–Einstein statistics: $n_1, n_2, \ldots, m_1, m_2, \ldots = 0, 1, 2, \ldots$.

Normal Products. The version of the classical Lagrangian and other observables of the field (energy, momentum, etc) suffer from an intrinsic ambiguity, because we must convert the products of classical quantities (which commute) into products of linear operators (in general, non-commuting). We can remove this ambiguity by defining, as we have just done, the quantum operator products in a way which ensures that their vacuum value is equal to zero. When we impose this condition, we speak of having *normal products* or *normal ordering.*

To formalise this condition, we observe that the field operators are sums of two components, characterised by the sign of the exponential in $t$ in the plane waves:

$$\begin{aligned} \phi^{(+)}(x) &\sim e^{-i(px)} \text{ (conventionally, positive frequency)} \\ \phi^{(-)}(x) &\sim e^{+i(px)} \text{ (conventionally, negative frequency).} \end{aligned} \tag{4.54}$$

The normal product of two operators is defined as what is obtained putting the positive frequency operators *to the right* of the expression and ignoring the result of eventual commutation. The normal product is commonly denoted with the symbol $N$ or, more simply enclosing the product between two colons $(: \cdots :)$. For example:

$$\begin{aligned} N(\phi(x)\phi^\dagger(y)) &=: \phi(x)\phi^\dagger(y) : \\ &=: (\phi^{(+)}(x) + \phi^{(-)}(x))((\phi^\dagger)^{(+)}(y) + (\phi^\dagger)^{(-)}(y)) : \\ &= \phi^{(+)}(x)(\phi^\dagger)^{(+)}(y) + (\phi^\dagger)^{(-)}(y)\phi^{(+)}(x) \\ &\quad + \phi^{(-)}(x)(\phi^\dagger)^{(+)}(y) + \phi^{(-)}(x)(\phi^\dagger)^{(-)}(y). \end{aligned} \tag{4.55}$$

In what follows, it will be understood that the Lagrangian, Hamiltonian and other observables should be constructed of normal products of the field. Correspondingly, the expression for energy is given by (4.53) without the infinite constant.

## 4.4 PROBLEMS FOR CHAPTER 4

### Sect. 4.1

1. Using the Klein–Gordon equation, show that the quantity

$$\int d^3x \, [f^\star_{\mathbf{k}} \cdot (\partial_t \phi) - (\partial_t f^\star_{\mathbf{k}}) \cdot \phi]$$

where $f_{\mathbf{k}}$ is given by eq. (4.12), is independent of time.

Sect. 4.3

**1.** Given the real scalar field $\phi(x)$,

- evaluate the commutator $[\phi(x), \phi(y)]$ for $x^0 \neq y^0$;
- show that each of the two terms contributing to the commutator is separately Lorentz invariant;
- show that the canonical commutation rule is recovered in the $x^0 \to y^0$ limit.

**2.** Find the momentum operator associated with the real scalar field $\phi$, defined as

$$P^i = \int d^3x \; T^{0i}$$

$T^{\mu\nu}$ being the energy, momentum tensor, and show that

$$[P^i, \phi] = i\frac{\partial \phi}{\partial x_i} \; .$$

**3.** Evaluate the vacuum expectation value

$$\langle 0|\phi(x_1)\phi(x_2)\phi(x_3)\phi(x_4)|0\rangle \; .$$

**4.** Consider the transformation

$$x^\mu \to x'^\mu = e^{-\alpha} x^\mu \quad , \quad \phi(x) \to \phi'(x') = e^{\alpha}\phi(x)$$

which leaves the Lagrangian density of a free massless scalar field invariant.

- obtain the conserved charge

$$D = \int d^3x j^0(x) \; ,$$

where $j^\mu$ is the corresponding Noether current;

- evaluate the commutators

$$[D, \phi(x)] \quad , \quad [D, \pi(x)]$$

where $\pi(x)$ is the field conjugate to $\phi(x)$.

**5.** Evaluate the commutator $[P^\mu, \phi(x)]$, where $P^\mu = T^{0\mu}$ and $T^{\mu\nu}$ is the energy-momentum tensor.

CHAPTER 5

# ELECTROMAGNETIC-FIELD QUANTISATION

## 5.1 MAXWELL'S EQUATIONS IN COVARIANT FORM

The equations which describe the behaviour of electric and magnetic fields in the presence of a given charge density, $\rho(\boldsymbol{x}, \boldsymbol{t})$, and current density, $\boldsymbol{J}(\boldsymbol{x}, \boldsymbol{t})$, are written in the following way:

$$\boldsymbol{\nabla} \cdot \boldsymbol{E} = \rho \tag{5.1}$$

$$\boldsymbol{\nabla} \cdot \boldsymbol{B} = 0 \tag{5.2}$$

$$\boldsymbol{\nabla} \times \boldsymbol{B} - \frac{1}{c}\frac{\partial \boldsymbol{E}}{\partial t} = \boldsymbol{J} \tag{5.3}$$

$$\boldsymbol{\nabla} \times \boldsymbol{E} + \frac{1}{c}\frac{\partial \boldsymbol{B}}{\partial t} = 0. \tag{5.4}$$

These equations can immediately be written in covariant form for Lorentz transformations [3].

We introduce the antisymmetric tensor $F^{\mu\nu}$, connected to the electric and magnetic fields by the equations:

$$F^{\mu\nu} = -F^{\nu\mu}$$

$$F^{0i} = E^i; \;\; F^{12} = B^3, \text{ and cyclic permutations,} \tag{5.5}$$

and the 4-vector for the charge-current density:

$$j^\mu = (\rho, \frac{1}{c}\mathbf{j}). \tag{5.6}$$

The inhomogeneous Maxwell's equations may then be written as:

$$\partial_\nu \, F^{\mu\nu} = j^\mu \tag{5.7}$$

while the homogeneous equations can be expressed as:

$$\partial_\mu F_{\nu\lambda} + \partial_\nu F_{\lambda\mu} + \partial_\lambda F_{\mu\nu} = 0; \qquad (\mu < \lambda < \nu). \tag{5.8}$$



DOI: 10.1201/9781003436263-5

The homogeneous equations take a more symmetric form if we express the three antisymmetric indices in terms of the Levi–Civita tensor and introduce the dual tensor:

$$\tilde{F}^{\mu\nu} = \epsilon^{\mu\nu\rho\sigma} F_{\rho\sigma} \tag{5.9}$$

The homogeneous equations (5.8) become:

$$\epsilon^{\mu\nu\rho\sigma} \partial_\nu F_{\rho\sigma} = 0. = \partial_\nu (\tilde{F}^{\mu\nu}) \tag{5.10}$$

Equation (5.10) expresses the fact that the *dual tensor does not contain sources*, unlike $F^{\mu\nu}$. Because the conversion from $F$ to $\tilde{F}$ requires the interchange of electric and magnetic fields, equation (5.10) implies the absence of *magnetic monopoles*, the magnetic analogue of electric charge.

Vector Potential. Equations (5.8) are constraints on the components of $F^{\mu\nu}$. Consequently, not all the six components of $\boldsymbol{E}$ and $\boldsymbol{B}$ are independent variables.

There are obviously four ways of choosing three different indices, each with four possible values, which is the number of independent homogeneous equations. In total, therefore, the electromagnetic field contains only *two dynamic variables.* A first step in isolating the independent components consists in observing that equations (5.8) are identically satisfied by the expression:

$$F^{\mu\nu} = \partial^\nu A^\mu - \partial^\mu A^\nu \tag{5.11}$$

as is verified immediately from (5.10). Equation (5.11) defines a new field, known as the *vector potential.* Explicitly, with $\Phi = A^0 =$ scalar potential:

$$F^{0i} = \partial^i A^0 - \partial^0 A^i = -\partial_i \Phi - \frac{1}{c}\frac{\partial}{\partial t} A^i; \tag{5.12}$$

$$\boldsymbol{B} = \boldsymbol{\nabla} \times \boldsymbol{A}. \tag{5.13}$$

If we use the components of the vector potential as dynamic variables, Maxwell's equations in a vacuum can be derived from an action principle, starting from the Maxwell Lagrangian:

$$\mathcal{L}_{e.m.} = -\frac{1}{4} F_{\mu\nu} F^{\mu\nu} = \frac{1}{2}(\boldsymbol{E}^2 - \boldsymbol{B}^2). \tag{5.14}$$

Gauge Invariance. The components of $A^\mu$ are four variables, so they still with some redundancy. If we carry out a *gauge transformation*

$$A^\mu \to A'^\mu = A^\mu + \partial^\mu f \tag{5.15}$$

the new vector potential gives rise to the same observables $\boldsymbol{E}$ and $\boldsymbol{B}$, *for any function f.*

It is possible and natural to use arbitrariness in the definition of $A^\mu$ to impose a covariance requirement on $A^\mu$. A condition often used is the *Lorenz gauge condition*[1]:

$$\partial_\mu A^\mu = 0. \tag{5.16}$$

It is easy to see that this condition can *always* be imposed. If we start from a given $A^\mu$ which does not satisfy (5.16), we can obtain an equivalent $A'^\mu$ which satisfies it by solving the equation:

$$0 = \partial_\mu A'^\mu = \partial_\mu A^\mu(x) + \Box f(x) \tag{5.17}$$

which has $f(\mathrm{x})$ as unknown. We can explicitly give a particular solution of this equation in terms of the inverse of the operator $\Box$. However, the solution is not unique, since the corresponding homogeneous equation:

$$\Box f = 0 \tag{5.18}$$

permits non-trivial solutions (as we have seen already in the case of the Klein–Gordon equation). This is in accord with a count of the degrees of freedom. Taking account of (5.16) we find three degrees of freedom for $A^\mu$, but the preceding count says that there should be two.

Unfortunately, the final condition cannot generally be given in covariant form. This is the origin of numerous problems, which will be confronted and resolved in what follows.

## 5.2 GREEN'S FUNCTIONS OF THE ELECTROMAGNETIC FIELD

In terms of the vector potential, Maxwell's inhomogeneous equations, (5.7), are written:

$$\partial_\nu F^{\mu\nu} = \Box A^\mu - \partial^\mu(\partial_\nu A^\nu) = J^\mu$$

which, if $A^\mu$ satisfies the Lorenz condition (5.16), reduces to the wave equation:

$$\partial_\nu \partial^\nu A^\mu = J^\mu \tag{5.19}$$

with the supplementary condition that the current should be conserved:

$$\partial_\mu J^\mu = 0 \to k_\mu \tilde{J}^\mu(k) = 0. \tag{5.20}$$

To characterise the solutions of (5.19), we can use the results of the preceding section, in the limit of zero mass, $\mu^2 \to 0$. As we have seen, the Green's functions of the Klein–Gordon equation contain singularities in the Fourier transform, localised at the points $k^0 = \pm\sqrt{\mu^2 + (\boldsymbol{k})^2}$. To discuss the electromagnetic field, it is desirable to begin with a fictitious mass, $\lambda$, which is small

[1]Proposed in 1867 by the Danish mathematical physicist, Ludvig Valentin Lorenz, not to be confused with H. A. Lorentz of the homonymous transformations.

but non-zero, to prevent singularities from coalescing at the same point when $\boldsymbol{k} \to 0$. The limit $\lambda \to 0$ can be taken at the end of the calculations.

From (5.19), we see that the Green's function for the vector potential satisfies the equation:

$$-\Box G^{\mu\nu}(x) = -g^{\mu\nu}\delta^{(4)}(x) \tag{5.21}$$

or, after the Fourier transform:

$$k^2 \tilde{G}^{\mu\nu}(k) = -g^{\mu\nu}. \tag{5.22}$$

The solution of (5.19) is therefore written (in the Lorenz gauge, with Feynman boundary conditions):

$$A^\mu(x) = lim_{\lambda\to 0}\left(\int \frac{d^4k}{(2\pi)^4}\frac{-g^{\mu\nu}}{k^2-\lambda^2+i\epsilon}\tilde{J}_\nu(k)\, e^{-i(k\cdot x)}\right) + A_0^\mu(x) \tag{5.23}$$

where $A_0^\mu$ satisfies the free wave equation.

As noted in the previous paragraph, the Lorenz condition does not completely determine the gauge and we can impose a further condition. To continue, we must identify an appropriate basis on which to project the four components of $A^\mu$.

We fix the 4-vector $k^\mu = (\omega(k), \boldsymbol{k})$ and, correspondingly, a space-time coordinate system identified by the following four vectors:

$$\begin{aligned} &\epsilon^\mu_{1,2} = (0, \boldsymbol{\epsilon}_{1,2}); \;\; \boldsymbol{k}\cdot\boldsymbol{\epsilon}_{(1,2)} = 0; \\ &\epsilon^\mu_3 = (0, \frac{\boldsymbol{k}}{|\boldsymbol{k}|}); \\ &\epsilon^\mu_0 = \eta^\mu = (1, \boldsymbol{0}). \end{aligned} \tag{5.24}$$

The normalisation conditions are:

$$\epsilon^\mu_\alpha \epsilon^\nu_\beta \, g_{\mu\nu} = g_{\alpha\beta} \tag{5.25}$$

and the completeness conditions are:

$$\begin{aligned} &\Sigma_{i=1,2}\, \epsilon^m_i \epsilon^n_i = (\delta^{mn} - \frac{k^m k^n}{|\boldsymbol{k}|}); \\ &\Sigma_\alpha\, g_{\alpha\alpha}\epsilon^\mu_\alpha \epsilon^\nu_\alpha = g^{\mu\nu}. \end{aligned} \tag{5.26}$$

These 4-vectors form a basis in which any other vector can be expanded. Using the second completeness equation from (5.26), we rewrite the Feynman Green's function in (5.23) in the following way (with the limit of zero mass understood):

$$\begin{aligned} \frac{-g^{\mu\nu}}{k^2+i\epsilon} &= \frac{-(\Sigma_\alpha\, g_{\alpha\alpha}\epsilon^\mu_\alpha\epsilon^\nu_\alpha)}{k^2+i\epsilon} \\ &= \frac{(\Sigma_{1,2}\,\epsilon^\mu_i\epsilon^\nu_i)}{k^2+i\epsilon} + \frac{(\epsilon^\mu_3\epsilon^\nu_3 - \eta^\mu\eta^\nu)}{k^2+i\epsilon}. \end{aligned} \tag{5.27}$$

Keeping in mind condition (5.20), we eliminate $\epsilon_3$ to leave $k^\mu$ and $\eta^\mu$:

$$\epsilon_3^\mu = \frac{k^\mu}{|\boldsymbol{k}|} - \frac{\omega}{|\boldsymbol{k}|}\eta^\mu \tag{5.28}$$

and consequently:

$$\epsilon_3^\mu \epsilon_3^\nu - \eta^\mu\eta^\nu = \frac{k^\mu k^\nu}{(\boldsymbol{k})^2} - \frac{\omega}{(\boldsymbol{k})^2}(k^\mu\eta^\nu + \eta^\mu k^\nu) - (1 - \frac{\omega^2}{(\boldsymbol{k})^2})\eta^\mu\eta^\nu. \tag{5.29}$$

Finally, we can rewrite the integrand in (5.23):

$$\begin{aligned}\frac{-g^{\mu\nu}}{k^2+i\epsilon}\tilde{J}_\nu(k)e^{-i(k\cdot x)} &= \frac{(\Sigma_{1,2}\,\epsilon_i^\mu\epsilon_i^\nu)}{k^2+i\epsilon}\;\tilde{J}_\nu(k)\;e^{-i(k\cdot x)}+ \\ &+\frac{1}{k^2+i\epsilon}\,[\frac{k^\mu k^\nu}{(\boldsymbol{k})^2} - \frac{\omega}{(\boldsymbol{k})^2}(k^\mu\eta^\nu+\eta^\mu k^\nu)]\;\tilde{J}_\nu(k)\;e^{-i(k\cdot x)} \\ &+\frac{1}{(\boldsymbol{k})^2}\eta^\mu\eta^\nu\tilde{J}_\nu(k)e^{-i(k\cdot x)}.\end{aligned} \tag{5.30}$$

We now analyse the different terms:

- The terms in the first line corresponds to the waves generated by the current, which are *transverse* waves with respect to the direction of propagation and represent the two degrees of freedom present in the field.

- In the second line, after integration, the terms proportional to $k^\mu$ contribute with terms of the type $\partial^\mu f$, which can be eliminated with a further (last) gauge transformation; the terms proportional to $k_\nu\tilde{J}^\nu$ vanish because of the conservation of the current; in total we can ignore the second line.

- In the third line, the Feynman propagator has been replaced by the Fourier transform of the Coulomb field; this term represents the electrostatic potential generated by the charge density in $J^\mu$.

Explicitly:

$$\begin{aligned}A^i(x) &= \int \frac{d^4k}{(2\pi)^4}\frac{1}{k^2+i\epsilon}(\delta^{ij} - \frac{k^i k^j}{|\boldsymbol{k}|^2})\tilde{J}^j(k)e^{-i(k\cdot x)}; \\ A^0(x) &= \int \frac{d^4k}{(2\pi)^4}\frac{1}{|\boldsymbol{k}|^2}\,\tilde{\rho}(k)e^{-i(k\cdot x)}.\end{aligned} \tag{5.31}$$

We note that:

$$\begin{aligned}\boldsymbol{\nabla}\cdot\boldsymbol{A}(x) &= 0; \\ -\boldsymbol{\nabla}\cdot\boldsymbol{\nabla}A^0(x) &= \rho(x)\end{aligned}$$

In general, we can fix the gauge so that the electric field is divided into transverse and longitudinal parts:

$$\boldsymbol{E} = \boldsymbol{E_L} + \boldsymbol{E_T};$$
$$\boldsymbol{\nabla} \cdot \boldsymbol{E_T} = 0; \qquad \boldsymbol{\nabla} \cdot \boldsymbol{E_L} = \rho. \tag{5.32}$$

From the preceding results we obtain explicitly:

$$\boldsymbol{E}_L(\mathbf{x}, t) = -\boldsymbol{\nabla} A^0 = -\boldsymbol{\nabla} \int d^3y \frac{1}{4\pi} \frac{1}{|\mathbf{x} - \mathbf{y}|} \rho(\mathbf{y}, t);$$
$$\boldsymbol{E}_T = -\frac{\partial}{\partial t} \boldsymbol{A} = -\frac{\partial}{\partial t} \int \frac{d^4k}{(2\pi)^4} \frac{1}{k^2 + i\epsilon} e^{-ikx} \, [\tilde{\boldsymbol{J}} - \frac{\boldsymbol{k}(\boldsymbol{k} \cdot \tilde{\boldsymbol{J}})}{|\boldsymbol{k}|^2}]. \tag{5.33}$$

## 5.3 THE MAXWELL–LORENTZ EQUATIONS

We consider the case in which the electromagnetic field is coupled to a pointlike particle with charge $q$. For the electron, $q = -e$, where $e$ is the elementary electric charge:

$$e = +1.60217653(14)10^{-19} \text{ C}. \tag{5.34}$$

For a pointlike particle:

$$j^\mu = (\rho, \frac{1}{c} \boldsymbol{j});$$
$$\rho = q \, \delta^{(3)}[x - x(t)]; \; \boldsymbol{j} = q \, \boldsymbol{v} \, \delta^{(3)}[x - x(t)]. \tag{5.35}$$

We therefore have:

$$\frac{1}{c} \frac{\partial}{\partial t} \rho = -q \frac{d\mathbf{x}(t)}{dt} \cdot \boldsymbol{\nabla} \, \delta^{(3)}[x - x(t)]$$
$$\frac{1}{c} \boldsymbol{\nabla} \cdot \mathbf{j} = q \boldsymbol{v}(t) \cdot \boldsymbol{\nabla} \, \delta^{(3)}[x - x(t)]. \tag{5.36}$$

As a consequence, $j^\mu$ satisfies the continuity equation:

$$\frac{1}{c} \frac{\partial}{\partial t} \rho + \frac{1}{c} \boldsymbol{\nabla} \cdot \boldsymbol{j} = 0. \tag{5.37}$$

The extension to the case of multiple charges is straightforward.

The overall action of the field+charge system is obtained from the equations of section 5.1, specifying the current according to (5.35) and adding the action due to the charge. Combining them, we find the Lagrangian density:

$$\mathcal{L}(x) = -\frac{1}{4} \, F_{\mu\nu} F^{\mu\nu} + \delta^{(3)}[x - x(t)](-mc^2 \sqrt{1 - \frac{\boldsymbol{v}^2}{c^2}} \,) - j_\mu \, A^\mu(x). \tag{5.38}$$

From this result, the field equations are obtained as before, in the form:

$$\partial_\nu F^{\mu\nu} = j^\mu = q \frac{1}{\gamma} \, u_\mu A^\mu[x(t)] \, \delta^{(3)}[x - x(t)] \tag{5.39}$$

where $u^\mu$ is the 4-velocity and, as usual, $\gamma = 1/\sqrt{1-\frac{v^2}{c^2}}$.

To derive the equations of motion of the charge, we calculate the conjugate momentum (where $\boldsymbol{L} = \int d^3x\mathcal{L}$):

$$\frac{\partial \boldsymbol{L}}{\partial \boldsymbol{v}} = m\gamma\, \boldsymbol{v} + q\boldsymbol{A}[\boldsymbol{x}(t),\, t] \tag{5.40}$$

from which:

$$\frac{d}{dt}\frac{\partial \boldsymbol{L}}{\partial \boldsymbol{v}} = \frac{d}{dt}(m\gamma\boldsymbol{v}) + q\frac{\partial \boldsymbol{A}}{\partial t} + q\frac{1}{c}(\boldsymbol{v}\cdot\boldsymbol{\nabla})\boldsymbol{A} \tag{5.41}$$

and the generalised force:

$$\frac{\partial \boldsymbol{L}}{\partial \boldsymbol{x}} = -q[\boldsymbol{\nabla} A^0 - \,\Sigma_i\, v^i\, \boldsymbol{\nabla} A^i]. \tag{5.42}$$

The equations of motion are therefore:

$$\begin{aligned} \frac{d}{dt}\frac{\partial \boldsymbol{L}}{\partial \boldsymbol{v}} &= \frac{\partial \boldsymbol{L}}{\partial \boldsymbol{x}}; \\ \frac{d}{dt}(m\gamma\boldsymbol{v}) &= \frac{d}{dt}(\boldsymbol{p}) = -q\frac{\partial \boldsymbol{A}}{\partial t} - q\boldsymbol{\nabla} A^0 + q\frac{1}{c}[-(\boldsymbol{v}\cdot\boldsymbol{\nabla})\boldsymbol{A} + \Sigma_i\, v^i(\boldsymbol{\nabla} A^i)] = \\ &= q\boldsymbol{E} + q\frac{1}{c}\,\boldsymbol{v}\times\nabla\times(\boldsymbol{A}) = q\boldsymbol{E} + q\frac{1}{c}\,\boldsymbol{v}\times\boldsymbol{B}. \end{aligned} \tag{5.43}$$

These are the equations for the spatial components of the 4-momentum, $\boldsymbol{p}$, of the charge. The equation for the time component is obtained by multiplying the previous equation by $\boldsymbol{p}$ and using the relations:

$$\begin{aligned} p^0 dp^0 &= \boldsymbol{p}\cdot d\boldsymbol{p}; \\ \boldsymbol{v}/c &= \boldsymbol{\beta} = \boldsymbol{p}/p^0. \end{aligned}$$

We therefore find:

$$\frac{d\epsilon}{dt} = c\frac{dp^0}{dt} = q\,\boldsymbol{v}\cdot\boldsymbol{E}. \tag{5.44}$$

Equation (5.44) expresses the conservation of energy: *the energy acquired by the particle in a unit of time is equal to the power provided by the electric field* (the Lorentz force, the second term in the second line of (5.43), does not perform work).

The preceding equations can be put in the covariant form, noting that:

$$\begin{aligned} \boldsymbol{v}\cdot\boldsymbol{E} &= \frac{1}{\gamma}\,F^{0\mu}u_\nu \\ \frac{dp^1}{dt} &= qF^{01} + \frac{q}{c}(v^2F^{12} - v^3F^{31}) = \frac{1}{c\gamma}F^{1\mu}u_\mu. \end{aligned}$$

We therefore obtain:

$$\frac{dp^\mu}{dt} = \frac{q}{c}\frac{1}{\gamma}F^{\mu\nu}u_\nu; \text{ or}$$

$$\frac{dp^\mu}{d\tau} = \frac{q}{c}F^{\mu\nu}u_\nu \tag{5.45}$$

in terms of the proper time. In the case of multiple particles, we obtain an equation of the type (5.45) for each particle, while in equations (5.39) we should include the contribution of each particle to the current.

With the boundary conditions that there should be no external fields and fields at infinity, the Maxwell–Lorentz equations (5.39, 5.45) describe the time evolution of a system of charged particles, each under the action of the field generated by itself and by the other particles.

Assuming that the other forces are negligible, this system of equations describes the behaviour of matter in terms of its elementary constituents. The Maxwell–Lorentz equations are the first example of a *Theory of Everything*, a theory which would describe all the phenomena found in Nature.

The systematic description of the properties of matter based on equations (5.45) was dealt with by Lorentz [4] at the start of the twentieth century, and represented a fundamental step forward in the understanding of the structure of matter.

For the behaviour of matter on the laboratory scale, the hypothesis that electromagnetic forces should dominate is completely adequate. On astronomical scales, it is necessary to take into account gravitational forces, which can be inserted into the picture by extending the principle of special relativity to Einstein's general relativity. The Maxwell–Lorentz–Einstein equations give an accurate description of phenomena on macroscopic scales, not yet superseded.

On a microscopic scale, at atomic and subatomic dimensions ($10^{-8}$ cm), the Maxwell–Lorentz picture must be replaced by quantum electrodynamics (QED). It is a very noteworthy fact that the Maxwell–Lorentz equations, once converted to the framework of quantum field theory, maintain their form essentially unaltered and are capable of describing the properties of condensed matter and of atoms extraordinarily accurately.

At nuclear and subnuclear levels (below $10^{-13}$ cm) other forces enter into play: the *nuclear forces* described for the first time in a covariant and quantum manner by Yukawa, at the beginning of the 1930s, and the *weak interactions*, identified by Fermi, also at the start of the 1930s, as responsible for the decay of the neutron and $\beta$ radioactivity.

Conservation of Energy and Momentum. Starting from the Maxwell–Lorentz Lagrangian, we can construct the conserved quantities corresponding to the energy and total momentum of a charged field.

It is natural to begin with the case of the electromagnetic field in the absence of charges (cf. Section 5.1):

$$\mathcal{L}_{e.m.} = -\frac{1}{4}F_{\mu\nu}F^{\mu\nu}. \tag{5.46}$$

The canonical energy-momentum tensor is given by:

$$\begin{aligned} T^{\mu,\nu}_{e.m.} &= \frac{\partial \mathcal{L}_{e.m.}}{\partial_\mu A_\beta}\partial^\nu A_\beta \; - \; g^{\mu\nu}\mathcal{L}_{e.m.} = \\ &= F^{\mu\beta}\partial^\nu A_\beta + \frac{g^{\mu\nu}}{4}F_{\alpha\beta}F^{\alpha\beta}. \end{aligned} \tag{5.47}$$

As well as $T^{\mu,\nu}_{e.m.}$ we consider the symmetric tensor, $\Theta^{\mu\nu}_{e.m.}$ obtained with the Belinfante–Rosenfeld procedure. $\Theta^{\mu\nu}_{e.m.}$ is obtained most simply by summing with (5.47) the conserved tensor:

$$S^{\mu,\nu} = -\partial_\beta(F^{\mu\beta}A^\nu) = -F^{\mu\beta}(\partial_\beta A^\nu)$$

since, in the absence of charges, $\partial_\beta F^{\mu\beta} = 0$. We note that:

$$\partial_\mu S^{\mu,\nu} = 0; \quad \int d^3x S^{0\nu} = 0.$$

In conclusion:

$$\Theta^{\mu\nu}_{e.m.} = -g_{\rho\sigma}F^{\rho\mu}F^{\sigma\nu} + \frac{g^{\mu\nu}}{4}F^{\alpha\beta}F_{\alpha\beta}. \tag{5.48}$$

The explicit forms of the energy and momentum densities obtained from (5.48) are:

$$\begin{aligned} E_{e.m.} &= \Theta^{00}_{e.m.} = \frac{1}{2}(\boldsymbol{E}\cdot\boldsymbol{E} + \boldsymbol{B}\cdot\boldsymbol{B}); \\ P^i_{e.m.} &= \Theta^{0i}_{e.m.} = (\boldsymbol{E}\times\boldsymbol{B})^i. \end{aligned} \tag{5.49}$$

We now consider the complete *Lagrangian density*, (5.38), which we rewrite for convenience:

$$\begin{aligned} \mathcal{L} &= \mathcal{L}_{e.m.} + \mathcal{L}_q + \mathcal{L}_{int}; \\ \mathcal{L}_q &= -mc^2\frac{1}{\gamma}\delta^{(3)}[\boldsymbol{x} - \boldsymbol{x}(t)]; \\ \mathcal{L}_{int} &= -j_\mu A^\mu = q\frac{1}{\gamma}\delta^{(3)}[\boldsymbol{x} - \boldsymbol{x}(t)]u_\mu A^\mu. \end{aligned}$$

The overall energy-momentum tensor is obtained by differentiating with respect to the degrees of freedom of the field and of the particle. We restrict ourselves to consideration of the spatial integral of the time components ($\boldsymbol{L}_q +$

$\boldsymbol{L}_{int} = \int d^3x(\mathcal{L}_q + \mathcal{L}_{int})$:

$$\begin{aligned}\boldsymbol{E}_{tot} &= \int d^3x T^{0,0}_{tot} \\ &= \int d^3x[\frac{\partial \mathcal{L}}{\partial_0 A_\beta}\partial^0 A_\beta - g^{00}\mathcal{L}_{e.m.}] + \frac{\partial(\boldsymbol{L}_q + \boldsymbol{L}_{int})}{\partial \boldsymbol{v}} \cdot \boldsymbol{v} - (\boldsymbol{L}_q + \boldsymbol{L}_{int}) \\ &= \int d^3x T^{0,0}_{e.m.} + mc^2\gamma + qA^0[\boldsymbol{x}(t), t] \end{aligned} \tag{5.50}$$

$$\begin{aligned}\boldsymbol{p}^i_{tot} &= \int d^3x T^{0i}_{e.m.} + \frac{\partial(\boldsymbol{L}_q + \boldsymbol{L}_{int})}{\partial v^i} \\ &= \int d^3x T^{0i}_{e.m.} + m\gamma v^i + qA^i[\boldsymbol{x}(t), t]. \end{aligned} \tag{5.51}$$

The tensor $T^{\mu,\nu}_{e.m.}$ is no longer conserved. Using the equation of motion for $F^{\mu\nu}$, (5.39), we find:

$$\begin{aligned}\frac{d\boldsymbol{p}^\nu_{e.m.}}{dt} &= \int d^3x \partial_0 T^{0,\nu}_{e.m.} = \int d^3x \partial_\mu T^{\mu,\nu}_{e.m.} = \\ &= -\int d^3x j_\beta \partial^\nu A^\beta + \int d^3x \frac{1}{2} F_{\mu\sigma}(\partial^\mu F^{\sigma\nu} + \partial^\sigma F^{\nu\mu} + \partial^\nu F^{\mu\sigma}).\end{aligned}$$

In the parentheses of the second term $\sigma \neq \mu$. If we also have $\nu \neq \sigma \neq \mu$, the bracketed term is zero for the homogeneous Maxwell's equations, (5.8). If instead $\nu = \sigma \neq \mu$, it vanishes because of the antisymmetry of $F^{\mu\nu}$. In any case, therefore:

$$\begin{aligned}\frac{d\boldsymbol{E}_{e.m.}}{dt} &= -\int d^3x\ j_\beta \partial^0 A^\beta = -q\partial_0 A^0 + q\boldsymbol{v} \cdot (\partial_0 \boldsymbol{A}); \\ \frac{d\boldsymbol{p}^i_{e.m.}}{dt} &= q\boldsymbol{v} \cdot (\partial_i \boldsymbol{A})\end{aligned}$$

from which:

$$\begin{aligned}\frac{d\boldsymbol{E}_{tot}}{dt} &= \frac{d\boldsymbol{E}_{e.m.}}{dt} + \frac{d\boldsymbol{E}_q}{dt} = \\ &= -q\partial_0 A^0 + q\boldsymbol{v} \cdot (\partial_0 \boldsymbol{A}) + \frac{d(mc^2\gamma)}{dt} + q\partial_0 A^0 + q\boldsymbol{v} \cdot \partial A^0 = \\ &= -q\boldsymbol{v} \cdot \boldsymbol{E} + \frac{\boldsymbol{d(mc^2\gamma)}}{\boldsymbol{dt}} = 0 \end{aligned} \tag{5.52}$$

as a consequence of the equation of motion of the particle, (5.44). Similarly, from the equation of motion, (5.43), we find:

$$\begin{aligned}\frac{d\boldsymbol{p}^i_{tot}}{dt} &= \frac{d\boldsymbol{p}^i_{e.m.}}{dt} + \frac{d\boldsymbol{p}^i_q}{dt} = \\ &= q\boldsymbol{v} \cdot (\partial_i \boldsymbol{A}) + \frac{dm\gamma v^i}{dt} + q\partial_0 A^i + \boldsymbol{v} \cdot \boldsymbol{\nabla} A^i = \\ &= \frac{dm\gamma v^i}{dt} - qE^i - q(\boldsymbol{v} \times \boldsymbol{B})^{\boldsymbol{i}} = 0. \end{aligned} \tag{5.53}$$

The evolution of the system in time consists of the continuous exchange of energy and momentum between field and particle, maintaining constant the values of the energy and the total momentum.

However, the definition of the energy and momentum associated with the particle or with the field are not unique at a given instant, since the quantities are not individually constants of the motion. A particularly simple description is obtained by eliminating $T^{\mu,\nu}_{e.m.}$ in favour of $\Theta^{\mu\nu}_{e.m.}$. Comparing (5.47) and (5.48) leads to:

$$T^{\mu,\nu} = \Theta^{\mu\nu} + F^{\mu\beta}\partial_\beta A^\nu = \Theta^{\mu\nu} + \partial_\beta(F^{\mu\beta}A^\nu) - j^\mu A^\nu.$$

The total derivative can be omitted and we can combine the last term with the expression for the energy-momentum of the particle. In this way we obtain:

$$\begin{aligned} \boldsymbol{E}_{tot} &= \frac{1}{2}\int d^3x(E^2 + B^2) + \frac{mc^2}{\sqrt{1-\frac{v^2}{c^2}}} \\ \boldsymbol{p}^i_{tot} &= \int d^3x(\boldsymbol{E}\times\boldsymbol{B})^i + \frac{mv^i}{\sqrt{1-\frac{v^2}{c^2}}}. \end{aligned} \tag{5.54}$$

In the expressions of (5.54) apparently, only the energy of the free particle appears. In the case of more than one particle, it can be asked where the energy from the electrostatic interaction between the particles has gone. The answer is that this energy is absorbed into the first term, as can be seen in the following way. We separate the electric field into longitudinal and transverse components, according to (5.32). We find:

$$\begin{aligned} \frac{1}{2}\int d^3x\; E^2 &= \frac{1}{2}\int d^3x(E_L^2 + E_T^2) = \\ &= \frac{1}{2}\int d^3x\; E_T^2 + \frac{1}{2}\int d^3x\;(\boldsymbol{\nabla}\Phi)\cdot(\boldsymbol{\nabla}\Phi) = \\ &= \frac{1}{2}\int d^3x\; E_T^2 - \frac{1}{2}\int d^3x\Phi(\boldsymbol{\nabla}\cdot\boldsymbol{\nabla}\Phi) = \\ &= \frac{1}{2}\int d^3x\; E_T^2 + \int d^3x\;\Phi\rho \end{aligned} \tag{5.55}$$

where $\Phi$ is the electrostatic potential. The second term in the final expression is the electrostatic energy which, for a system of pointlike particles, leads to the expression:

$$V_{Coul} = \frac{1}{2}\Sigma_{ij}\frac{1}{4\pi}\frac{q_i\, q_j}{|\boldsymbol{x}_i(t) - \boldsymbol{x}_j(t)|}\;. \tag{5.56}$$

We find:

$$E_{tot} = \frac{1}{2}\int d^3x(E_T^2 + B^2) + \frac{1}{2}\Sigma_{ij}\frac{1}{4\pi}\frac{q_i\, q_j}{|\boldsymbol{x}_i - \boldsymbol{x}_j|} + \frac{mc^2}{\sqrt{1-\frac{v^2}{c^2}}}. \tag{5.57}$$

The first term contains only the degrees of freedom of the radiation field, while the second and third contain the degrees of freedom of the particles.

Starting from (5.54) the unintegrated energy-momentum tensor can finally be derived, in the form:

$$\Theta^{\mu\nu} = \Theta^{\mu\nu}_{e.m.} + \theta^{\mu\nu};$$
$$\theta^{\mu\nu} = mc^2 \frac{1}{\gamma}\, u^\mu u^\nu\, \delta^{(3)}[\boldsymbol{x} - \boldsymbol{x}(t)].$$

Comment. For a pointlike particle we have introduced the Lagrangian *density*, current and energy-momentum in the form:

$$\mathcal{L} = -mc^2 \frac{1}{\gamma}\, \delta[\boldsymbol{x} - \boldsymbol{x}(t)];$$
$$j^\mu = q\frac{1}{\gamma}\, u^\mu\, \delta[\boldsymbol{x} - \boldsymbol{x}(t)];$$
$$\theta^{\mu\nu} = mc^2 \frac{1}{\gamma}\, u^\mu u^\nu\, \delta[\boldsymbol{x} - \boldsymbol{x}(t)];$$

It is interesting to observe that the factor $1/\gamma$ is essential to compensate the non-covariance of the $\delta$-function and produce the apppropriate covariant quantities, as seen from the following argument. We multiply $j^\mu$ by an explicitly covariant 4-density and by a finite but small 4-volume around $(x(\boldsymbol{t}), \boldsymbol{t})$:

$$\Delta(jA) = \Delta^3 x \Delta t (j_\mu A^\mu) = q\frac{\Delta t}{\gamma} u_\mu A^\mu(\mathbf{x}(t), t) = q\Delta\tau\, [u_\mu A^\mu(\mathbf{x}(\tau), \tau)] \quad (5.58)$$

where $\tau$ is the proper time of the particle. We have obtained an invariant result *for any* 4-vector $A^\mu$, therefore $j^\mu$ must transform like a 4-vector. Similar reasoning holds for the other densities.

## 5.4 HAMILTON FORMALISM AND MINIMAL SUBSTITUTION

The conjugate momentum of the particle was already calculated in (5.51):

$$\boldsymbol{p} = m\gamma\mathbf{v} + q\boldsymbol{A}[\boldsymbol{x}(t), t]. \quad (5.59)$$

From here we should derive $\boldsymbol{v}$ as a function of $\boldsymbol{p}$. We find:

$$\boldsymbol{v} = \frac{c\boldsymbol{\pi}}{\sqrt{|\boldsymbol{\pi}|^2 + (mc)^2}};$$
$$\boldsymbol{\pi} = \boldsymbol{p} - q\boldsymbol{A}.$$

The charged particle Hamiltonian is obtained from:

$$\begin{aligned}\boldsymbol{H}_p &= \boldsymbol{p}\cdot\boldsymbol{v} - (\boldsymbol{L}_p + \boldsymbol{L}_{int}) = \\ &= \boldsymbol{p}\cdot\boldsymbol{v} + qA^0 - q\boldsymbol{v}\cdot\boldsymbol{A} + mc^2\sqrt{1-\frac{v^2}{c^2}} = \\ &= qA^0 + \boldsymbol{v}\cdot\boldsymbol{\pi} + mc^2\sqrt{1-\frac{v^2}{c^2}}.\end{aligned}$$

Eliminating $\boldsymbol{v}$ we find, finally:

$$\boldsymbol{H}_p = qA^0 + \sqrt{(mc^2)^2 + c^2(\boldsymbol{p} - q\boldsymbol{A})^2}. \tag{5.60}$$

If we compare with the free particle Hamiltonian:

$$\boldsymbol{H}_p = \sqrt{(mc^2)^2 + (c\boldsymbol{p})^2} \tag{5.61}$$

we see that the interaction of a free-charged particle in a known electromagnetic field is introduced with the *minimal substitution*:

$$\begin{aligned}&\boldsymbol{H}_p \to \boldsymbol{H}_p - qA^0; \\ &\boldsymbol{p} \to \boldsymbol{p} - q\boldsymbol{A}\end{aligned} \tag{5.62}$$

or, in covariant terms:

$$p^\mu \to p^\mu - qA^\mu. \tag{5.63}$$

We conclude with the Hamiltonian of the electromagnetic field. Starting from the definition of the conjugate momentum of $A^\mu$:

$$\Pi^\mu = \frac{\partial L}{\partial\partial_0 A_\nu} = F^{0\mu}. \tag{5.64}$$

We note that $\Pi^0 = F^{00} = 0$, consistent with the fact that $A^0$ is not a real variable of the electromagnetic field. Moreover, as we saw in equations (5.31) and after, we can choose a gauge in which the vector potential is transverse, $\partial_i A^i = 0$. In this gauge:

$$\Pi^i = -\partial_0 A^i = E_T^i \tag{5.65}$$

The Hamiltonian of the field is given by:

$$\begin{aligned}\boldsymbol{H}_{e.m.} &= \int d^3x\{\boldsymbol{\pi}\cdot(\partial_0\boldsymbol{A}) - L_{e.m.}\} = \int d^3x\{E_T^2 - \frac{1}{2}(E^2 - B^2)\} = \\ &= \frac{1}{2}\int d^3x\{(E_T^2 + B^2)\} - \frac{1}{2}\int d^3x E_L^2.\end{aligned}$$

We must add this result to the Hamiltonian of the particle, (5.60). Considering the case of several charged particles, we find (with $A_i^\mu = A^\mu[\mathbf{x}_i(t), t]$):

$$\begin{aligned} \boldsymbol{H}_{tot} &= \frac{1}{2}\int d^3x(E_T^2 + B^2) - \\ &\quad - \frac{1}{2}\int d^3x\ E_L^2 + \Sigma_i q A_i^0 + \Sigma_i \sqrt{(m_i c^2)^2 + c^2(\boldsymbol{p}_i - \boldsymbol{A}_i)^2}\} = \\ &= \frac{1}{2}\int d^3x(E_T^2 + B^2) + \frac{1}{2}\Sigma_{ij}\frac{q_i q_j}{4\pi|\mathbf{x}_i - \mathbf{x}_j|} + \\ &\quad + \Sigma_i \sqrt{(m_i c^2)^2 + c^2(\boldsymbol{p}_i - \boldsymbol{A}_i)^2}. \end{aligned} \tag{5.66}$$

In the non-relativistic limit we find the well known result:

$$\boldsymbol{H}_{tot} = \frac{1}{2}\int d^3x(E_T^2 + B^2) + \frac{1}{2}\Sigma_{ij}\frac{q_i q_j}{4\pi|\mathbf{x}_i - \mathbf{x}_j|} + \Sigma_i \frac{(\boldsymbol{p}_i - q_i \boldsymbol{A}_i)^2}{2m_i^2}. \tag{5.67}$$

The Classical Zeeman Effect. The minimal substitution leads to the Lorentz force and numerous examples of classical electromagnetic phenomena therefore confirm its validity.

A characteristic example is the *Zeeman effect*, the splitting of spectral lines emitted or absorbed by atoms in a magnetic field, considered in classical theory by Lorentz [4]. We consider an electron in an atomic system described the Hamiltonian:

$$\boldsymbol{H} = \frac{\boldsymbol{p}^2}{2m} + \cdots + V(\mathbf{x}, \ldots) \tag{5.68}$$

where $\mathbf{x}, \mathbf{p}$ are the canonical variables of the electrons and the dots $(\ldots)$ indicate further terms in the kinetic energy or the potential from other degrees of freedom of the system. We suppose in addition that $V$ has a spherical symmetry, $V = V(r^2, \ldots)$ with $r^2 = x^2 + y^2 + z^2$. Hamilton's equations are:

$$\begin{aligned} \frac{d\mathbf{x}}{dt} &= \frac{\partial \boldsymbol{H}}{\partial \boldsymbol{p}} = \frac{\boldsymbol{p}}{m}; \\ \frac{d\boldsymbol{p}}{dt} &= -\frac{\partial \boldsymbol{H}}{\partial \boldsymbol{x}} = -\frac{\partial V}{\partial \boldsymbol{x}} = -\frac{\partial V}{\partial r^2} 2\boldsymbol{x}. \end{aligned} \tag{5.69}$$

Now we introduce a constant magnetic field directed along the $z$-axis, generated by the vector potential:

$$\boldsymbol{A} = \frac{B}{2}(-y, x, 0); \quad \nabla \times \boldsymbol{A} = \boldsymbol{B}.$$

According to the minimal substitution, the new Hamiltonian is:

$$\boldsymbol{H} = \frac{(\boldsymbol{p} + e\boldsymbol{A})^2}{2m} + \ldots + V(r^2, \ldots). \tag{5.70}$$

The spherical symmetry is reduced to an axial symmetry around the direction of $\boldsymbol{B}$, so we must treat the variables $x, p_x$ and $y, p_y$ separately from $z, p_z$. We replace the former with the complex variables:

$$\zeta = x + iy; \qquad p = p_x + ip_y \tag{5.71}$$

and in addition:

$$A = A_x + iA_y = i\frac{B}{2}\zeta. \tag{5.72}$$

Hamilton's equations for the coordinates transverse to $B$ are:

$$\frac{d\zeta}{dt} = \frac{\partial \boldsymbol{H}}{\partial p_x} + i\frac{\partial \boldsymbol{H}}{\partial p_y} = \frac{p}{m} + i\omega_L\zeta$$

where we have introduced the *Larmor frequency*:

$$\omega_L = \frac{eB}{2m}. \tag{5.73}$$

We define a new variable $\chi(t)$, setting:

$$\zeta(t) = \chi(t)e^{i\omega_L t}. \tag{5.74}$$

The previous equation gives:

$$\frac{d\chi}{dt}e^{i\omega_L t} = \frac{p}{m} \tag{5.75}$$

which suggests to redefine the conjugate momentum also, by setting:

$$p(t) = \pi(t)e^{i\omega_L t}. \tag{5.76}$$

Hamilton's equation for $p$ is then written;

$$\begin{aligned} \frac{dp}{dt} &= (\frac{d\pi}{dt} + i\omega_L\pi)e^{i\omega_L t} = -(\frac{\partial \boldsymbol{H}}{\partial x} + i\frac{\partial \boldsymbol{H}}{\partial y}) = \\ &= -\frac{\partial V}{\partial r^2}2\zeta + i\omega_L p = (-\frac{\partial V}{\partial r^2}2\chi + i\omega_L\pi)e^{i\omega_L t} \end{aligned} \tag{5.77}$$

where we have neglected terms quadratic in the magnetic field, which is negligible in normal experimental situations.

The conclusion obtained from equations (5.73) and (5.77) is that the variables $\chi$ and $\pi$ obey the same equations of motion as the unperturbed atom. From (5.74) we see that a precession around the magnetic field direction at the Larmor frequency is superimposed onto the unperturbed motion. The motion along $z$ is not affected by the field (the $z$-component of the Lorentz force is zero).

If we assume that the unperturbed motion is harmonic with frequency $\omega_0$, we have:

$$\chi(t) = ae^{+i\omega_0 t} + be^{-i\omega_0 t}, \; z = ce^{i\omega_0 t}$$

for the solutions of (5.69), while:

$$\zeta(t) = ae^{+i(\omega_0+\omega_L)t} + be^{-i(\omega_0-\omega_L)t}, \; z = ce^{i\omega_0 t} \tag{5.78}$$

for the motion in the magnetic field.

A spectral line of frequency $\omega_0$ absorbed or emitted by the atom in normal conditions separates into three components with frequency $\omega_0 \pm \omega_L$, $\omega_0$. In addition, light which travels in the direction of the magnetic field, which cannot contribute to the motion along $z$, contains only the first two components.

The classical predictions are observed in a certain number of cases (*normal* Zeeman effect). In other cases, the line structure in the magnetic field is more complicated. The *anomalous* Zeeman effect can only be explained by taking the magnetic moment of the electron spin into account, which we will return in Section 6.1.5.

## 5.5 QUANTISATION OF THE ELECTROMAGNETIC FIELD IN VACUUM

Despite the fact that the observation of electromagnetic phenomena was at the origin of the development of classical field theory, quantisation of the electromagnetic field in the canonical formalism presents non-trivial problems.

The starting point is the Lagrangian density (5.14)

$$\mathcal{L}_{e.m.} = -\frac{1}{4}F^{\mu\nu}F_{\mu\nu} = -\frac{1}{2}\left(\partial^\nu A^\mu - \partial^\mu A^\nu\right)\partial_\nu A_\mu = \frac{1}{2}\left(|\boldsymbol{E}|^2 - |\boldsymbol{B}|^2\right) \tag{5.79}$$

from which Maxwell's equations in the absence of charges and currents are derived, through the action principle. The canonical variables conjugate to the components of the vector potential $A^\mu$, which are obtained from:

$$\pi^\mu = \frac{\partial \mathcal{L}}{\partial \dot{A}_\mu} = F^{0\mu} \tag{5.80}$$

are $(i = 1, 2, 3)$:

$$\pi^0 = F^{00} = 0, \qquad \pi^i = F^{0i} = \partial^i A^0 - \partial^0 A^i = E^i. \tag{5.81}$$

From the corresponding expression for the Hamiltonian density:

$$\begin{aligned}\mathcal{H} = \sum_{i=1}^{3} \pi^i \dot{A}_i - \mathcal{L} = \sum_{i=1}^{3} E^i \dot{A}_i - \frac{1}{2}\left(|\boldsymbol{E}|^2 - |\boldsymbol{B}|^2\right) = \\ = \frac{1}{2}\left(|\boldsymbol{E}|^2 + |\boldsymbol{B}|^2\right) + \boldsymbol{E}\cdot\boldsymbol{\nabla}\phi\,,\end{aligned} \tag{5.82}$$

with $\phi = A^0$, it follows that (comparing with (5.54)):

$$H = \frac{1}{2}\int d^3x \left(|\boldsymbol{E}|^2 + |\boldsymbol{B}|^2\right)\,, \tag{5.83}$$

as is easily seen by integrating by parts and using $\boldsymbol{\nabla} \cdot \boldsymbol{E} = 0$.

The canonical commutation rules for $A^\mu$ and $\pi^i$ $(i, k = 1, 2, 3)$ are:

$$[A^\mu(\mathbf{x}, t), A^\nu(\mathbf{x}', t)] = 0 \tag{5.84}$$

$$\left[\pi^i(\mathbf{x}, t), \pi^k(\mathbf{x}', t)\right] = 0 \tag{5.85}$$

$$\left[\pi^k(\mathbf{x}, t), A^0(\mathbf{x}', t)\right] = 0 \tag{5.86}$$

showing that, because of the vanishing of the conjugate momentum $\pi^0$, the component $A^0$ of the vector potential, unlike the components $A^i$, commutes with all $A^\mu$ and $\pi^i$, and can therefore be described by a number, instead of an operator. This then limits application of the quantum formalism to the components $A^i$ only, and requires $A^0$ to be treated as a classical field.

This procedure, originally followed by Dirac [5], while being of great utility in many applications, has the disadvantage of not being manifestly covariant, since the components of the four-vector $A^\mu$ are treated in an asymmetric manner. The covariant quantisation of the electromagnetic field presents non-trivial technical problems and will be discussed elsewhere.

Energy and Momentum of the Electromagnetic Field. In the gauge defined by the condition:

$$\boldsymbol{\nabla} \cdot \boldsymbol{A} = 0 , \tag{5.87}$$

the vector potential satisfies:

$$\Box \boldsymbol{A}(x) = \boldsymbol{j}(x) - \boldsymbol{\nabla} \frac{\partial \phi(x)}{\partial t} , \tag{5.88}$$

as follows from (5.3), while (5.1) implies that the scalar potential $A^0 = \phi$, the solution of Poisson's equation:

$$\nabla^2 \phi(x) = -\rho(x) , \tag{5.89}$$

is the Coulomb potential generated by the charge distribution $\rho(x)$:

$$\phi(x) = \int d^3x' \, \frac{\rho(\mathbf{x}', t)}{|\mathbf{x} - \mathbf{x}'|} . \tag{5.90}$$

For this reason, the gauge (5.87) is called the Coulomb gauge.

In principle, the term containing $\phi$ on the right-hand side of (5.88) can be obtained from (5.90). However, separating the longitudinal and transverse components of the current, by writing:

$$\boldsymbol{j}(x) = \boldsymbol{j}_T(x) + \mathbf{j}_L(x) \tag{5.91}$$

with

$$\boldsymbol{\nabla} \times \boldsymbol{j}_L = 0, \quad \boldsymbol{\nabla} \cdot \boldsymbol{j}_T = 0 \tag{5.92}$$

and using the identity $\boldsymbol{\nabla} \times (\boldsymbol{\nabla} \times \boldsymbol{j}) = \boldsymbol{\nabla}(\boldsymbol{\nabla} \cdot \boldsymbol{j}) - \nabla^2 \boldsymbol{j}$, it is easily shown that

$$\boldsymbol{\nabla} \frac{\partial \phi(x)}{\partial t} = \boldsymbol{j}_L(x) \,, \tag{5.93}$$

which means that the source term in the equation which determines the vector potential $\boldsymbol{A}$ is reduced to the transverse component of the current, $\boldsymbol{j}_T(x)$.

In the absence of charges and currents, $\phi(x) \equiv 0$ and (5.88) becomes:

$$\Box \boldsymbol{A} = 0 \,. \tag{5.94}$$

A particular solution to (5.94) which satisfies periodic boundary conditions on a cubic box, as discussed in Section 4.1, can be written in the form:

$$\boldsymbol{u}_{\boldsymbol{k}}(\boldsymbol{x}, t) = \frac{1}{V} \, \boldsymbol{\epsilon}_{\boldsymbol{k}} \, \mathrm{e}^{-i(\omega_k t - \boldsymbol{k} \cdot \boldsymbol{x})} \,, \tag{5.95}$$

where $V$ is the normalisation volume and

$$\omega_k = |\boldsymbol{k}| \,. \tag{5.96}$$

The gauge condition (5.87) implies that the polarisation vector has the property:

$$\boldsymbol{k} \cdot \boldsymbol{\epsilon}_{\boldsymbol{k}} = 0 \,, \tag{5.97}$$

that is, for every wave vector $\boldsymbol{k}$, $\boldsymbol{\epsilon}_{\boldsymbol{k}}$ lies on the plane perpendicular to $\boldsymbol{k}$. Therefore we can define two real unit vectors $\boldsymbol{\epsilon}_{\mathbf{k}1}$ and $\boldsymbol{\epsilon}_{\mathbf{k}2}$, such that $(r, r' = 1, 2)$

$$\boldsymbol{\epsilon}_{\boldsymbol{k}r} \cdot \boldsymbol{\epsilon}_{\boldsymbol{k}r'} = \delta_{rr'} \tag{5.98}$$

and

$$\boldsymbol{\epsilon}_{\boldsymbol{k}r} \cdot \boldsymbol{k} = 0 \,. \tag{5.99}$$

The general solution of (5.94) which satisfies the gauge condition (5.87) is obtained from a linear combination of $\boldsymbol{u}_{\boldsymbol{k}}(\boldsymbol{x}, t)$ (we recall that $\boldsymbol{A}(\mathbf{x}, t)$ is a real function)

$$\begin{aligned} \boldsymbol{A}(\mathbf{x}, t) &= \sum_{\boldsymbol{k}} \sum_{r=1}^{2} \left[ c_{\boldsymbol{k}r} \boldsymbol{u}_{\boldsymbol{k}r}(\boldsymbol{x}, t) + c^*_{\mathbf{k}r} \boldsymbol{u}^*_{\boldsymbol{k}r}(\boldsymbol{x}, t) \right] \\ &= \sum_{\boldsymbol{k}} \sum_{r=1}^{2} \left[ c_{\boldsymbol{k}r}(t) \boldsymbol{u}_{\boldsymbol{k}r}(\mathbf{x}) + c^*_{\boldsymbol{k}r}(t) \boldsymbol{u}^*_{\boldsymbol{k}r}(\boldsymbol{x}) \right] \,, \end{aligned} \tag{5.100}$$

where we have defined

$$c_{\boldsymbol{k}r}(t) = c_{\boldsymbol{k}r} \mathrm{e}^{-i\omega_k t} \tag{5.101}$$

and the functions

$$\boldsymbol{u}_{\boldsymbol{k}r}(\boldsymbol{x}) = \frac{1}{\sqrt{V}} \, \boldsymbol{\epsilon}_{\boldsymbol{k}r} \, \mathrm{e}^{i\boldsymbol{k} \cdot \boldsymbol{x}} \tag{5.102}$$

satisfy the orthonormality conditions:

$$\int d^3x\ \boldsymbol{u}^*_{\boldsymbol{k}r}(\boldsymbol{x})\boldsymbol{u}_{\boldsymbol{k}'r'}(\boldsymbol{x}) = \frac{\boldsymbol{\epsilon}_{\boldsymbol{k}r}\cdot\boldsymbol{\epsilon}_{\boldsymbol{k}'r'}}{V}\int d^3x\ \mathrm{e}^{i(\boldsymbol{k}-\boldsymbol{k}')\cdot\boldsymbol{x}} = \delta_{rr'}\delta_{\boldsymbol{k}\boldsymbol{k}'}\ . \tag{5.103}$$

Substituting into the equation for the energy of the electromagnetic field (comparing with (5.54)):

$$H = \frac{1}{2}\int d^3x\ \left(\ |\boldsymbol{B}|^2 + |\boldsymbol{E}|^2\ \right) = \frac{1}{2}\int d^3x\ \left(\ |\boldsymbol{\nabla}\times\boldsymbol{A}|^2\ + \left|\frac{\partial\boldsymbol{A}}{\partial t}\right|^2\ \right) \tag{5.104}$$

the Fourier expansion (5.100) gives

$$H = \frac{1}{2}\sum_{\boldsymbol{k}r}\sum_{\boldsymbol{k}'r'}\int d^3x\ \Bigg\{[c_{\boldsymbol{k}r}(t)\ \boldsymbol{\nabla}\times\boldsymbol{u}_{\boldsymbol{k}r}(\boldsymbol{x})+c.c.]\cdot[c^*_{\boldsymbol{k}'r'}(t)\ \boldsymbol{\nabla}\times\boldsymbol{u}^*_{\boldsymbol{k}'r'}(\boldsymbol{x})+c.c.]$$
$$+\left[\frac{\partial c_{\boldsymbol{k}r}}{\partial t}\ \boldsymbol{u}_{\boldsymbol{k}r}(\boldsymbol{x}) + c.c.\right]\cdot\left[\frac{\partial c^*_{\boldsymbol{k}'r'}}{\partial t}\ \boldsymbol{u}^*_{\boldsymbol{k}'r'}(\boldsymbol{x}) + c.c.\right]\Bigg\}. \tag{5.105}$$

The calculation of the contribution of the magnetic field requires integrations of the type:

$$\begin{aligned}
&\int d^3x\ (\boldsymbol{\nabla}\times\boldsymbol{u}_{\boldsymbol{k}r}(\mathbf{x}))\cdot(\boldsymbol{\nabla}\times\boldsymbol{u}^*_{\boldsymbol{k}'r'}(\boldsymbol{x}))\\
&= \int d^3x\ \boldsymbol{\nabla}\cdot[\boldsymbol{u}_{\boldsymbol{k}r}(\mathbf{x})\times(\boldsymbol{\nabla}\times\boldsymbol{u}^*_{\boldsymbol{k}'r'}(\boldsymbol{x}))]+\\
&\qquad+\int d^3x\ \boldsymbol{u}_{\boldsymbol{k}r}(\mathbf{x})\cdot[\boldsymbol{\nabla}\times(\boldsymbol{\nabla}\times\boldsymbol{u}^*_{\boldsymbol{k}'r'}(\boldsymbol{x}))]\\
&= -\int d^3x\ \boldsymbol{u}_{\boldsymbol{k}r}(\mathbf{x})\nabla^2\boldsymbol{u}^*_{\boldsymbol{k}'r'}(\boldsymbol{x})\\
&= \frac{|\boldsymbol{k}'|^2}{V}\ \boldsymbol{\epsilon}_{\boldsymbol{k}r}\cdot\boldsymbol{\epsilon}_{\boldsymbol{k}'r'}\int d^3x\ \mathrm{e}^{i(\boldsymbol{k}-\boldsymbol{k}')\cdot\boldsymbol{x}} = \omega^2_{k'}\ \delta_{rr'}\delta_{\boldsymbol{k}\boldsymbol{k}'}\ ,
\end{aligned}$$

which are easily carried out using the periodic boundary conditions at the edges of the volume of integration and the identities

$(\boldsymbol{\nabla}\times\boldsymbol{u})\cdot(\boldsymbol{\nabla}\times\boldsymbol{v}) = \boldsymbol{\nabla}\cdot[\boldsymbol{u}\times(\boldsymbol{\nabla}\times\boldsymbol{v})] + \boldsymbol{u}\cdot[\boldsymbol{\nabla}\times(\boldsymbol{\nabla}\times\boldsymbol{v})]$

and

$\boldsymbol{\nabla}\times(\boldsymbol{\nabla}\times\boldsymbol{u}) = \boldsymbol{\nabla}\cdot(\boldsymbol{\nabla}\cdot\boldsymbol{u}) - \nabla^2\boldsymbol{u}.$

The corresponding integrals for the contribution of the electric field are instead of the type:

$$\int d^3x\ \frac{\partial c_{\boldsymbol{k}r}}{\partial t}\boldsymbol{u}_{\boldsymbol{k}r}(\boldsymbol{x})\cdot\frac{\partial c^*_{\boldsymbol{k}'r'}}{\partial t}\boldsymbol{u}^*_{\boldsymbol{k}'r'}(\boldsymbol{x}) = \omega_k\omega_{k'}\ \delta_{rr'}\delta_{\boldsymbol{k}\boldsymbol{k}'}\ c_{\boldsymbol{k}r}(t)c^*_{\boldsymbol{k}'r'}(t)\ .$$

The final result is:

$$H = \sum_{\boldsymbol{k}r}\omega^2_k\ [c_{\boldsymbol{k}r}(t)c^*_{\boldsymbol{k}r}(t) + c^*_{\boldsymbol{k}r}(t)c_{\boldsymbol{k}r}(t)]\ , \tag{5.106}$$

where the functions $c_{\boldsymbol{k}r}(t)$ satisfy the differential equation:

$$\ddot{c}_{\boldsymbol{k}r}(t) = -\omega_k^2 \, c_{\boldsymbol{k}r}(t) , \tag{5.107}$$

which is the equation of motion of a classical harmonic oscillator of angular frequency $\omega_k$ and unit mass. We note that, if the functions $c_{\boldsymbol{k}r}(t)$ are complex numbers, $[c_{\boldsymbol{k}r}(t)c^*_{\boldsymbol{k}r}(t) + c^*_{\boldsymbol{k}r}(t)c_{\boldsymbol{k}r}(t)] = 2c_{\boldsymbol{k}r}(t)c^*_{\boldsymbol{k}r}(t)$. The reason we have written $H$ in the form (5.106) will be made clear shortly.

The energy of a system of classical, unit mass oscillators, vibrating in the directions defined by the wave vectors $\boldsymbol{k}$ with angular frequency $\omega_k$ is:

$$H_{osc} = \sum_{\boldsymbol{k}r} \frac{1}{2} \left(p^2_{\boldsymbol{k}r} + \omega_k^2 x^2_{\boldsymbol{k}r}\right) , \tag{5.108}$$

where $x_{\boldsymbol{k}r}$ and $p_{\boldsymbol{k}r}$ are the canonical classical variables, which satisfy the equations of motion:

$$\ddot{x}_{\boldsymbol{k}r}(t) = -\omega_k^2 \, x_{\boldsymbol{k}r}(t), \quad \ddot{p}_{\boldsymbol{k}r}(t) = -\omega_k^2 \, p_{\boldsymbol{k}r}(t) . \tag{5.109}$$

Comparing (5.106) with (5.108) we see immediately that $H$ coincides with $H_{osc}$ if

$$c_{\boldsymbol{k}r} = \frac{1}{2\omega_k}(\omega_k x_{\boldsymbol{k}r} + i p_{\boldsymbol{k}r}) , \qquad c^*_{\boldsymbol{k}r} = \frac{1}{2\omega_k}(\omega_k x_{\boldsymbol{k}r} - i p_{\boldsymbol{k}r}) , \tag{5.110}$$

or if

$$x_{\boldsymbol{k}r} = c_{\boldsymbol{k}r} + c^*_{\boldsymbol{k}r} , \qquad p_{\boldsymbol{k}r} = -i\omega_k(c_{\boldsymbol{k}r} - c^*_{\boldsymbol{k}r}) . \tag{5.111}$$

This result suggests to interpret $H$ as the energy of a set of classical oscillators. The path to the quantum case is straightforward. The energy of a system of quantum oscillators has the form:

$$H = \sum_{\boldsymbol{k}r} \frac{\omega_k}{2} \left(a^\dagger_{\boldsymbol{k}r} a_{\boldsymbol{k}r} + a_{\boldsymbol{k}r} a^\dagger_{\boldsymbol{k}r}\right) , \tag{5.112}$$

which agrees with the energy of the electromagnetic field (5.106) if the coefficients of the expansion of $\boldsymbol{A}$ in a Fourier series are interpreted as quantum operators defined by the equations:

$$c_{\boldsymbol{k}r} = \frac{1}{\sqrt{2\omega_k}} \, a_{\boldsymbol{k}r} , \qquad c^*_{\boldsymbol{k}r} = \frac{1}{\sqrt{2\omega_k}} \, a^\dagger_{\boldsymbol{k}r} . \tag{5.113}$$

The vector potential $\boldsymbol{A}$ can therefore be expressed in terms of the creation and destruction operators $a^\dagger_{\boldsymbol{k}r}$ and $a_{\boldsymbol{k}r}$

$$\boldsymbol{A}(x) = \sum_{\boldsymbol{k}r} \frac{1}{\sqrt{2V\omega_k}} \boldsymbol{\epsilon}_{\boldsymbol{k}r} \left[a_{\boldsymbol{k}r}(t) \mathrm{e}^{-i\boldsymbol{k}\cdot\boldsymbol{x}} + a^\dagger_{\boldsymbol{k}r}(t) \mathrm{e}^{i\boldsymbol{k}\cdot\boldsymbol{x}}\right] \tag{5.114}$$

$$= \sum_{\boldsymbol{k}r} \frac{1}{\sqrt{2V\omega_k}} \boldsymbol{\epsilon}_{\boldsymbol{k}r} \left[a_{\boldsymbol{k}r} \mathrm{e}^{-i(\omega_k t - \boldsymbol{k}\cdot\boldsymbol{x})} + a^\dagger_{\boldsymbol{k}r} \mathrm{e}^{i(\omega_k t - \boldsymbol{k}\cdot\boldsymbol{x})}\right]$$

$$= \boldsymbol{A}^+(x) + \boldsymbol{A}^-(x) , \tag{5.115}$$

and is itself an operator in the Hilbert space whose state vectors are:

$$|n_{\boldsymbol{k}_1 r_1}, n_{\boldsymbol{k}_2 r_2}, \ldots n_{\boldsymbol{k}_n r_n} \ldots\rangle, \tag{5.116}$$

where $n_{\boldsymbol{k}r}$ is the quantum number of the mode of oscillation characterised by the wave vector $\boldsymbol{k}$ and by the polarisation vector $\boldsymbol{\epsilon}_{\boldsymbol{k}r}$. The state (5.116) can be obtained starting from the vacuum state, which corresponds to $n_{\boldsymbol{k}r} \equiv 0$, using

$$|n_{\boldsymbol{k}_1 r_1}, n_{\boldsymbol{k}_2 r_2}, \ldots\rangle = \prod_{\boldsymbol{k}_i r_i} \frac{(a^\dagger_{\boldsymbol{k}_i r_i})^{n_{\boldsymbol{k}_i r_i}}}{\sqrt{n_{\boldsymbol{k}_i r_i}!}} |0\rangle \ . \tag{5.117}$$

Notice that in (5.115) we have separated the contribution of the terms containing creation operators ($\boldsymbol{A}^-(x)$) from those of annihilation ($\boldsymbol{A}^+(x)$). Evidently:

$$\boldsymbol{A}^+(x)|0\rangle = 0 \ . \tag{5.118}$$

Choosing $\langle 0|H|0\rangle$ as the zero of the energy scale, the Hamiltonian of the electromagnetic field becomes:

$$H = \sum_{\boldsymbol{k}r} \omega_k \ N_{\boldsymbol{k}r} \ , \tag{5.119}$$

with $N_{\boldsymbol{k}r} = a^\dagger_{\boldsymbol{k}r} a_{\boldsymbol{k}r}$, and satisfying the eigenvalue equations:

$$H|n_{\boldsymbol{k}_1 r_1}, n_{\boldsymbol{k}_2 r_2}, \ldots\rangle = \sum_{\boldsymbol{k}_i r_i} n_{\boldsymbol{k}_i r_i} \omega_{\boldsymbol{k}_i} \ |n_{\boldsymbol{k}_1 r_1}, n_{\boldsymbol{k}_2 r_2}, \ldots\rangle \ . \tag{5.120}$$

Because $\boldsymbol{A}(x)$ is linear in the creation and destruction operators, from the commutation rules

$$[a_{\boldsymbol{k}r}, N_{\boldsymbol{k}'r'}] = \left[a_{\boldsymbol{k}r}, a^\dagger_{\boldsymbol{k}'r'} a_{\boldsymbol{k}'r'}\right] = \delta_{rr'} \delta_{\boldsymbol{k}\boldsymbol{k}'} a_{\boldsymbol{k}r} \tag{5.121}$$

and

$$\left[a^\dagger_{\boldsymbol{k}r}, N_{\boldsymbol{k}'r'}\right] = \left[a^\dagger_{\boldsymbol{k}r}, a^\dagger_{\boldsymbol{k}'r'} a_{\boldsymbol{k}'r'}\right] = \delta_{rr'} \delta_{\boldsymbol{k}\boldsymbol{k}'} a^\dagger_{\boldsymbol{k}r} \ , \tag{5.122}$$

it follows that the operator $N_{\boldsymbol{k}r}$ does not commute with $\boldsymbol{A}(x)$ nor, therefore, with the electric and magnetic fields $\boldsymbol{E}$ and $\boldsymbol{B}$. This result implies that the values of $\boldsymbol{E}$ and $\boldsymbol{B}$ and $n_{\boldsymbol{k}r}$ cannot be simultaneously measured with arbitrary precision. Moreover, from the linearity of $\boldsymbol{A}(x)$ in $a_{\boldsymbol{k}r}$ and $a^\dagger_{\boldsymbol{k}r}$, it also follows that the expectation values $\langle \boldsymbol{E}\rangle$ and $\langle \boldsymbol{B}\rangle$ in the states (5.116) are all zero.

Classically the momentum of an electromagnetic wave is given by the Poynting vector (compare with equation (5.54))

$$\boldsymbol{p} = \int d^3x \ \boldsymbol{E} \times \boldsymbol{B} = -\int d^3x \ \frac{\partial \boldsymbol{A}}{\partial t} \times (\boldsymbol{\nabla} \times \boldsymbol{A}) \ . \tag{5.123}$$

Substituting (5.114) in (5.123) we find

$$\begin{aligned} \boldsymbol{p} &= \sum_{\boldsymbol{k}r} \boldsymbol{k}\, \frac{1}{2}\left(a^\dagger_{\boldsymbol{k}r} a_{\boldsymbol{k}r} + a_{\boldsymbol{k}r} a^\dagger_{\boldsymbol{k}r}\right) \\ &= \sum_{\boldsymbol{k}r} \boldsymbol{k} \left(a^\dagger_{\boldsymbol{k}r} a_{\boldsymbol{k}r} + \frac{1}{2}\right) \\ &= \sum_{\boldsymbol{k}r} \boldsymbol{k}\, N_{\boldsymbol{k}r} \ . \end{aligned} \tag{5.124}$$

To obtain (5.124) it is sufficient to rewrite (5.123) in the form

$$\boldsymbol{p} = -\frac{1}{2}\int d^3x\ (\boldsymbol{E}\times\boldsymbol{B} - \boldsymbol{B}\times\boldsymbol{E}) \ . \tag{5.125}$$

In this way it is immediately seen that the terms containing two creation or two destruction operators do not contribute. For the calculation of the other terms we use

$$\begin{aligned} &\int d^3x \sum_{\boldsymbol{k}r}\sum_{\boldsymbol{k}'r'} \frac{1}{2V}\frac{1}{\sqrt{\omega_k\omega_{k'}}} \boldsymbol{\epsilon}_{\boldsymbol{k}r}\times(\boldsymbol{k}\times\boldsymbol{\epsilon}_{\boldsymbol{k}'r'})\omega_k \times \\ &\qquad\qquad \left[a_{\boldsymbol{k}r}a^\dagger_{\boldsymbol{k}'r'}e^{-i(k-k')x} + a^\dagger_{\boldsymbol{k}r}a_{\boldsymbol{k}'r'}e^{i(k-k')x}\right] \\ &= \sum_{\boldsymbol{k}r}\sum_{\boldsymbol{k}'r'} \frac{1}{2}\frac{\omega_k}{\sqrt{\omega_k\omega_{k'}}} \boldsymbol{\epsilon}_{\boldsymbol{k}r}\times(\boldsymbol{k}\times\boldsymbol{\epsilon}_{\boldsymbol{k}'r'}) \left(a_{\boldsymbol{k}r}a^\dagger_{\boldsymbol{k}'r'} + a^\dagger_{\boldsymbol{k}r}a_{\boldsymbol{k}'r'}\right)\delta_{\boldsymbol{k}\boldsymbol{k}'} \\ &= \sum_{\boldsymbol{k}r} \boldsymbol{k} \left(a^\dagger_{\boldsymbol{k}r}a_{\boldsymbol{k}r} + \frac{1}{2}\right) \end{aligned} \tag{5.126}$$

where, clearly, $\sum_{\boldsymbol{k}} \boldsymbol{k} = 0$.

Equation (5.124) shows that the quantum of the electromagnetic field with energy $\omega_k = |\boldsymbol{k}|$ has momentum $\boldsymbol{k}$, and can therefore be identified with a particle of zero mass (as follows from $\omega_k^2 - |\boldsymbol{k}^2| = 0$), the photon. The corpuscular nature of electromagnetic radiation was confirmed in 1922 by the observation of the Compton effect, which demonstrates the conservation of momentum and energy in an elastic collision between photons and atomic electrons.

## 5.6 THE SPIN OF THE PHOTON

The polarisation state of the photon is determined by the projection of its angular momentum along the quantisation axis, which we can take to be $x^3$. We apply the definition of the canonical tensor for the angular momentum from Chapter 3 (equations (3.90) and (3.91)), to the case of the electromagnetic field. Because this is a vector field (which transforms like a 4-vector under Lorentz transformations) the indices $MN$ which appear in the definition of $\Sigma^{\alpha\beta}_{MN}$ (equation(3.87)) are Lorentz indices in this case. In this way, we find:

$$\Sigma^{\alpha\beta}_{\mu\nu} = g^{\alpha\mu}g^{\beta}_{\nu} - g^{\mu}_{\beta}g^{\nu\alpha} \ , \tag{5.127}$$

that, substituted in the equation analogous to (3.91) for $J^3$, gives the result

$$J^3 = \int d^3x \left[ \sum_{i=1}^{3} \dot{A}^i \left( x^1 \frac{\partial}{\partial x_2} - x^2 \frac{\partial}{\partial x_1} \right) A^i - (\dot{A}^1 A^2 - \dot{A}^2 A^1) \right] . \quad (5.128)$$

Now substituting (5.114) into (5.128) we see that $J^3$ satisfies the commutation rules

$$\left[ J^3, a^\dagger_{\boldsymbol{k}r} \right] = i \left( \epsilon^1_{\boldsymbol{k}r} a^\dagger_{\boldsymbol{k}2} - \epsilon^2_{\boldsymbol{k}r} a^\dagger_{\boldsymbol{k}1} \right) , \quad (5.129)$$

where we have chosen the axis $x^3$ along the direction of the wave vector $\boldsymbol{k}$. Note that only the second term in the integrand of (5.128), which we can interpret as spin angular momentum, gives a non-zero contribution to the commutator (5.129).

We now define two new operators

$$a^\dagger_{\boldsymbol{k}R} = \frac{1}{\sqrt{2}} \left( a^\dagger_{\boldsymbol{k}1} + i a^\dagger_{\boldsymbol{k}2} \right), \quad a^\dagger_{\boldsymbol{k}L} = \frac{1}{\sqrt{2}} \left( a^\dagger_{\boldsymbol{k}1} - i a^\dagger_{\boldsymbol{k}2} \right) , \quad (5.130)$$

which create *circularly* polarised photons; that is, photons whose state of polarisation is described by the vectors

$$\boldsymbol{\epsilon}_{\boldsymbol{k}R} = \frac{1}{\sqrt{2}} \left( \boldsymbol{\epsilon}_{\boldsymbol{k}1} + i \boldsymbol{\epsilon}_{\boldsymbol{k}2} \right), \quad \boldsymbol{\epsilon}_{\boldsymbol{k}L} = \frac{1}{\sqrt{2}} \left( \boldsymbol{\epsilon}_{\boldsymbol{k}1} - i \boldsymbol{\epsilon}_{\boldsymbol{k}2} \right) . \quad (5.131)$$

Setting $\boldsymbol{\epsilon}_{\boldsymbol{k}1} \equiv (1,0,0)$ and $\boldsymbol{\epsilon}_{\boldsymbol{k}2} \equiv (0,1,0)$ and rewriting (5.129) in terms of the new operators we obtain the commutation rules

$$\left[ J^3, a^\dagger_{\boldsymbol{k}R} \right] = a^\dagger_{\boldsymbol{k}R}, \quad \left[ J^3, a^\dagger_{\boldsymbol{k}L} \right] = -a^\dagger_{\boldsymbol{k}L} , \quad (5.132)$$

from which it follows that the third component of the angular momentum of the state with a photon is given by

$$J^3 a^\dagger_{\boldsymbol{k}R} |0\rangle = \left[ J^3, a^\dagger_{\boldsymbol{k}R} \right] |0\rangle = a^\dagger_{\boldsymbol{k}R} |0\rangle , \quad (5.133)$$

$$J^3 a^\dagger_{\boldsymbol{k}L} |0\rangle = \left[ J^3, a^\dagger_{\boldsymbol{k}L} \right] |0\rangle = -a^\dagger_{\boldsymbol{k}L} |0\rangle . \quad (5.134)$$

This result shows that the photon has spin $|\boldsymbol{J}| = 1$, as required by the vector nature of the electromagnetic field, and that the two projections $J_3 = \pm 1$ corresponds to the circularly polarised states. The absence of the state with $J_3 = 0$ is a consequence of the transversality condition (5.87), which reduces the number of degrees of freedom by one unit.

## 5.7 PROBLEMS FOR CHAPTER 5

### Sect. 5.5

**1.** Consider a radiation field in the Coulomb gauge: $\boldsymbol{\nabla} \cdot \boldsymbol{A} = 0$. Using the plane wave expansions of the fields $A_i$ $(i = 1, \, ,2, \, 3)$ and of the corresponding conjugate variables $\pi_i$, show that

$$[A_i(t, \boldsymbol{x}), \pi_j(t, \boldsymbol{x}'))] = i\delta^{\perp}_{ij}(\boldsymbol{x} - \boldsymbol{x}')$$

where the *transverse δ-function* is defined as

$$\delta^{\perp}_{ij}(\boldsymbol{x} - \boldsymbol{x}') = \sum_{\boldsymbol{k}} \left( \delta_{ij} - \frac{k_i k_j}{\boldsymbol{k}^2} \right) \frac{1}{\sqrt{V}} e^{\boldsymbol{k} \cdot (\boldsymbol{x} - \boldsymbol{x}')} \, .$$

**2.** In the Coulomb gauge, Section 5.5, the interactions of a collection of atomic electrons with the electromagnetic field are described by the Hamiltonian [see eq. (5.67)]

$$H_{\text{int}} = \sum_{i=1}^{Z} \frac{e}{m} (\boldsymbol{p}_i \cdot \boldsymbol{A})$$

where $e$ and $m$ are the electron charge and mass, while $\boldsymbol{A}$ is the radiation field.

Using the dipole approximation, which amounts to replacing $e^{i\boldsymbol{k} \cdot \boldsymbol{x}} \to 1$ in the plane-wave expansion of the field $\boldsymbol{A}$, evaluate the transition matrix element associated with photon emission and absorption

$$M_{i \to f} = \langle f | \; H_{\text{int}} \; | i \rangle$$

where

$$|i\rangle = |A, \; n_{\boldsymbol{k}r}\rangle \quad , \quad |f\rangle = |B, \; n_{\boldsymbol{k}r} \pm 1\rangle \, .$$

In the above equations, $n_{\boldsymbol{k}r}$ is the number of photons in the state of momentum $\boldsymbol{k}$ and polarisation specified by the index $r$, whereas $A$ and $B$ denote the quantum numbers of the atomic initial and final states, respectively.

**3.** Obtain the commutation rules satisfied by the components of the electric and magnetic fields $(i = 1, \; 2, \; 3)$

$$[E^i(x), E^j(y)], \quad [B^i(x), B^j(y)], \quad [E^i(x), B^j(y)]$$

for $x_0 = y_0$.

**4.** The basis of eigenstates of the electromagnetic Hamiltonian, specified by the number of photons with a given wave number, **k**, and polarisation, $r$, provides a useful representation when the number of photons is small. The states produced by a classical light source, however, are best described using coherent states.

Consider the coherent state associated with a single oscillation mode of the electromagnetic field

$$|\alpha\rangle = e^{-|\alpha|^2} \sum_{n=0}^{\infty} \frac{a^{\dagger n}}{\sqrt{n!}} |n\rangle$$

where $\alpha$ is an arbitrary complex number, $a^\dagger$ is the creation operator and $|n\rangle$ is the $n$-photon state.

- Show that $\alpha$ is an eigenstate of the annihilation operator, satisfying the eigenvalue equation

$$a|\alpha\rangle = \alpha|\alpha\rangle \ .$$

- Using the above result and the relations

$$a = \frac{1}{\sqrt{2\hbar\omega}}(\omega x + ip), \quad a^\dagger = \frac{1}{\sqrt{2\hbar\omega}}(\omega x - ip)$$

show that the $|\alpha\rangle$ is a minimum uncertainty state, i.e. that

$$\Delta x \Delta p = \frac{\hbar}{2} \ .$$

Hint: use $(\Delta x)^2 = \langle x^2\rangle - \langle x\rangle^2$ and $(\Delta p)^2 = \langle p^2\rangle - \langle p\rangle^2$.

**5.** The Lagrangian density

$$\mathcal{L} = -\frac{1}{4}F_{\mu\nu}F^{\mu\nu} + \frac{1}{2}m^2 A_\mu A^\mu - j_\mu A^\mu$$

with

$$F^{\mu\nu} = \partial^\nu A^\mu - \partial^\mu A^\nu \ ,$$

describes a massive vector field interacting with the current $j_\mu$.

- Derive the equations of motion of the field $A^\mu$.
- Using the obtained result, determine which condition must be fulfilled to satisfy the relation

$$\partial_\mu A^\mu = 0 \ .$$

- Calculate the variation of the action, defined as

$$\delta S = \int d^4x \ \mathcal{L}(A'_\mu, \partial_\nu A'_\mu) - \int d^4x \ \mathcal{L}(A_\mu, \partial_\nu A_\mu)$$

associated with the gauge transformation

$$A_\mu \to A_\mu{}' = A_\mu - \partial_\mu \Lambda \ .$$

**6.** Starting from the transformation rules

$$A^{\mu'}(x') = \Lambda^{\mu}_{\nu} A^{\nu}(x)$$

calculate the angular momentum matrices $(\Sigma^{\alpha\beta})^{\mu}_{\nu}$, introduced in Section 3.6, and show that the symmetric energy-momentum tensor, $\theta^{\mu\nu}_{e.m.}$ of Eq. (5.48), can be obtained from the canonical energy-momentum tensor $T^{\mu,\nu}$ using the method of Belinfante and Rosenfeld.

CHAPTER 6

# THE DIRAC EQUATION

From this chapter onward we will use natural units, in which $\hbar = c = 1$.

According to non-relativistic quantum mechanics, the evolution of the wave function of a free particle of mass $m$ and momentum $\mathbf{p}$ is described by the Schrödinger equation:

$$i\frac{\partial \psi}{\partial t} = \boldsymbol{H}\psi \ , \tag{6.1}$$

with the Hamiltonian operator $\boldsymbol{H} = -\boldsymbol{\nabla}^2/2m$ obtained from the expression for energy

$$E = \frac{\boldsymbol{p}^2}{2m} \ , \tag{6.2}$$

using the substitution

$$E \to i\frac{\partial}{\partial t}, \quad \boldsymbol{p} \to -i\boldsymbol{\nabla} \ . \tag{6.3}$$

Schrödinger himself [6] first suggested a generalisation of equation (6.1) based on the use of the relativistic energy equation:

$$E^2 = \boldsymbol{p}^2 + m^2 \ . \tag{6.4}$$

The outcome of this process is the Klein–Gordon equation:

$$(\square + m^2)\psi = 0 \ , \tag{6.5}$$

Multiplying by $\psi^*$ and subtracting the product of $\psi$ and the complex conjugate of (6.5) from the result, the continuity equation is obtained

$$\frac{\partial \rho}{\partial t} = \boldsymbol{\nabla j} \ , \tag{6.6}$$

with

$$\rho = \psi^* \frac{\partial \psi}{\partial t} - \psi\frac{\partial \psi^*}{\partial t} \tag{6.7}$$


DOI: 10.1201/9781003436263-6

and

$$j = -\nabla \left(\psi^* \nabla \psi - \psi \nabla \psi^*\right) . \qquad (6.8)$$

However, the $\rho$ which appears in equation (6.6) cannot be identified with the probability density, in analogy with the similar quantity obtained from the Schrödinger equation, because it does not have the required property of being always positive-definite. To be convinced of this it is sufficient to substitute $i\partial/\partial t \to E$ in (6.7). The result,

$$\rho = E|\psi|^2 ,$$

shows that $\rho$ can be either positive or negative, following from the fact that the Klein–Gordon equation has solutions for the energy with both signs

$$E = \pm\sqrt{\boldsymbol{p}^2 + m^2} ,$$

Note that in the non-relativistic limit $E \approx m > 0$, and the familiar result $\rho \propto |\psi|^2$ is recovered.

In addition, the presence of the second-time derivative conflicts with the postulate of quantum mechanics stating that the wave function contains all the information on the state of a physical system, which must therefore be completely determined by its initial value.

Because of these problems the Klein–Gordon equation was initially abandoned, until Pauli and Weisskopf [7] suggested that the solution should be interpreted as a quantum field, instead of the wave function of a particle.

## 6.1 FORM AND PROPERTIES OF THE DIRAC EQUATION

If, at a given instant, the wave function should contain all the information on the state, the wave equation must be of first order with respect to time. Because the relativistic treatment requires that the time and the spatial coordinates should be treated symmetrically, this implies that the spatial derivatives must also be of first order. Moreover, the solutions should be compatible with the Klein–Gordon equation, which is obtained directly from the relativistic expression for energy (6.4).

To satisfy all these conditions, Dirac proposed [9] to write the wave equation in the form

$$i\frac{\partial \psi}{\partial t} = (-i\boldsymbol{\alpha} \cdot \nabla + \beta m)\psi , \qquad (6.9)$$

where $\psi$ is a vector with $N$ components

$$\psi = \begin{pmatrix} \psi_1 \\ \psi_1 \\ \cdots \\ \psi_N \end{pmatrix} \qquad (6.10)$$

and $\alpha_i$ $(i = 1, 2, 3)$ and $\beta$ are $N \times N$, matrices with $N$ to be determined.

Note that wave functions of the type (6.10) are encountered in non-relativistic quantum mechanics. For example, the wave function of a particle of spin $\frac{1}{2}$ is a two-component vector.

Because the Hamiltonian is a hermitian operator, it must be true that $\boldsymbol{\alpha} = \boldsymbol{\alpha}^\dagger$ and $\beta = \beta^\dagger$. Moreover, from the requirement that (6.9) should be compatible with the Klein–Gordon equation, or

$$(\boldsymbol{\alpha} \cdot \mathbf{p} + \beta m)(\boldsymbol{\alpha} \cdot \mathbf{p} + \beta m) = E^2 = \mathbf{p}^2 + m^2,$$

it follows that:

$$\begin{aligned}
&\alpha_i \alpha_j p_i p_j + m(\alpha_i \beta + \beta \alpha_i) p_i + \beta^2 m^2 \\
&= \frac{1}{2} \left( \{\alpha_i, \alpha_j\} + [\alpha_i, \alpha_j] \right) p_i p_j + m \{\alpha_i, \beta\} p_i + \beta^2 m^2 \\
&= p_i p_j \delta_{ij} + m^2.
\end{aligned}$$

Note that $p_i p_j$ is a symmetric tensor, whose contraction with the antisymmetric tensor $[\alpha_i, \alpha_j]$ is zero. Therefore we find:

$$\{\alpha_i, \alpha_j\} = 2\delta_{ij} \tag{6.11}$$

$$\{\alpha_i, \beta\} = 0 \tag{6.12}$$

$$\alpha_i^2 = \beta^2 = 1 \; . \tag{6.13}$$

It is helpful to introduce a new set of matrices $\gamma^\mu$ ($\mu = 0, 1, 2, 3$) defined as ($i = 1, 2, 3$)

$$\gamma^0 = \beta, \quad \gamma^i = \beta \alpha_i = \gamma^0 \alpha_i \; , \tag{6.14}$$

which satisfies the anticommutation rules

$$\{\gamma^\mu, \gamma^\nu\} = 2g^{\mu\nu}, \tag{6.15}$$

and have the properties

$$(\gamma^0)^2 = 1, \quad (\gamma^i)^2 = -1 \; , \tag{6.16}$$

$$\gamma^{\mu\dagger} = \gamma^0 \gamma^\mu \gamma^0 \; . \tag{6.17}$$

Using the $\gamma^\mu$ matrices the Dirac equation can be rewritten in the form given by Feynman. Multiplying (6.9) on the left by $\beta = \gamma^0$ gives

$$i\beta \frac{\partial \psi}{\partial t} = i\gamma^0 \frac{\partial \psi}{\partial t} = (\beta \alpha_i p_i + \beta^2 m)\psi = \left( -i\gamma^i \frac{\partial}{\partial x^i} + m \right) \psi \; ,$$

which is

$$(i\gamma^\mu \partial_\mu - m)\psi = 0 \; . \tag{6.18}$$

Finally, introducing the notation $\partial\!\!\!/ = \gamma^\mu \partial_\mu$, equation (6.18) can be put in the form

$$(i\partial\!\!\!/ - m)\psi = 0 \; . \tag{6.19}$$

Properties of the Dirac Matrices. The anticommutation rules (6.11) – $\alpha_i$ and $\beta$ should satisfy. Equation (6.13) implies that the eigenvalues of the matrix $\alpha_i$ and $\beta$ are all equal to $\pm 1$, while from (6.11) it follows, for $j \neq i$, that

$$\begin{aligned} \alpha_i \alpha_j \alpha_j = \alpha_i &= -\alpha_j \alpha_i \alpha_j \\ Tr(\alpha_i) = -Tr(\alpha_j \alpha_i \alpha_j) &= -Tr(\alpha_j \alpha_j \alpha_i) = -Tr(\alpha_i) \ , \end{aligned}$$

thus

$$Tr(\alpha_i) = 0 \ . \tag{6.20}$$

Using the same procedure it can also be shown that from (6.11) it follows that

$$Tr(\beta) = 0 \ . \tag{6.21}$$

An $N \times N$ matrix whose eigenvalues are all equal to $\pm 1$ can have a zero trace only if $N$ is an even number. Therefore the possible dimensions of the Dirac matrices are $N = 2, 4, \ldots$.

We can immediately exclude the case $N = 2$. The Pauli matrices

$$\sigma_1 = \begin{pmatrix} 0 & 1 \\ 1 & 0 \end{pmatrix} , \quad \sigma_2 = \begin{pmatrix} 0 & -i \\ i & 0 \end{pmatrix} , \quad \sigma_3 = \begin{pmatrix} 1 & 0 \\ 0 & -1 \end{pmatrix} , \tag{6.22}$$

used in non-relativistic quantum mechanics to describe particles of spin $\frac{1}{2}$ satisfy the anticommutation rules

$$\{\sigma_i, \sigma_j\} = 2\delta_{ij} \ , \tag{6.23}$$

similar to (6.11). However, it is impossible to find a fourth independent matrix which anticommutes with the $\sigma_i$. In fact the Pauli matrices, together with the unit matrix $\mathbb{1}$, form a basis for $2 \times 2$ matrices, from which any $2 \times 2$ matrix, $M$, can be constructed according to

$$M = M_0 \mathbb{1} + M_i \sigma_i$$

with

$$\mathbb{1} = \begin{pmatrix} 1 & 0 \\ 0 & 1 \end{pmatrix} .$$

A matrix $M$ that is independent of $\sigma_i$ must have $M_0 \neq 0$, but in this case it obviously cannot anticommute with the $\sigma_i$. The smallest dimension which the matrices $\alpha_i$ and $\beta$ can have is $N = 4$. It can easily be verified that the $4 \times 4$ matrices (written in terms of $2 \times 2$ blocks)

$$\alpha_i = \begin{pmatrix} 0 & \sigma_i \\ \sigma_i & 0 \end{pmatrix} , \quad \beta = \begin{pmatrix} 1 & 0 \\ 0 & -1 \end{pmatrix} , \tag{6.24}$$

satisfy (6.11)–(6.13).

The representation of the $\gamma^\mu$ matrices obtained from (6.24), known as the Pauli representation, is not unique. Given a non-singular matrix $S$, the new matrices

$$\widetilde{\gamma}^\mu = S^{-1}\gamma^\mu S \tag{6.25}$$

satisfy the same anticommutation rules as $\gamma^\mu$:

$$\{\widetilde{\gamma}^\mu, \widetilde{\gamma}^\nu\} = S^{-1}\gamma^\mu S S^{-1}\gamma^\nu S + S^{-1}\gamma^\nu S S^{-1}\gamma^\mu S = S^{-1}\{\gamma^\mu, \gamma^\nu\} S = 2g^{\mu\nu} .$$

It can be shown that, given two sets of $\gamma^\mu$ and $\widetilde{\gamma}^\mu$ matrices which satisfy the anticommutation rules (6.15), they are always connected by a transformation of the type (6.25), with a particular non-singular matrix $S$.

### 6.1.1 Spin

The Hamiltonian operator for the Dirac equation

$$H = \boldsymbol{\alpha}\cdot\boldsymbol{p} + \beta m \tag{6.26}$$

does not commute with the orbital angular momentum

$$\boldsymbol{L} = \boldsymbol{x}\times\boldsymbol{p} . \tag{6.27}$$

Using the commutation rules $[p_i, x_j] = -i\delta_{ij}$ gives, for example:

$$\begin{aligned} [H, L_3] &= [\alpha_i p_i, x_1 p_2 - p_2 x_1] \\ &= \alpha_1 p_2 [p_1, x_1] - \alpha_2 p_1 [p_2, x_2] \\ &= -i(\alpha_1 p_2 - \alpha_2 p_1) \neq 0 . \end{aligned}$$

The constants of the motion associated with the invariance under rotation of (6.9) are the components of the *total angular momentum*, defined as

$$\begin{aligned} \boldsymbol{J} &= \boldsymbol{L} + \frac{1}{2}\boldsymbol{\Sigma}; \\ \boldsymbol{\Sigma} &= \begin{pmatrix} \boldsymbol{\sigma} & 0 \\ 0 & \boldsymbol{\sigma} \end{pmatrix} , \end{aligned} \tag{6.28}$$

which do commute with the Hamiltonian (6.26). To see this, we define the antisymmetric tensor

$$\sigma^{\mu\nu} = \frac{i}{2}[\gamma^\mu, \gamma^\nu] , \tag{6.29}$$

whose components $\sigma^{ij}$ $(i, j = 1, 2, 3)$ are

$$\sigma^{ij} = -\frac{i}{2}\begin{pmatrix} [\sigma^i, \sigma^j] & 0 \\ 0 & [\sigma^i, \sigma^j] \end{pmatrix} = \epsilon^{ijk}\begin{pmatrix} \sigma_k & 0 \\ 0 & \sigma_k \end{pmatrix} = \epsilon^{ijk}\,\Sigma_k .$$

Thus we obtain

$$\Sigma_3 = \frac{i}{2}(\gamma_1\gamma_2 - \gamma_2\gamma_1) = -\frac{i}{2}(\alpha_1\alpha_2 - \alpha_2\alpha_1) .$$

In this form we can easily calculate the commutator of $H$ with $\Sigma_3$:

$$\frac{1}{2}\left[H, \Sigma_3\right] = -\frac{i}{4}\left[\alpha_i p_i, \alpha_1\alpha_2 - \alpha_2\alpha_1\right] = -i(\alpha_2 p_1 - \alpha_1 p_2) = -\left[H, L_3\right] ,$$

implying

$$[H, J_3] = \left[H, L_3 + \frac{1}{2}\Sigma_3\right] = 0 .$$

The eigenvalues of $\Sigma_3$ are $\pm 1$. Equation (6.28) shows that the Dirac equation describes particle with spin $\frac{1}{2}$.

When the momentum of the particle is non-zero, while the projection of the spin along an arbitrary axis is not conserved, as we have just seen, the projection of the spin along the direction of motion commutes with the Hamiltonian. This quantity is given the name *helicity*, and is described by the operator

$$\sigma_p = \frac{(\boldsymbol{\Sigma} \cdot \boldsymbol{p})}{|\boldsymbol{p}|} . \tag{6.30}$$

Comment. The appearance of spin explains why the wave function which satisfies the Dirac equation must be a multidimensional vector. However for spin $\frac{1}{2}$ a two-component wave function was expected, while the minimum dimension of the Dirac matrices is $N = 4$. The doubling of the components is due to the necessity of the presence of an antiparticle, as we shall see later.

### 6.1.2 Relativistic Invariance

We would like to show that, if $\psi(x)$ satisfies the Dirac equation in a given frame of reference $O$, the wave function determined by an observer in another system $O'$ satisfies the Dirac equation in $O'$.

This is similar to what happens in the case of the electromagnetic field tensor: the components of $\boldsymbol{E}$ and $\boldsymbol{B}$ change from one frame to the other, but the form of Maxwell's equations remains invariant.

We consider the homogeneous Lorentz transformation from $O$ to $O'$

$$x'^{\mu} = \Lambda^{\mu}{}_{\nu} x^{\nu} .$$

Correspondingly, the components of $\psi$ should transform linearly, to respect the superposition principle, with a matrix which depends on the transformation $\Lambda$

$$\psi'(x') = S(\Lambda)\psi(x). \tag{6.31}$$

The dependence of $S$ on $\Lambda$ must be such as to respect the rule for combining Lorentz transformations, at least for those transformations close to the identity:

$$S(\Lambda_1\Lambda_2) = S(\Lambda_1)S(\Lambda_2). \tag{6.32}$$

Note that we do not know *a priori* the form of $S(\Lambda)$. The relativistic invariance of the Dirac equation requires that *it should be possible to determine* $S(\Lambda)$ so that:

- the transformations are in accord with the combination rule (6.32),
- they lead to a $\psi'$ which satisfies the Dirac equation in $O'$ if $\psi$ satisfies it in $O$.

We now consider the Dirac equation in $O$:

$$\begin{aligned} 0 &= \left(i\gamma^\lambda \frac{\partial}{\partial x^\lambda} - m\right)\psi(x) = \left(i\gamma^\lambda \frac{\partial}{\partial x^\lambda} - m\right) S^{-1}(\Lambda)\psi'(x') = \\ &= \left(i\gamma^\lambda \frac{\partial x'^\mu}{\partial x^\lambda}\frac{\partial}{\partial x'^\mu} - m\right) S^{-1}(\Lambda)\psi'(x') = \\ &= \left(i\gamma^\lambda \Lambda^\mu{}_\lambda \frac{\partial}{\partial x'^\mu} - m\right) S^{-1}(\Lambda)\psi'(x'). \end{aligned} \tag{6.33}$$

Multiplying by the matrix $S(\Lambda)$, we obtain:

$$\begin{aligned} &\left(i\widetilde{\gamma}^\mu \frac{\partial}{\partial x'^\mu} - m\right)\psi'(x') = 0; \\ &\widetilde{\gamma}^\mu = \Lambda^\mu{}_\nu\, S(\Lambda)\gamma^\nu S^{-1}(\Lambda). \end{aligned} \tag{6.34}$$

Equation (6.34) agrees with the Dirac equation in system $O'$ if the matrices $\widetilde{\gamma}^\mu$ are identical to $\gamma^\mu$, or if $S(\Lambda)$ satisfies the relation:

$$S^{-1}(\Lambda)\gamma^\mu S(\Lambda) = \Lambda^\mu{}_\nu \gamma^\nu. \tag{6.35}$$

To solve (6.35), we restrict ourselves to infinitesimal transformations, which we have seen in (3.85) to be of the form (cf. Section 3.6):

$$\Lambda^\mu{}_\nu = \delta^\mu{}_\nu + \epsilon^\mu{}_\nu\,,$$

with

$$\begin{aligned} \epsilon^\mu{}_\nu &= g^{\mu\alpha}\epsilon_{\alpha\nu}; \\ \epsilon_{\alpha\beta} &= -\epsilon_{\beta\alpha}\,. \end{aligned} \tag{6.36}$$

Now we write

$$S = 1 + T\,, \quad S^{-1} = 1 - T$$

with $T$ infinitesimal:

$$T = \frac{1}{2}\epsilon_{\alpha\beta}T^{\alpha\beta}\,, \tag{6.37}$$

and $T^{\alpha\beta}$ antisymmetric.

Substituting in equation (6.35) we obtain (to first order in $\epsilon$)

$$\Lambda^\mu{}_\nu\gamma^\nu = \gamma^\mu + \epsilon^\mu{}_\nu\gamma^\nu = S^{-1}\gamma^\mu S = \gamma^\mu + (\gamma^\mu T - T\gamma^\mu)\,.$$

which gives (using (6.37) and (6.36))

$$\left(g^{\beta\mu}\gamma^{\alpha} - g^{\alpha\mu}\gamma^{\beta}\right) = \left[\gamma^{\mu}, T^{\alpha\beta}\right] . \tag{6.38}$$

Equation (6.38) has as solution the antisymmetric tensor (compare with (6.29))

$$T^{\alpha\beta} = \frac{1}{4}\left[\gamma^{\alpha}, \gamma^{\beta}\right] = -\frac{i}{2}\sigma^{\alpha\beta} ,$$

The transformation $S$ that we seek is therefore:

$$S = 1 - \frac{i}{4}\,\epsilon_{\alpha\beta}\sigma^{\alpha\beta} , \tag{6.39}$$

and we note the property

$$\gamma^0 S^{\dagger}\gamma^0 = S^{-1} . \tag{6.40}$$

which is shown using

$$\sigma^{\mu\nu\dagger} = \gamma^0\,\sigma^{\mu\nu}\gamma^0 , \tag{6.41}$$

and which implies

$$\gamma^0 S^{\dagger}\gamma^0 = 1 + \frac{i}{4}\,\epsilon_{\mu\nu}\gamma^0\sigma^{\mu\nu\dagger}\gamma^0 = 1 + \frac{i}{4}\,\epsilon_{\mu\nu}\sigma^{\mu\nu} = S^{-1} . \tag{6.42}$$

Note that (6.40) is valid for any proper Lorentz transformation which can be obtained as a product of infinitesimal transformations. To demonstrate this we consider the case in which

$$S = S_1 S_2 , \quad S^{-1} = S_2^{-1}S_1^{-1} ,$$

with $S_1$ and $S_2$ infinitesimal transformations. It follows from (6.40) that

$$\gamma^0 S^{\dagger}\gamma^0 = \gamma^0 S_2^{\dagger}S_1^{\dagger}\gamma^0 = \gamma^0 S_2^{\dagger}\gamma^0\gamma^0 S_1^{\dagger}\gamma^0 = S_2^{-1}S_1^{-1} = S^{-1} .$$

The Adjoint Spinor. In general $\psi$ is a complex wave function. Alongside $\psi$ we can introduce the complex conjugate spinor $\psi^*$. If we consider $\psi$ as a column vector, see (6.10), we can introduce the spinor $\psi^{\dagger}$, a row vector which has the elements of $\psi^*$ as components:

$$\psi^{\dagger} = (\psi^*)^T . \tag{6.43}$$

We note that $\psi^{\dagger}\psi$ is not invariant under Lorentz transformations, since the matrix $S(\Lambda)$ is not unitary. An invariant can be constructed by considering the *adjoint spinor*

$$\bar{\psi} = \psi^{\dagger}\gamma^0 . \tag{6.44}$$

Using equation (6.42) it can be seen that

$$\bar{\psi}'(x') = \psi^{\dagger}(x)S^{\dagger}\gamma^0 = \bar{\psi}(x)S^{-1} \tag{6.45}$$

so that:

$$(\bar{\psi}\psi)'(x') = \bar{\psi}(x)S^{-1}S\psi(x) = (\bar{\psi}\psi)(x) . \tag{6.46}$$

Covariant Bilinear Forms. Multiplying together two or more $\gamma$ matrices generates a matrix algebra. Because the symmetric product of two $\gamma$ matrices is $\pm I$, we can limit our considerations to the products which are antisymmetric in their Lorentz indices. Thus fifteen $4 \times 4$ matrices which are linearly independent are found (the four $\gamma$ matrices, six products of two $\gamma$, four of three and one of four) which together with the identity matrix make up the Dirac algebra

$$\Gamma^S = I \ , \ \Gamma^V_\mu = \gamma_\mu \ , \ \Gamma^T_{\mu\nu} = \sigma_{\mu\nu} \ ,$$

$$\Gamma^P = \gamma_5 = \gamma^5 = i\gamma^0\gamma^1\gamma^2\gamma^3 \ , \ \Gamma^A_\mu = \gamma_\mu\gamma_5 \ .$$

We note that $\gamma_5$ is hermitian while relations similar to (6.17) hold for the other matrices of the algebra

$$\Gamma = \gamma^0\Gamma^\dagger\gamma^0 \ . \tag{6.47}$$

Using the results of the previous paragraph one easily obtains transformation rules for the bilinear forms of the type

$$\psi^\dagger\gamma^0\Gamma\psi = \bar{\psi}\Gamma\psi \ , \tag{6.48}$$

in terms of the adjoint spinor $\bar{\psi}$, which transforms according to (6.45). The covariant bilinear forms (6.48) perform an important role, because they have definite transformation properties under Lorentz transformations. They are the basic ingredients with which to construct observable quantities and invariant Lagrangian densities.

We consider, for example, the continuity equation (6.6) which is obtained from the Dirac equation. It is easily shown that

$$\rho = \psi^\dagger\psi = \bar{\psi}\gamma^0\psi, \qquad \boldsymbol{j} = \psi^\dagger\boldsymbol{\alpha}\psi = \bar{\psi}\boldsymbol{\gamma}\psi \ .$$

The bilinear form $\bar{\psi}\gamma^\mu\psi$ is a 4-current associated with the particle described by the Dirac equation and $j^0 = \rho$ is the probability density, whose volume integral is a conserved quantity.

We can now proceed to determine the transformation rules for the bilinear forms.

- $\bar{\psi}\Gamma_S\psi = \bar{\psi}\psi$ transforms like a scalar, because

$$\bar{\psi}'\psi' = \bar{\psi}S^{-1}S\psi = \bar{\psi}\psi \ .$$

- $\bar{\psi}\Gamma^\mu_V\psi = \bar{\psi}\gamma^\mu\psi$ transforms like the components of a covariant four-vector, because

$$\bar{\psi}'\gamma^\mu\psi' = \bar{\psi}S^{-1}\gamma^\mu S\psi = \Lambda^\mu{}_\nu\bar{\psi}\gamma^\nu\psi \ .$$

- $\bar{\psi}\Gamma^{\mu\nu}_T\psi = \bar{\psi}\sigma^{\mu\nu}\psi = \bar{\psi}\ i[\gamma^\mu,\gamma^\nu]/2\ \psi$ transforms like the elements of an antisymmetric tensor, because

$$\bar{\psi}'\gamma^\mu\gamma^\nu\psi' = \bar{\psi}S^{-1}\gamma^\mu SS^{-1}\gamma^\nu S\psi = \Lambda^\mu{}_\alpha\Lambda^\nu{}_\beta\bar{\psi}\gamma^\alpha\gamma^\beta\psi \ .$$

- $\bar{\psi}\Gamma_P\psi = \bar{\psi}\gamma^5\psi$ transforms according to

$$\bar{\psi}'\gamma^5\psi' = \det(\Lambda)\ \bar{\psi}\gamma^5\psi\ ,$$

that is, it transforms like a scalar in the case of proper Lorentz transformations ($\det(\Lambda) = 1$) but changes sign in the case of parity transformations $(x_0, x_i) \to (x_0, -x_i)$ whose determinant equals $-1$. Therefore $\bar{\psi}\gamma^5\psi$ is a *pseudoscalar* density.

The transformation rule is obtained using the definition

$$\gamma^5 = i\gamma^0\gamma^1\gamma^2\gamma^3 = \frac{i}{4!}\epsilon_{\mu\nu\alpha\beta}\gamma^\mu\gamma^\nu\gamma^\alpha\gamma^\beta\ ,$$

where $\epsilon_{\mu\nu\alpha\beta}$ is the unit antisymmetric tensor with four indices. Thus, one finds

$$\begin{aligned} S^{-1}\gamma^5 S &= \frac{i}{4!}\epsilon_{\mu\nu\alpha\beta}\Lambda^\mu{}_\delta\Lambda^\nu{}_\lambda\Lambda^\alpha{}_\sigma\Lambda^\beta{}_\rho\gamma^\delta\gamma^\lambda\gamma^\sigma\gamma^\rho \\ &= \frac{i}{4!}\det(\Lambda)\epsilon_{\delta\lambda\sigma\rho}\gamma^\delta\gamma^\lambda\gamma^\sigma\gamma^\rho = \det(\Lambda)\gamma^5\ . \end{aligned}$$

- $\bar{\psi}\Gamma_A^{\mu\nu}\psi = \bar{\psi}\gamma^5\gamma^\mu\psi$ transforms according to

$$\bar{\psi}'\gamma^5\gamma^\mu\psi' = \det(\Lambda)\ \Lambda^\mu{}_\nu\bar{\psi}\gamma^5\gamma^\nu\psi\ ,$$

that is, like the components of a four-vector in the case of proper Lorentz transformations and in the opposite way under parity transformations.

Comment 1. A correspondence $\Lambda \to S(\Lambda)$ between the elements of a group and a set of matrices which are subject to the same composition rule as the group defines a *representation* of the group. Equations (6.31) and (6.32) define a representation of the Lorentz group, to be added to the irreducible tensors already discussed in Section 3.3. According to the classification considered there, the four-dimensional representation of the Dirac spinors corresponds to $(0, 1/2) \oplus (1/2, 0)$. This is a *reducible* representation, where a non-trivial matrix ($\gamma_5$) exists which commutes with all the generators of the group. The tensor product of an odd number of spinor representations generates a new series of representations which, from the point of view of rotations, contains representations of half-integer spin.

Comment 2. According to (6.40) the matrices $S(\Lambda)$ are *pseudounitary*, but not unitary. In fact it can be shown that representations of the Lorentz group by unitary operators are necessarily infinite-dimensional. This is due to the fact that $L_+^\uparrow$ is a *non-compact* group: the spaces of parameters which describe the transformations make up a non-compact set, unlike the rotations. For these, the matrix elements are bounded functions of the rotation angles, within the interval $(0, 2\pi)$. Conversely, the parameter $\gamma = 1/\sqrt{1-\beta^2}$, which appears in $\Lambda^\mu_\nu$, is unbounded.

Comment 3. According to quantum mechanics, the squared modulus of the product of two states, $|<B|A>|^2$, represents the probability of finding the state $|B>$ as the outcome of a measurement on state $|A>$. This probability must be the same in all frames of reference, so if $|\Lambda A>$ and $|\Lambda B>$ are the kets which represent the transformed states, we must have, for any $|A>$ and $|B>$:

$$|<\Lambda B|\Lambda A>|^2 = |<B|A>|^2 . \tag{6.49}$$

Again, according to quantum mechanics, the transformed states are obtained by applying a linear operator $U(\Lambda)$, which represents the Lorentz transformation $\Lambda$. Therefore it must be true that:

$$|<B|U(\Lambda)^\dagger U(\Lambda)|A>|^2 = |<B|A>|^2 . \tag{6.50}$$

Wigner showed that (10.39) has two possible solutions:

$$<B|U(\Lambda)^\dagger U(\Lambda)|A>=<B|A> \text{ or } <B|U(\Lambda)^\dagger U(\Lambda)|A>=<B|A>^* . \tag{6.51}$$

We can exclude the second case by continuity: when $\Lambda$ tends towards the unity transformation, $|\Lambda A>$ must tend to $|A>$, so $U(\Lambda)$ tends to unity, satisfying the first condition and not the second. Finally

- *Lorentz transformations are represented by quantum states of unitary operators.*

Comment 4. Given the non-unitary nature of the $S(\Lambda)$ matrices, it is interesting to ask in what way the scalar products between states which are solutions of the Dirac equation can provide a unitary representation of the Lorentz group, as required by the considerations of the previous comment. In Dirac's theory, for two states $|A>$ and $|B>$, we have:

$$<A|B>=\int d^3x\ \psi_A^*(\boldsymbol{x},t)\psi_B(\boldsymbol{x},t) .$$

The density inside the integral is not invariant. Instead, the form of $S(\Lambda)$ implies that it should be the time component of a 4-current, conserved by the Dirac equation:

$$\psi_A^*(\boldsymbol{x},t)\psi_B(\boldsymbol{x},t) = \bar{\psi}_A(\boldsymbol{x},t)\gamma^0\psi_B(\boldsymbol{x},t) = J^0_{A,B};$$
$$\partial_\mu J^\mu_{A,B} = 0 .$$

But as we saw in Section 3.5, the space integral of the time component of a conserved 4-current is a relativistic invariant. Therefore, for any Lorentz transformation:

$$<\Lambda A|\Lambda B>=<A|B>$$

which is exactly the invariance condition for the scalar product. The non-invariance of $\psi^*\psi$ is in the proportion required to balance exactly the non-invariance of the size of $d^3x$, so that the $S(\Lambda)$ matrices impose on the physical states a unitary representation of the Lorentz group. This result will be derived explicitly in Section 7.4.

### 6.1.3 Boost

This expression denotes a specific Lorentz transformation, which corresponds to the passage from the given frame of reference, $\mathcal{O}$, to a frame $\mathcal{O}'$ moving along the positive x-axis at a velocity $\beta$ with respect to $\mathcal{O}$.

The transformation rule between the coordinates of $\mathcal{O}$ and $\mathcal{O}'$ involves only $x^0$ and $x^1$, and is written:

$$x^{0,\prime} = \Lambda^0_0 x^0 + \Lambda^0_1 x^1 = \gamma(t - \beta x)$$
$$x^{1,\prime} = \Lambda^1_0 x^0 + \Lambda^1_1 x^1 = \gamma(-\beta t + x)$$

where:

$$\gamma = (1 - \beta^2)^{-1/2} = \cosh\theta \tag{6.52}$$

and $\theta$ is the rapidity, with:

$$\beta = \tanh\theta. \tag{6.53}$$

The origin of $\mathcal{O}'$, $x'$=0, moves according to $x = \beta t$ as it should. Setting $\beta$ infinitesimal:

$$\beta = \delta\theta$$

and defining the infinitesimal parameters of the transformation according to (6.36) we obtain

$$\epsilon_{1,0} = -\epsilon_{0,1} = \delta\theta\ .$$

The matrix S($\Lambda$), which determines the transformation between $\psi$(x) and $\psi'$(x$'$) is therefore, according to (6.39):

$$S(\Lambda) = 1 - \frac{i}{4}\ \epsilon_{\alpha\beta}\sigma^{\alpha\beta} =$$
$$= 1 - \frac{i\delta\theta}{2}\sigma^{10}\ . \tag{6.54}$$

Using the definition:

$$\sigma^{10} = \frac{i}{2}[\gamma^1, \gamma^0] = i\gamma^1\gamma^0 = -i\alpha_1$$

we find:

$$S(\Lambda) = 1 - \frac{\delta\theta}{2}\alpha_1 \ . \tag{6.55}$$

Transformations with finite rapidity, $\theta = \tanh^{-1}\beta$, are obtained by combining infinitesimal transformations, which corresponds to multiplying together the relevant matrices (6.55). We put:

$$\delta\theta = \frac{\theta}{N}; \quad (N\ large)$$

and find:

$$S(\Lambda) = \left(1 - \frac{\theta}{2N}\alpha_1\right)^N \rightarrow e^{-\frac{\theta}{2}\alpha_1} \ . \tag{6.56}$$

The exponential with $\alpha_1$ is easily expressed in elementary terms, because $(\alpha_1)^2 = 1$. Expanding in series, we obtain:

$$\begin{aligned} e^{-\frac{\theta}{2}\alpha_1} &= \sum_{n=0}^{n=\infty} \frac{1}{n!}\left(-\frac{\theta}{2}\alpha_1\right)^n = \\ &= \sum_{k=0}^{k=\infty} \frac{1}{(2k)!}\left(-\frac{\theta}{2}\right)^{2k} + \alpha_1 \sum_{k=0}^{k=\infty} \frac{1}{(2k+1)!}\left(-\frac{\theta}{2}\right)^{2k+1} \\ &= \cosh\frac{\theta}{2} - \alpha_1 \sinh\frac{\theta}{2} \end{aligned} \tag{6.57}$$

and finally using the well-known hyperbolic trigonometric relationships:

$$\cosh^2\frac{\theta}{2} = \frac{\cosh\theta + 1}{2}; \qquad \sinh^2\frac{\theta}{2} = \frac{\cosh\theta - 1}{2} \tag{6.58}$$

to obtain:

$$\begin{aligned} S(\Lambda) &= \cosh\frac{\theta}{2}\begin{bmatrix} 1 & -\tanh\frac{\theta}{2}\sigma_1 \\ -\tanh\frac{\theta}{2}\sigma_1 & 1 \end{bmatrix} = \\ &= \sqrt{\frac{\gamma+1}{2}}\begin{bmatrix} 1 & -\frac{\beta\gamma}{\gamma+1}\sigma_1 \\ -\frac{\beta\gamma}{\gamma+1}\sigma_1 & 1 \end{bmatrix} . \end{aligned} \tag{6.59}$$

Equation (6.59) can be generalised immediately to the case in which the velocity is directed along a general vector, $\vec{n}$:

$$S(\Lambda) = \sqrt{\frac{\gamma+1}{2}}\begin{bmatrix} 1 & -\frac{\beta\gamma}{\gamma+1}(\vec{n}\cdot\vec{\sigma}) \\ -\frac{\beta\gamma}{\gamma+1}(\vec{n}\cdot\vec{\sigma}) & 1 \end{bmatrix} . \tag{6.60}$$

### 6.1.4 Solutions of the Dirac Equation for a Free Particle

The Dirac equation, in the form (6.9) or (6.18), has solutions in the form of relativistic plane waves. We write

$$\psi(x) = u(p)e^{-i(px)};\ p^\mu = (E, \vec{p})\ . \tag{6.61}$$

The equation for $u(p)$ takes the form:

$$(\not{p} - m)u(p) = (p_\mu \gamma^\mu - m)u(p) = 0 \tag{6.62}$$

where $u$ is a four-component spinor.

To begin, we consider solutions of the Dirac equation for a particle at rest. In this case (6.62) reduces to

$$(\gamma^0 E - m)u(p) = 0\ , \tag{6.63}$$

or, explicitly:

$$\begin{pmatrix} E-m & 0 & 0 & 0 \\ 0 & E-m & 0 & 0 \\ 0 & 0 & -E-m & 0 \\ 0 & 0 & 0 & -E-m \end{pmatrix} \begin{pmatrix} u_1 \\ u_2 \\ u_3 \\ u_4 \end{pmatrix} = 0\ . \tag{6.64}$$

Equation (6.64) has eigenvalues $E^{(1)} = E^{(2)} = m$ and $E^{(3)} = E^{(4)} = -m$, whose corresponding eigenvectors are:

$$u^{(1)} = \begin{pmatrix} 1 \\ 0 \\ 0 \\ 0 \end{pmatrix},\quad u^{(2)} = \begin{pmatrix} 0 \\ 1 \\ 0 \\ 0 \end{pmatrix},\quad u^{(3)} = \begin{pmatrix} 0 \\ 0 \\ 1 \\ 0 \end{pmatrix},\quad u^{(4)} = \begin{pmatrix} 0 \\ 0 \\ 0 \\ 1 \end{pmatrix}. \tag{6.65}$$

Like the Klein–Gordon equation, the Dirac equation also has negative energy, as well as positive energy, solutions. The two positive energy solutions correspond to the two possible states of a particle with spin $\frac{1}{2}$. We will later discuss the meaning of the solutions with negative energy.

In the general case of $\vec{p} \neq 0$ we write the 4-spinor $u^{(r)}(p)$ in terms of two two-component spinors $u_A$ and $u_B$

$$u^{(r)}(\boldsymbol{p}) = \begin{pmatrix} u_A \\ u_B \end{pmatrix}. \tag{6.66}$$

From (6.62) we obtain the matrix equation

$$\begin{pmatrix} E-m & -\boldsymbol{\sigma}\cdot\boldsymbol{p} \\ \boldsymbol{\sigma}\cdot\boldsymbol{p} & -E-m \end{pmatrix} \begin{pmatrix} u_A \\ u_B \end{pmatrix} = 0\ , \tag{6.67}$$

which has non-trivial solutions for

$$E^2 - m^2 = (\boldsymbol{\sigma}\cdot\boldsymbol{p})^2 = \boldsymbol{p}^2\ . \tag{6.68}$$

Rewriting (6.67) in the form

$$u_A = \frac{(\boldsymbol{\sigma} \cdot \boldsymbol{p})}{E - m} u_B \tag{6.69}$$

$$u_B = \frac{(\boldsymbol{\sigma} \cdot \boldsymbol{p})}{E + m} u_A , \tag{6.70}$$

it is immediately seen that the positive energy solutions can be obtained from (6.70), by choosing $u_A$ equal to one of the Pauli two-component spinors

$$\chi_1 = \begin{pmatrix} 1 \\ 0 \end{pmatrix} , \quad \chi_2 = \begin{pmatrix} 0 \\ 1 \end{pmatrix} . \tag{6.71}$$

The negative energy solutions are obtained in a similar way starting from (6.69). We can therefore write the four solutions to (6.62) in the form

$$u_r^{(+)}(\boldsymbol{p}) = \begin{pmatrix} \chi_r \\ \frac{(\boldsymbol{\sigma}\cdot\boldsymbol{p})}{E_p+m}\chi_r \end{pmatrix} , \quad u_r^{(-)}(\boldsymbol{p}) = \begin{pmatrix} -\frac{(\boldsymbol{\sigma}\cdot\boldsymbol{p})}{E_p+m}\chi_r \\ \chi_r \end{pmatrix} , \tag{6.72}$$

with $r = 1, 2$ and $E_p = +\sqrt{\boldsymbol{p}^2 + m^2} = |E|$. Obviously, for $\boldsymbol{p} \to 0$, (6.72) reduces to (6.65).

To describe the negative energy states, it is helpful to introduce two new 4-spinors $v_r(\boldsymbol{p})$ $(r = 1, 2)$, defined by

$$v_1(\boldsymbol{p}) = u_2^{(-)}(-\boldsymbol{p}) , \quad v_2(\boldsymbol{p}) = -u_1^{(-)}(-\boldsymbol{p}) . \tag{6.73}$$

or:

$$v_r(\boldsymbol{p}) = \epsilon^{rs} \; u_s^{(-)}(-\boldsymbol{p}), \quad (r, \; s = 1, \; 2) \tag{6.74}$$

where $\epsilon^{rs}$ is the completely antisymmetric two-index tensor[1] and the sum over repeated indices is understood. Furthermore we put

$$u_r(\boldsymbol{p}) = u_r^{(+)}(\boldsymbol{p}) . \tag{6.75}$$

It immediately follows from the definitions that $u_r$ and $v_r$ satisfy the equations:

$$(\not{p} - m)u_r = 0 , \quad (\not{p} + m)v_r = 0 . \tag{6.76}$$

The normalisation of the 4-spinors is defined by writing the solution of (6.62) in the form

$$\psi(x) = N \left(\frac{m}{VE_p}\right)^{1/2} \times \begin{cases} u_r(\boldsymbol{p})e^{-ipx} & \text{(positive energy)} \\ v_r(\boldsymbol{p})e^{ipx} & \text{(negative energy)} \end{cases} \tag{6.77}$$

and requiring that

$$\int d^3x \; \psi^\dagger(x)\psi(x) = 1 , \tag{6.78}$$

[1] $\epsilon^{12} = -\epsilon^{21} = 1$, $\epsilon^{11} = \epsilon^{22} = 0$.

which gives

$$u_r^\dagger(\boldsymbol{p})u_r(\boldsymbol{p}) = v_r^\dagger(\boldsymbol{p})v_r(\boldsymbol{p}) = \frac{E_p}{m} \ . \tag{6.79}$$

In this way the result $N = [(E_p + m)/2m]^{1/2}$ is obtained.

With this choice of normalisation, the orthonormality relations for the 4-spinors are:

$$\begin{aligned} u_r^\dagger(\boldsymbol{p})u_s(\boldsymbol{p}) &= v_r^\dagger(\boldsymbol{p})v_s(\boldsymbol{p}) = \delta_{rs}\,\frac{E_p}{m} \\ u_r^\dagger(\boldsymbol{p})v_s(-\boldsymbol{p}) &= u_r^{(+)\dagger}(\boldsymbol{p})u_s^{(-)}(\boldsymbol{p}) = 0 \end{aligned} \tag{6.80}$$

and, with simple formal manipulation[2] we obtain from (6.76) the relations

$$\begin{aligned} &m\ \bar{u}(\boldsymbol{p})\gamma^\mu u(\boldsymbol{p}) = p^\mu\ \bar{u}(\boldsymbol{p})u(\boldsymbol{p}) \\ &\bar{u}(\boldsymbol{p})v(\boldsymbol{p}) = 0 \ . \end{aligned} \tag{6.81}$$

The orthonormality conditions for $u$ and $v$ with respect to multiplication by the adjoint spinors follow from the first of these

$$\begin{aligned} &\bar{u}_r(\boldsymbol{p})u_s(\boldsymbol{p}) = -\bar{v}_r(\boldsymbol{p})v_s(\boldsymbol{p}) = \delta_{rs} \ , \\ &\bar{u}_r(\boldsymbol{p})v_s(\boldsymbol{p}) = \bar{v}_r(\boldsymbol{p})u_s(\boldsymbol{p}) = 0 \ . \end{aligned} \tag{6.82}$$

The completeness of the set of solutions of the Dirac equation is expressed by the relation

$$\sum_r [(u_r)_\alpha(\boldsymbol{p})(\bar{u}_r)_\beta(\boldsymbol{p}) - (v_r)_\alpha(\boldsymbol{p})(\bar{v}_r)_\beta(\boldsymbol{p})] = \delta_{\alpha\beta} \ , \tag{6.83}$$

where $(u_r)_\alpha(\boldsymbol{p})$ is the $\alpha$ component of the 4-spinor. Equation (6.83) is easily obtained from the 4-spinor definition, which implies

$$\sum_r (u_r)_\alpha(\boldsymbol{p})(\bar{u}_r)_\beta(\boldsymbol{p}) = \left(\Lambda_p^+\right)_{\alpha\beta} = \left(\frac{\not p + m}{2m}\right)_{\alpha\beta} \tag{6.84}$$

$$-\sum_r (v_r)_\alpha(\boldsymbol{p})(\bar{v}_r)_\beta(\boldsymbol{p}) = \left(\Lambda_p^-\right)_{\alpha\beta} = -\left(\frac{\not p - m}{2m}\right)_{\alpha\beta} \ . \tag{6.85}$$

The operators $\Lambda_p^+$ and $\Lambda_p^-$ are projectors of, respectively, the positive and negative energy states, i.e.

$$\Lambda_p^+ u_r = u_r \quad , \quad \Lambda_p^- v_r = v_r \ , \tag{6.86}$$

$$\Lambda_p^+ v_r = \Lambda_p^- u_r = 0 \tag{6.87}$$

[2] The first line of (6.81) is obtained by multiplying (6.76) on the left by $\bar{u}(\boldsymbol{p})\gamma^\mu$ and summing with the adjoint of the same equation multiplied from the right by $\gamma^\mu u(\boldsymbol{p})$; the second is the outcome of a similar rearrangement of the equation for $v(\boldsymbol{p})$ in (6.76).

Starting from the complete basis of solutions to the Dirac equation, we can express any solution, according to the expansion

$$\psi(x) = \sum_{\boldsymbol{p},r} \sqrt{\frac{m}{E(p)V}} [a_r(\boldsymbol{p})u_r(\boldsymbol{p})e^{-i(px)} + (b_r(\boldsymbol{p}))^* v_r(\boldsymbol{p})e^{i(px)}] \,. \tag{6.88}$$

For later use, we give here the expressions for normalisation of the wave function and the expectation values of the energy and momentum in terms of the amplitudes of the normal modes of oscillation which appear in (6.88).

To carry out the calculations, we treat the components of the spinors $u$ and $v$ as commuting numbers, but avoid exchanging the order in which the coefficients $a$ and $b$ appear. With the help of the orthogonality conditions (6.80) we find:

$$\begin{aligned} N &= \int d^3x\, \psi^\dagger(\boldsymbol{x},t)\psi(\boldsymbol{x},t) = \\ &= \sum_{\boldsymbol{p},r} [a_r(\boldsymbol{p})(a_r(\boldsymbol{p}))^* + (b_r(\boldsymbol{p}))^* b_r(\boldsymbol{p})]; \end{aligned} \tag{6.89}$$

$$\begin{aligned} < E > &= \int d^3x\, \psi^\dagger(\boldsymbol{x},t)[-i\boldsymbol{\alpha}\cdot\boldsymbol{\nabla} + \beta m]\psi(\boldsymbol{x},t) = \\ &= \sum_{\boldsymbol{p},r} E(p)[a_r(\boldsymbol{p})(a_r(\boldsymbol{p}))^* - (b_r(\boldsymbol{p}))^* b_r(\boldsymbol{p})]; \end{aligned} \tag{6.90}$$

$$\begin{aligned} < \boldsymbol{P} > &= \int d^3x\, \psi^\dagger(\boldsymbol{x},t)(-i\boldsymbol{\nabla})\psi(\boldsymbol{x},t) = \\ &= \sum_{\boldsymbol{p},r} \boldsymbol{p}[a_r(\boldsymbol{p})(a_r(\boldsymbol{p}))^* - (b_r(\boldsymbol{p}))^* b_r(\boldsymbol{p})]. \end{aligned} \tag{6.91}$$

From (6.90) we see that the normalisation factor $\sqrt{m/EV}$ has been chosen to ensure that the energy of the oscillation mode $\boldsymbol{p}, r$ is equal to $E(p)$ for unit oscillation amplitude.

**Solutions with $\boldsymbol{p} \neq 0$ and Lorentz Boost.** The solutions for $\boldsymbol{p} \neq 0$ can be obtained with a Lorentz transformation starting from those for a particle at rest, using the representation for the Lorentz boost defined in Section 6.1.3.

We denote the spinor in frame $\mathcal{O}'$, introduced in Section 6.1.3, with:

$$\psi'(x') = e^{-i(p'x')}\, u'(p') \tag{6.92}$$

We then have

$$\psi(x) = e^{-i(px)}\, u(p) = S^{-1}(\Lambda)\psi'(x') = e^{-i(p'x')} S^{-1}(\Lambda) u'(p') \tag{6.93}$$

or, given that $(p'x') = (px)$:

$$\begin{aligned} u(p) &= S^{-1}(\Lambda)u'(p'); \\ S^{-1}(\Lambda) &= \sqrt{\frac{\gamma+1}{2}} \begin{bmatrix} 1 & \frac{\beta\gamma\sigma_1}{\gamma+1} \\ \frac{\beta\gamma\sigma_1}{\gamma+1} & 1 \end{bmatrix} \,. \end{aligned} \tag{6.94}$$

If the particle is at rest in $\mathcal{O}'$, the energy and momentum in $\mathcal{O}$ are given by

$$p = \pm m\beta\gamma; \quad p^0 = \pm E = \pm m\gamma \tag{6.95}$$

for the positive and negative energy solutions. Therefore, for the first we find (compare with the definitions (6.65)–(6.72)):

$$\begin{aligned} u_r^{(+)}(p) = u(p) &= \sqrt{\frac{E+m}{2m}} \begin{bmatrix} 1 & \frac{p\sigma_1}{E+m} \\ \frac{p\sigma_1}{E+m} & 1 \end{bmatrix} \begin{pmatrix} \chi_r \\ 0 \end{pmatrix} = \\ &= \begin{pmatrix} \chi_r \\ \frac{p\sigma_1}{E+m}\chi_r \end{pmatrix} \quad (r = 1,\ 2)\ , \end{aligned} \tag{6.96}$$

and for the negative energy solutions:

$$\begin{aligned} u_r^{(-)}(p) &= \sqrt{\frac{E+m}{2m}} \begin{bmatrix} 1 & \frac{-p\sigma_1}{E+m} \\ \frac{-p\sigma_1}{E+m} & 1 \end{bmatrix} \begin{pmatrix} 0 \\ \chi_r \end{pmatrix} = \\ &= \begin{pmatrix} \frac{-p\sigma_1}{E+m}\chi \\ \chi_r \end{pmatrix} \\ v_r(p) = \epsilon^{rs} u_s^{(-)}(-p) &= \epsilon^{rs} \begin{pmatrix} \frac{p\sigma_1}{E+m}\chi_s \\ \chi_s \end{pmatrix} \quad (r,\ s = 1,\ 2)\ . \end{aligned} \tag{6.97}$$

The solutions corresponding to positive and negative energies do not mix under a Lorentz transformation.

Note. As equation (6.96) shows, for small velocities the lower components of $u(p)$ are of the order of $\beta = v/c = p/m$ with respect to the upper components. In this limit, upper and lower components are referred to as the *big* and *small* components of $u(p)$, respectively. The roles are exchanged for $v(p)$, as shown by equation (6.97).

### 6.1.5 The Magnetic Moment of the Electron

We consider an electron in a known electromagnetic field $A^\mu \equiv (\phi, \boldsymbol{A})$. The Dirac equation in the presence of the field is obtained with the minimal substitution discussed in Section 5.4:

$$i\partial^\mu \to i\partial^\mu + eA^\mu\ , \tag{6.98}$$

where $e$ is the electronic charge. We obtain:

$$(i\not{\partial} + e\not{A} - m)\psi = 0\ , \tag{6.99}$$

or

$$\left[\beta\left(i\frac{\partial}{\partial t} + e\phi\right) - \beta\boldsymbol{\alpha}\cdot(\boldsymbol{p} + e\boldsymbol{A}) - m\right]\psi = 0\ . \tag{6.100}$$

If we restrict ourselves to solutions with positive energy, equation (6.98) provides an extraordinarily accurate description of the behaviour of the electron in a given electromagnetic field, both for the electric field generated by an atomic nucleus and for a classical external field. We want now to study equation (6.100) for the latter case, in the non-relativistic limit, for

$$E = \sqrt{\boldsymbol{p}^2 + m^2} \approx m\left(1 + \frac{\boldsymbol{p}^2}{2m^2}\right) = m + \frac{\boldsymbol{p}^2}{2m}\,, \tag{6.101}$$

with $m \gg \boldsymbol{p}^2/2m$. In these conditions it is helpful to isolate the rapidly varying phase factor which corresponds to the rest energy, and rewrite the solution to (6.100) in the form

$$\psi = \widetilde{\psi}\mathrm{e}^{-imt}\,, \tag{6.102}$$

where $\widetilde{\psi}$ oscillates much more slowly than $e^{-imt}$.

Substituting (6.102) in (6.100) and multiplying on the left by $\beta e^{imt}$ gives

$$\left(i\frac{\partial}{\partial t} + m\right)\widetilde{\psi} = \left[\boldsymbol{\alpha}\cdot(\boldsymbol{p} + e\boldsymbol{A}) + \beta m - e\phi\right]\widetilde{\psi}\,. \tag{6.103}$$

$\widetilde{\psi}$ is a four-component spinor which we can write in the form

$$\widetilde{\psi} = \begin{pmatrix} \varphi \\ \eta \end{pmatrix}, \tag{6.104}$$

with $\varphi$ and $\eta$ two-component spinors. From

$$\boldsymbol{\alpha}\widetilde{\psi} = \begin{pmatrix} 0 & \boldsymbol{\sigma} \\ \boldsymbol{\sigma} & 0 \end{pmatrix}\begin{pmatrix} \varphi \\ \eta \end{pmatrix} = \begin{pmatrix} \boldsymbol{\sigma}\eta \\ \boldsymbol{\sigma}\varphi \end{pmatrix} \tag{6.105}$$

$$\beta\widetilde{\psi} = \begin{pmatrix} I & 0 \\ 0 & -I \end{pmatrix}\begin{pmatrix} \varphi \\ \eta \end{pmatrix} = \begin{pmatrix} \varphi \\ -\eta \end{pmatrix}, \tag{6.106}$$

a set of coupled equations for $\varphi$ and $\eta$ is obtained

$$\left(i\frac{\partial}{\partial t} + e\phi\right)\varphi = \boldsymbol{\sigma}\cdot(\boldsymbol{p} + e\boldsymbol{A})\eta \tag{6.107}$$

$$\left(i\frac{\partial}{\partial t} + e\phi + 2m\right)\eta = \boldsymbol{\sigma}\cdot(\boldsymbol{p} + e\boldsymbol{A})\varphi\,. \tag{6.108}$$

Now we again use the non-relativistic approximation, for which

$$\left(i\frac{\partial}{\partial t} - e\phi + 2m\right)\eta \approx 2m\eta \tag{6.109}$$

and thus

$$\eta \approx \frac{\boldsymbol{\sigma}\cdot(\boldsymbol{p} + e\boldsymbol{A})}{2m}\varphi\,. \tag{6.110}$$

Equation (6.110) shows that $\eta$ is the *small component* of $\widetilde{\psi}$, being of order $\boldsymbol{p}/m$ with respect to $\varphi$.

Substituting (6.110) into (6.107) gives the equation for $\varphi$

$$\left(i\frac{\partial}{\partial t} + e\phi\right)\varphi = \frac{[\boldsymbol{\sigma}\cdot(\boldsymbol{p}+e\boldsymbol{A})]^2}{2m}\varphi\ , \tag{6.111}$$

which can be rewritten in a more familiar form using the relationship

$$\begin{aligned}
[\boldsymbol{\sigma}\cdot(\boldsymbol{p}+e\boldsymbol{A})]^2 &= (\boldsymbol{p}+e\boldsymbol{A})^2 + i\boldsymbol{\sigma}\cdot[(\boldsymbol{p}+e\boldsymbol{A})\times(\boldsymbol{p}+e\boldsymbol{A})] \\
&= (\boldsymbol{p}+e\boldsymbol{A})^2 + ie\boldsymbol{\sigma}\cdot(\boldsymbol{p}\times\boldsymbol{A}+\boldsymbol{A}\times\boldsymbol{p}) \\
&= (\boldsymbol{p}+e\boldsymbol{A})^2 + e\boldsymbol{\sigma}\cdot(\boldsymbol{\nabla}\times\boldsymbol{A}+\boldsymbol{A}\times\boldsymbol{\nabla}) \\
&= (\boldsymbol{p}+e\boldsymbol{A})^2 + e\boldsymbol{\sigma}\cdot(\boldsymbol{\nabla}\times\boldsymbol{A}) \\
&= (\boldsymbol{p}+e\boldsymbol{A})^2 + e\boldsymbol{\sigma}\cdot\boldsymbol{B}\ ,
\end{aligned}$$

where $\boldsymbol{B} = (\boldsymbol{\nabla}\times\boldsymbol{A})$ is the magnetic field. This then gives

$$\left(i\frac{\partial}{\partial t} + e\phi\right)\varphi = \left[\frac{1}{2m}(\boldsymbol{p}+e\boldsymbol{A})^2 + \frac{e}{2m}\boldsymbol{\sigma}\cdot\boldsymbol{B}\right]\varphi\ , \tag{6.112}$$

which is the Schrödinger equation for a particle with electric charge $-e$ and spin $\boldsymbol{s} = \boldsymbol{\sigma}/2$, interacting with an electromagnetic field described by the potentials $\phi$ and $\boldsymbol{A}$.

The last term on the right hand side of equation (6.112) is the interaction energy between the magnetic field $\boldsymbol{B}$ and a magnetic dipole moment

$$\boldsymbol{\mu} = \frac{-e}{2m}\boldsymbol{\sigma} = g\frac{-e}{2m}\boldsymbol{s}\ . \tag{6.113}$$

The coefficient $g$ is known as the *gyromagnetic ratio* and expresses the relationship between the magnetic moment, given in units of Bohr magnetons, and the corresponding angular momentum.

The complete interaction with the magnetic field is obtained by inserting into (6.112) the vector potential corresponding to a constant field along the $z$-axis:

$$\boldsymbol{A} = \frac{B}{2}(-y, x, 0);\quad \boldsymbol{\nabla}\cdot\boldsymbol{A} = 0;\quad \nabla\times\boldsymbol{A} = \boldsymbol{B}\ .$$

Neglecting terms of order $B^2$, we obtain:

$$\begin{aligned}
\boldsymbol{H} &= \frac{\boldsymbol{p}^2}{2m} + \frac{e}{2m}(\boldsymbol{p}\cdot\boldsymbol{A}+\boldsymbol{A}\cdot\boldsymbol{p}) + \frac{e}{2m}\boldsymbol{\sigma}\cdot\boldsymbol{B} = \\
&= \frac{\boldsymbol{p}^2}{2m} + \frac{eB}{2m}(xp_y - yp_x) + \frac{e}{2m}\boldsymbol{\sigma}\cdot\boldsymbol{B} = \\
&= \frac{\boldsymbol{p}^2}{2m} + \frac{e}{2m}(\boldsymbol{L}+2\boldsymbol{S})\cdot\boldsymbol{B}\ .
\end{aligned} \tag{6.114}$$

The term which contains the orbital angular momentum provides the quantum explanation of the normal Zeeman effect. The emission or absorption of a photon obeys the selection rule $\Delta L_z = \pm 1, 0$. For atoms in which $L_z$ is a good quantum number, in the presence of a magnetic field the spectral line splits into three components separated by $\pm eB/2m, 0$, which coincides exactly with the Larmor frequency (5.73). However, in complex atoms, $L_z$ and $S_z$ are not individually diagonal and the difference between the levels involved in the transition is also a function of the gyromagnetic spin ratio [9]. Equation (6.113) correctly describes the behaviour of the electron; one finds $g = 2$ as originally hypothesised by Goudsmit and Uhlenbeck to explain the anomalous Zeeman effect.

The result (6.114) constitutes an extraordinary success for Dirac's theory, in which the spin-magnetic field term, which in non-relativistic quantum mechanics must be added *ad hoc* to the Schrödinger equation, emerges in a natural way from the quantum minimal substitution applied to the Dirac equation of the free particle.

## 6.2 THE RELATIVISTIC HYDROGEN ATOM

In this section we discuss the spectrum of the hydrogen atom starting from the Dirac equation (following the original treatment by Dirac [9]; see also Schiff [10]). The calculation in the non-relativistic limit, based on the Schrödinger equation, is given in Appendix B. As in that case, the starting point is the factorisation of the Hamiltonian in polar coordinates.

### 6.2.1 Factorisation of the Dirac Equation in Polar Coordinates

The $\alpha$ matrices, in the Dirac theory, describe the velocity of the electron.

$$\frac{d\boldsymbol{x}}{dt} = -i\,[\boldsymbol{x}, H] = -i\,[\boldsymbol{\alpha} \cdot \boldsymbol{p} + \beta m] = \boldsymbol{\alpha}\ . \tag{6.115}$$

We can therefore define the radial velocity as:

$$\alpha_r = \frac{1}{r}\boldsymbol{x} \cdot \boldsymbol{\alpha} \tag{6.116}$$

and the radial momentum, $p_r$ (see Appendix B), as:

$$p_r = \frac{1}{r}(\boldsymbol{x} \cdot \boldsymbol{p} - i) = -i(\frac{\partial}{\partial r} + \frac{1}{r})\ . \tag{6.117}$$

The factorisation needed is obtained starting from the product:

$$\alpha_r(\boldsymbol{\alpha} \cdot \boldsymbol{p}) = \frac{1}{r}x_i(\alpha_i\alpha_j)p_j = \frac{1}{r}x_i\left(\delta_{ij} + \frac{1}{2}\left[\alpha_i, \alpha_j\right]\right)p_j\ . \tag{6.118}$$

The commutator of two $\alpha$ matrices uses the matrices $\Sigma_i = 2s_i$, introduced in equation (6.28), which describe the spin of the electron, for example:

$$[\alpha_1, \alpha_2] = 2\alpha_1\alpha_2 = 2i\begin{pmatrix} \sigma_3 & 0 \\ 0 & \sigma_3 \end{pmatrix} = 2i\Sigma_3\ . \tag{6.119}$$

In general:

$$\frac{1}{2}\left[\alpha_i, \alpha_j\right] = i\epsilon_{ijk}\Sigma_k \tag{6.120}$$

from which, using the definition of $p_r$, equation (6.117), and the orbital angular momentum, $\boldsymbol{L} = \boldsymbol{x} \times \boldsymbol{p}$, we can rewrite equation (6.118) as:

$$\alpha_r(\boldsymbol{\alpha} \cdot \boldsymbol{p}) = p_r + \frac{i}{r}(\boldsymbol{L} \cdot \boldsymbol{\Sigma} + 1) = p_r + \beta\frac{i}{r}k \tag{6.121}$$

where we have defined the hermitian operator $k$:

$$k = \beta(\boldsymbol{L} \cdot \boldsymbol{\Sigma} + 1) \ . \tag{6.122}$$

The Dirac Hamiltonian can be rewritten in the new variables as:

$$\boldsymbol{\alpha} \cdot \boldsymbol{p} + \beta m = \alpha_r p_r + i\alpha_r\beta(r)^{-1}k + \beta m \ . \tag{6.123}$$

The physical meaning of $k$ becomes clear by calculating the square:

$$\begin{aligned} k^2 &= (\boldsymbol{L} \cdot \boldsymbol{\Sigma} + 1)^2 = L_i(\delta_{ij} + \frac{1}{2}\left[\Sigma_i, \Sigma_j\right])L_j + 2\boldsymbol{L} \cdot \boldsymbol{\Sigma} + 1 = \\ &= L^2 + i\epsilon_{ijk}\Sigma_k L_i L_j + 2\boldsymbol{L} \cdot \boldsymbol{\Sigma} + 1 = L^2 + \boldsymbol{L} \cdot \boldsymbol{\Sigma} + \frac{3}{4} + \frac{1}{4} = \\ &= (\boldsymbol{L} + \frac{1}{2}\boldsymbol{\Sigma})^2 + \frac{1}{4} = J^2 + \frac{1}{4} \ . \end{aligned} \tag{6.124}$$

The square of $k$ is a function of the square of the total angular momentum and its eigenvalues are equal to $j(j+1) + 1/4 = (j + 1/2)^2$. Obviously $p_r$, $\alpha_r$ and $r$ commute with $J$, therefore with $k$ and also $\beta$. It follows that $k$ is a constant of the motion and that $H$ and $k$ are simultaneously diagonalisable. We note the anticommutation relation:

$$\{\alpha_r, \beta\} = 0, \ \ \alpha_r^2 = \beta^2 = 1 \ . \tag{6.125}$$

The overall Hamiltonian for the electron in the Coulomb field of the proton is written:

$$H = \alpha_r p_r + i\alpha_r\beta(r)^{-1}k + \beta m - \frac{\alpha}{r} \tag{6.126}$$

with $\alpha \simeq 1/137$ the fine structure constant.

### 6.2.2 Separation of Variables

To achieve the separation of variables, we must construct the Dirac spinors which correspond to the eigenvalues of the total angular momentum, $j$, and the component along the $z$-axis, $j_z$. These are similar to the spherical harmonics introduced in Appendix B, equation (B.12) for the particle of spin zero (for a complete treatment, see [11]).

**Spinors with Definite Angular Momentum.** For a spin $\frac{1}{2}$ particle, a given value of the total angular momentum can be obtained starting from two values of the orbital angular momentum:

$$l = j \pm 1/2, \; l' = j \mp 1/2 \,. \tag{6.127}$$

Even if $L_z$ is not conserved, see equation (6.28), the value of the total orbital angular momentum is a constant of the motion, since it is related to the behaviour of the wave function under parity. For inversion of the axes, the state with total orbital angular momentum $l$ has a factor $(-1)^l$, which assumes opposite values for the orbital momenta in (6.127).

We take the value of $l$ which corresponds to the plus sign in (6.127). The *two-dimensional* spinor with angular momentum $j$, $j_z$, is easily constructed combining the spherical harmonics $Y_l^m$ and $Y_l^{m+1}$ with the *up* and *down* spinors:

$$\begin{aligned}
&\chi^{(+)}(j = l - 1/2, j_z = m + 1/2; \theta, \phi) = \\
&= aY_l^m(\theta, \phi)\, \chi(\uparrow) + bY_l^{m+1}(\theta, \phi)\, \chi(\downarrow) = \\
&= \begin{pmatrix} aY_l^m(\theta, \phi) \\ bY_l^{m+1}(\theta, \phi) \end{pmatrix}
\end{aligned} \tag{6.128}$$

where $a$ and $b$ are coefficients[3] which characterise the combination corresponding to the total angular momentum $j = l - 1/2$. In detail, see [11], one finds:

$$\begin{aligned}
a &= \sqrt{\frac{l - m}{2l + 1}} \\
b &= -\sqrt{\frac{l + m + 1}{2l + 1}} \,.
\end{aligned} \tag{6.129}$$

Similarly, we define the spinor corresponding to the minus sign in (6.127):

$$\begin{aligned}
&\chi^{(-)}(j = l + 1/2, j_z = m + 1/2; \theta, \phi) = \\
&= cY_l^m(\theta, \phi)\, \chi(\uparrow) + dY_l^{m+1}(\theta, \phi)\, \chi(\downarrow) = \\
&= \begin{pmatrix} cY_l^m(\theta, \phi) \\ dY_l^{m+1}(\theta, \phi) \end{pmatrix}
\end{aligned} \tag{6.130}$$

which gives:

$$\begin{aligned}
c &= \sqrt{\frac{l + m + 1}{2l + 1}} \\
d &= \sqrt{\frac{l - m}{2l + 1}} \,.
\end{aligned} \tag{6.131}$$

[3] Known as Clebsch–Gordan coefficients; see [13].

To construct the Dirac spinors which correspond to specific values of $j$ and $j_z$, we anticipate from Section 12.1 that, in the theory of Dirac, parity is represented by the matrix $\gamma_0$=diag$(1,1,-1,-1)$. Therefore, the two-dimensional spinors which represent the first two and the second two rows must have opposite parity. From this, two possibilities follow:

$$u(+)_{j,j_z} = \begin{pmatrix} \frac{1}{r}F(r)\chi^{(+)} \\ \frac{1}{r}G(r)\chi^{(-)} \end{pmatrix}$$
$$u(-)_{j,j_z} = \begin{pmatrix} \frac{1}{r}F(r)\chi^{(-)} \\ \frac{1}{r}G(r)\chi^{(+)} \end{pmatrix} \tag{6.132}$$

where $r^{-1}F$ and $r^{-1}G$ are radial wave functions. As well as $j$ and $j_z$, we can characterise the spinor $u$ with the value $l$ of the orbital momentum of the upper component, which in the hydrogen atom is dominant, knowing that the lower component will be related to the orbital momentum $l'$ given by (6.127), and write simply:

$$u_{j,l,j_z} = \begin{pmatrix} \frac{1}{r}F(r)\chi(l) \\ \frac{1}{r}G(r)\chi(l') \end{pmatrix}. \tag{6.133}$$

In spectroscopic notation, see Appendix B, the two states which correspond to $j = 1/2$ are written:

$$\begin{aligned} &S_{1/2}, \text{ corresponding to } u_{j,l=0,j_z}: \\ &\qquad \chi(l=0) = \chi^{(-)}, \quad \chi(l'=1) = \chi^{(+)} \\ &P_{1/2}, \text{ corresponding to } u_{j,l=1,j_z}: \\ &\qquad \chi(l=1) = \chi^{(+)}, \quad \chi(l'=0) = \chi^{(-)}. \end{aligned} \tag{6.134}$$

Radial Equations. The two spinors, $\chi^{(\pm)}$ are connected by the operator $\sigma_r = r^{-1}\boldsymbol{\sigma}\cdot\boldsymbol{x}$. The relation follows from the fact that the operator $\sigma_r$ is rotationally invariant, so cannot change the total angular momentum, and has negative parity, therefore must exchange $l$ and $l'$:

$$\sigma_r(\theta,\phi)\chi^{(-)}(\theta,\phi) = \eta\chi^{(+)}(\theta,\phi) \tag{6.135}$$

with $\eta$ a complex number whose value is determined by setting $\theta = 0$ in (6.135) and applying $\sigma_r(0,\phi) = \sigma_3$ to the spinor with $m = 0$. We find[4] $\eta = 1$.

With this equation we can determine the action of the operator $\alpha_r$, equation (6.126), on the spinors (6.132) by choosing for $\alpha_r$ the representation[5]:

$$\alpha_r = \begin{pmatrix} 0 & -i\sigma_r \\ i\sigma_r & 0 \end{pmatrix} \tag{6.136}$$

[4]Using the relation $Y_l^0(\cos\theta = 1,\phi) = \sqrt{\frac{2l+1}{4\pi}}$.

[5]The reader can show that the new representation of the $\boldsymbol{\alpha}$ matrices is connected to the Dirac representation, equation (6.24), by a unitary transformation.

With this choice:

$$i\alpha_r \begin{pmatrix} \frac{1}{r}F(r)\chi^{(+)} \\ \frac{1}{r}G(r)\chi^{(-)} \end{pmatrix} = \begin{pmatrix} \frac{1}{r}G(r)\chi^{(+)} \\ -\frac{1}{r}F(r)\chi^{(-)} \end{pmatrix} \tag{6.137}$$

while:

$$\beta \begin{pmatrix} \frac{1}{r}F(r)\chi^{(+)} \\ \frac{1}{r}G(r)\chi^{(-)} \end{pmatrix} = \begin{pmatrix} \frac{1}{r}F(r)\chi^{(+)} \\ -\frac{1}{r}G(r)\chi^{(-)} \end{pmatrix} \tag{6.138}$$

and $k$ is simply a multiple of the identity:

$$k \begin{pmatrix} \frac{1}{r}F(r)\chi^{(+)} \\ \frac{1}{r}G(r)\chi^{(-)} \end{pmatrix} = k(j) \begin{pmatrix} \frac{1}{r}F(r)\chi^{(+)} \\ \frac{1}{r}G(r)\chi^{(-)} \end{pmatrix} \tag{6.139}$$

where $k(j)$, the eigenvalue of $k$, has the values:

$$k(j) = \sqrt{(j+1/2)^2} = \pm 1, \pm 2, \cdots \tag{6.140}$$

Consequently we can reduce the Dirac equation to a two-dimensional equation on the vector $(\frac{1}{r}F(r), \frac{1}{r}G(r))$ introducing the two-dimensional representation:

$$\beta = \begin{pmatrix} 1 & 0 \\ 0 & -1 \end{pmatrix}; \quad \alpha_r = \begin{pmatrix} 0 & -i \\ i & 0 \end{pmatrix} \tag{6.141}$$

which clearly satisfies the anticommutation relations (6.125). In this representation the Hamiltonian (6.126) gives rise to two radial equations:

$$\begin{aligned} G'(r) + \frac{k(j)}{r}G(r) + (\frac{\alpha}{r} - m + E)F(r) &= 0 \\ -F'(r) + \frac{k(j)}{r}F(r) + (\frac{\alpha}{r} + m + E)G(r) &= 0\,. \end{aligned} \tag{6.142}$$

**Why Does $\beta$ Appear in the Definition of $k$?** The spin-angular momentum coupling operator which appears in $k$, equation (6.122), can be applied to the two-dimensional spinors introduced in (6.128) and (6.130). We find:

$$\begin{aligned} (\boldsymbol{\sigma}\cdot\boldsymbol{L}+1)\chi(l) &= [j(j+1) - l(l+1) + \frac{1}{4}]\chi(l) = \\ &= -(j+1/2)\chi(l), \text{ for } l = j+1/2 \\ &= (j+1/2)\chi(l), \text{ for } l = j-1/2 \end{aligned} \tag{6.143}$$

in agreement with (6.140), while:

$$\begin{aligned} (\boldsymbol{\sigma}\cdot\boldsymbol{L}+1)\chi(l') &= [j(j+1) - l(l+1) + \frac{1}{4}]\chi(l') = \\ &= (j+1/2)\chi(l'), \text{ for } l' = j-1/2 \\ &= -(j+1/2)\chi(l'), \text{ for } l' = j+1/2. \end{aligned} \tag{6.144}$$

Therefore, the spinors (6.132) are *not* eigenstates of the spin-orbit operator, $\boldsymbol{\Sigma} \cdot \boldsymbol{L}$, but they are eigenstates of $k$, owing to the multiplication by $\beta$, which makes the eigenvalue of $\chi(l')$ equal to that of $\chi(l)$. We also note the relations used in the literature:

$$k(j) = -l, \text{for } l = j + 1/2; \quad k(j) = l + 1, \text{for } l = j - 1/2. \tag{6.145}$$

Boundary Conditions. For $r \to +\infty$, the radial equations reduce to:

$$\begin{aligned} &G'(r) - (m - E)F(r) = 0 \\ &-F'(r) + (m + E)G(r) = 0 \ . \end{aligned} \tag{6.146}$$

We define:

$$\epsilon_{1,2} = (m \pm E) > 0; \quad \epsilon = \sqrt{\epsilon_1 \epsilon_2} \ . \tag{6.147}$$

Taking the derivative of the second equation and substituting $G'$ from the first, we find:

$$F''(r) = -\epsilon^2 F(r) \tag{6.148}$$

from which the general solution follows:

$$F = Ae^{-\epsilon r} + Be^{+\epsilon r} \quad (r \to +\infty). \tag{6.149}$$

The same holds for $G$. In conclusion, we must require:

- $F(r), G(r) \to 0$ per $r \to 0^+$;
- $F(r), G(r) \sim e^{-\epsilon r}$ per $r \to +\infty$ .

### 6.2.3 Eigenvalues of the Hamiltonian

With obvious variations, the procedure follows the one used for the non-relativistic case of Appendix B. We write

$$\begin{aligned} F(r) &= e^{-\epsilon r}(a_0 r^s + a_1 r^{s+1} + \cdots + a_\nu r^{s+\nu} + \cdots) \\ G(r) &= e^{-\epsilon r}(b_0 r^s + a_1 r^{s+1} + \cdots + b_\nu r^{s+\nu} + \cdots). \end{aligned} \tag{6.150}$$

$\nu = 0$. The series should not contain terms in $r^{s-1}$. Substituting in (6.142), we find:

$$\begin{aligned} a_{-1} &= [-s + k(j)]a_0 + \alpha b_0 \\ b_{-1} &= \alpha a_0 + [s + k(j)]b_0 \ . \end{aligned} \tag{6.151}$$

To ensure $a_{-1} = b_{-1} = 0$ with $a_0$ and $b_0$ non-zero, the determinant of (6.151) must vanish:

$$\begin{aligned} &s^2 = k(j)^2 - \alpha^2 \ \text{ or} : s = +\sqrt{(j + 1/2)^2 - \alpha^2} \\ &\frac{b_0}{a_0} = \frac{-\alpha}{s + k(j)} \ . \end{aligned} \tag{6.152}$$

We now consider the general term of the series, of order $\nu-1$. Substituting in (6.142) we find (this time it is necessary to differentiate the exponentials as well):

$$\begin{aligned}[s+\nu-k(j)]a_\nu - \alpha b_\nu &= \epsilon a_{\nu-1} + \epsilon_1 b_{\nu-1} \\ \alpha a_\nu + [s+\nu+k(j)]b_\nu &= \epsilon_2 a_{\nu-1} + \epsilon b_{\nu-1}\ .\end{aligned} \tag{6.153}$$

The set on the right-hand side has determinant zero, therefore we can eliminate $a_{\nu-1}$ and $b_{\nu-1}$ and obtain the relation:

$$[\epsilon(s+\nu-k(j)) - \epsilon_1\alpha]a_\nu = [\epsilon_1(s+\nu+k(j)) + \epsilon\alpha]b_\nu\ . \tag{6.154}$$

General $\nu$. For general values of $\epsilon$ the recurrence relations (B.27) give rise to the series (6.150) starting from $a_0$ and $b_0$. For $\nu \to \infty$, we obtain:

$$\begin{aligned}\nu a_\nu &= \epsilon a_{\nu-1} + \epsilon_1 b_{\nu-1} = (2\epsilon)a_{\nu-1} \\ \nu b_\nu &= \epsilon_2 a_{\nu-1} + \epsilon b_{\nu-1} = (2\epsilon)b_{\nu-1}\end{aligned} \tag{6.155}$$

making use of (6.154). Summing the series, we obtain (the same holds for $G$):

$$F(r) \sim e^{-\epsilon r}e^{+2\epsilon r} = e^{+\epsilon r} \tag{6.156}$$

which is not acceptable; the series must terminate at a certain value $\nu = \bar{\nu}$. Setting $\nu = \bar{\nu}+1$ in (B.27) and making the first term vanish, we find:

$$b_{\bar{\nu}} = \frac{-\epsilon}{\epsilon_1}a_{\bar{\nu}} \tag{6.157}$$

and, substituting in (6.154):

$$E\alpha = \epsilon(s+\bar{\nu}) \tag{6.158}$$

which shows that $E > 0$. Squaring:

$$E^2 = \frac{m^2}{1+\frac{\alpha^2}{(s+\bar{\nu})^2}} \tag{6.159}$$

and, finally:

$$E = \frac{m}{\sqrt{1+\frac{\alpha^2}{(\bar{\nu}+\sqrt{(j+1/2)^2-\alpha^2})^2}}}\ . \tag{6.160}$$

Special consideration is required for the states with $\bar{\nu} = 0$, where there could be a conflict between equation (6.152):

$$\frac{b_0}{a_0} = \frac{-\alpha}{s+k(j)} \tag{6.161}$$

Table 6.1 Levels of the hydrogen atom in the Dirac equation for $n = \bar{\nu} + j + \frac{1}{2} \leq 4$. The energies increase from the bottom towards the top and from left to right. Levels with the same values of $j$ and $\bar{\nu}$ are degenerate. In the last column, the number of states in the row, equal to $2n^2$, is given (see Eq. (B.35)).

| | $\bar{\nu}$ | 3 | 2 | 1 | 0 | number of states |
|---|---|---|---|---|---|---|
| $n = \bar{\nu} + j + \frac{1}{2}$ | | | | | | |
| 4 | | $S_{1/2}, P_{1/2}$ | $P_{3/2}, D_{3/2}$ | $D_{5/2}, F_{5/2}$ | $F_{7/2}$ | $2 \cdot 16$ |
| 3 | | -- | $S_{1/2}, P_{1/2}$ | $P_{3/2}, D_{3/2}$ | $D_{5/2}$ | $2 \cdot 9$ |
| 2 | | -- | -- | $S_{1/2}, P_{1/2}$ | $P_{3/2}$ | $2 \cdot 4$ |
| 1 | | -- | -- | -- | $S_{1/2}$ | $2 \cdot 1$ |

and equation (6.154), which for $\bar{\nu} = 0$ reads:

$$\epsilon a_0 + \epsilon_1 b_0 = 0. \tag{6.162}$$

The latter equation clearly requires that $a_0$ and $b_0$ should have opposite signs, which can be obtained from the preceding equation only if:

$$k(j) = +(j + 1/2) \tag{6.163}$$

which in turn, given equation (6.144), implies the selection rule:

$$\bar{\nu} = 0: \text{ only } l = j - 1/2. \tag{6.164}$$

To the lowest order in $\alpha$, equation (6.160) reduces to the non-relativistic result, with the principal quantum number $n = j + 1/2 + \bar{\nu}$.

The resulting arrangement of the states can be represented by a series ordered according to the quantum numbers $n$, $l$, $j$. Table 6.1 shows the levels with $n \leq 4$, in spectroscopic notation (Appendix B):

The noteworthy feature of the result (6.160) is that the eigenvalues depend only on $\bar{\nu}$ and $j$. Dirac's theory largely resolves the degeneracy in $l$ of the non-relativistic result, giving a satisfactory explanation of the so-called *fine structure* of the levels, but a residual degeneracy remains between those pairs of states which have equal values of $\bar{\nu}$ and $j$ and which differ by one unit in orbital angular momentum.

The most celebrated example is that of the $2S_{1/2}$ and $2P_{1/2}$ states. In 1947, Lamb and Retherford observed a small energy difference between these states, the so-called *Lamb shift*. This difference was interpreted by Bethe as caused by a higher-order electrodynamic correction, owing to the interaction between the electron and the electrodynamic fluctuations of the vacuum. The calculation of Bethe and Schwinger's calculation of the difference of the magnetic moment of the electron from the value in Dirac's theory ($g = 2$), marked the beginning of modern quantum electrodynamics, QED.

## 6.3 TRACES OF THE $\gamma$ MATRICES

The traces of $\gamma$ matrix products enter into practically all field theory calculations. Here we give the rules for calculating these traces, and explicit results in some simpler cases. Today there are programs capable of numerically or symbolically computing traces of products up to very high levels, but it is nevertheless useful to acquire some familiarity with the properties of the traces and the calculation methods. The definitions of the $\gamma$ matrices and of $\gamma_5$ is given in Section 6.1. The point of departure is the elementary properties already discussed:

$$Tr(\gamma^\mu) = 0; \tag{6.165}$$
$$Tr(\gamma^\mu \gamma^\nu) = 4g^{\mu\nu} \ . \tag{6.166}$$

$\gamma_5$ The matrix $\gamma_5$ is defined as:

$$\gamma_5 = i\gamma^0\gamma^1\gamma^2\gamma^3 \ . \tag{6.167}$$

It follows from this that[6]:

- the trace of $\gamma_5$ with a number $\leq 3$ of $\gamma$ matrices is zero,
- $$Tr\left(\gamma^\mu\gamma^\nu\gamma^\rho\gamma^\sigma\gamma_5\right) = +4i\epsilon^{\mu\nu\rho\sigma} \ . \tag{6.168}$$

Odd Number of $\gamma$ Matrices. The rule (6.165) implies that the $\gamma$ matrices have an even number of dimensions and an equal number of $+1$ and $-1$ eigenvalues. Its generalisation is that

- the trace of an odd number of $\gamma$ matrices $= 0$.

Even Number of $\gamma$ Matrices. We first consider the case of four $\gamma$ matrices:

$$Tr\left(\gamma^\mu\gamma^\nu\gamma^\rho\gamma^\sigma\right).$$

We can advance the first matrix by one place using the anticommutation rule:

$$Tr\left(\gamma^\mu\gamma^\nu\gamma^\rho\gamma^\sigma\right) = -Tr\left(\gamma^\nu\gamma^\mu\gamma^\rho\gamma^\sigma\right) + 2g^{\mu\nu}Tr\left(\gamma^\rho\gamma^\sigma\right).$$

[6] We recall that the completely antisymmetric Levi–Civita tensor is defined by $\epsilon_{0123} = +1 = -\epsilon^{0123}$, equation (3.40).

The second term contains two fewer $\gamma$ matrices and is elementary. If we continue to advance using the anticommutation rule, we obtain:

$$Tr\,(\gamma^\mu\gamma^\nu\gamma^\rho\gamma^\sigma) = \\ = +8g^{\mu\nu}g^{\rho\sigma} - 8g^{\mu\rho}g^{\nu\sigma} + 8g^{\mu\sigma}g^{\nu\rho} - Tr\,(\gamma^\nu\gamma^\rho\gamma^\sigma\gamma^\mu)\ .$$

The final trace is equal to the initial one, because of the cyclic property; therefore, taking it to the other side of the equation, we find:

$$Tr\,(\gamma^\mu\gamma^\nu\gamma^\rho\gamma^\sigma) = +4\,(g^{\mu\nu}g^{\rho\sigma} - g^{\mu\rho}g^{\nu\sigma} + g^{\mu\sigma}g^{\nu\rho})\ . \tag{6.169}$$

The rule can be generalised, since, with an even number of $\gamma$ matrices, if we move the first matrix forward until it reaches the last place we always end up with the starting trace multiplied by $-1$. Therefore, we can reduce the trace of $2n$ $\gamma$ matrices to a combination with alternating signs of traces of $2n-2$ $\gamma$ matrices. Iterating, we reduce to $n = 2+$.

Of course, the number of terms in the expansion of the traces increases very rapidly, as $(2n-1)!!$, making the use of computer programs necessary.

As an alternative to the previous method, we can reduce the number of matrices in a trace by using the relation which reduces the product of three matrices to a combination of $\gamma^\alpha$ and $\gamma^\alpha\gamma_5$, which follows from the completeness of the basis of matrices introduced in equation (6.47):

$$\gamma^\mu\gamma^\nu\gamma^\lambda = g^{\mu\nu}\gamma^\lambda + g^{\nu\lambda}\gamma^\mu - g^{\mu\lambda}\gamma^\nu - i\epsilon^{\mu\nu\lambda\rho}\gamma_\rho\gamma_5\ . \tag{6.170}$$

## 6.4 PROBLEMS FOR CHAPTER 6

### Sect. 6.1

**1.** For a massless particle, the Dirac equation (6.9) involves only three Dirac matrices:

$$i\frac{\partial}{\partial t}\psi = \boldsymbol{\gamma}\cdot\boldsymbol{p}\psi\ .$$

Show that in this case, to recover the energy-momentum relation: $E^2 = |\mathbf{p}|^2$, we need only three anticommuting matrices, which we may take as the Pauli matrices $\sigma_i$, $i = 1, 2, 3$. The above equation is known as the Weyl equation and it describes the *two-component* neutrino, a particle with only two states, the *antineutrino*, with helicity $h = +1$, a candidate for the particle emitted with the electron in neutron's $\beta$-decay, Sect. 15.1, and its antiparticle, the *neutrino* with helicity $h = -1$ a candidate for the particle emitted in the solar fusion reactions, Sect. 2.4.

**2.** Prove that the four matrices $\tilde{\gamma}^\mu$:

$$\tilde{\gamma}^0 = \alpha_2 = \begin{pmatrix} 0 & \sigma_2 \\ \sigma_2 & 0 \end{pmatrix};\ \tilde{\gamma}^1 = -i\Sigma^3 = -i\begin{pmatrix} \sigma_3 & 0 \\ 0 & \sigma_3 \end{pmatrix}$$

$$\tilde{\gamma}^2 = \gamma^2;\ \tilde{\gamma}^3 = i\Sigma^1 = i\begin{pmatrix} \sigma_1 & 0 \\ 0 & \sigma_1 \end{pmatrix}$$

satisfy the Dirac anticommutation relations (6.15) and therefore provide an acceptable representation of the Dirac matrices known as the *Majorana representation*, to be discussed in Section 13.2.

**3.** Determine the unitary transformation that brings the matrices $\widetilde{\gamma}^\mu$ of Problem **2** into the standard Pauli representation (6.24).

**4.** The matrices $\widetilde{\gamma}^\mu$ of Problem **2** are all imaginary so that the Dirac equation in this representation has only real coefficients. What can you deduce from this fact? Draw your conclusions and then consult Sect. 13.2.

**5.** In the Weyl, or chiral, representation, the $\gamma$-matrices are written in the form

$$\gamma^0 = \begin{pmatrix} 0 & \mathbf{1} \\ \mathbf{1} & 0 \end{pmatrix}, \gamma^i = \begin{pmatrix} 0 & \sigma^i \\ -\sigma^i & 0 \end{pmatrix}.$$

- Verify that the above matrices fullfill the anti-commutation rule

$$\{\gamma^\mu, \gamma^\nu\} = 2g^{\mu\nu} .$$

- Derive the explicit expressions of the operators $P_\pm = (1 \pm \gamma_5)$, and verify that they are projection operators.

**6.** Given the Dirac spinor $\psi(x)$, and the transformation

$$\psi'(x') = S(\Lambda)\psi(x)$$
$$x' = \Lambda x$$

determine the form of the matrix $S(\Lambda)$ corresponding to rotation by an angle $\phi$ around the $z$-axis. Consider the cases of:

- infinitesimal transformation;
- finite transformation, discussing the result obtained for $\phi = 2\pi$.

**7.** Show that the operators

$$\Pi^\pm = \frac{1}{2}\left(1 \pm \frac{\boldsymbol{\Sigma} \cdot \boldsymbol{p}}{|\boldsymbol{p}|}\right)$$

where

$$\Sigma_i = \begin{pmatrix} \sigma_i & 0 \\ 0 & \sigma_i \end{pmatrix}$$

- are projection operators;
- satisfy the relations:

$$\Pi^+ u_r(\boldsymbol{p}) = \delta_{r1} u_r(\boldsymbol{p}), \quad \Pi^+ v_r(\boldsymbol{p}) = \delta_{r2} v_r(\boldsymbol{p})$$
$$\Pi^- u_r(\boldsymbol{p}) = \delta_{r2} u_r(\boldsymbol{p}), \quad \Pi^- v_r(\boldsymbol{p}) = \delta_{r1} v_r(\boldsymbol{p})$$

where $u_r(\boldsymbol{p})$ and $v_r(\boldsymbol{p})$ ($r = 1,\ 2$), are four-component spinors, solutions of the Dirac equation describing a free particle with momentum $\boldsymbol{p}$ along the $z$-axis.

**8.** Under a finite, i.e. not infinitesimal, Lorentz transformation $\Lambda$, a four-spinor transforms according to

$$\psi \to S(\Lambda)\psi \; ,$$

with

$$S[\Lambda(\omega)] = e^{-\frac{i}{4}\omega_{\mu\nu}\sigma^{\mu\nu}}$$

and

$$\sigma^{\mu\nu} = \frac{i}{2}[\gamma^\mu, \gamma^\nu] \; .$$

The non-vanishing components of the antisymmetric tensor $\omega_{\mu\nu}$ correspond to a boost along the direction of the $i$-axis ($\omega_{0i}$, with $i = 1, 2, 3$) and to rotations around the $k$-axis, perpendicular to the $(i, j)$-plane ($\omega_{ij}$).

- compute the matrix $S$ corresponding to a boost along the direction of the 1-axis;
- verify that

$$S(\Lambda)\gamma^\mu S(\Lambda)^{-1} = (\Lambda^{-1})^\mu{}_\nu \gamma^\nu \; .$$

Sect. 6.3

**1.** Prove that the trace of an odd number of gamma matrices vanishes.

CHAPTER 7

# QUANTISATION OF THE DIRAC FIELD

## 7.1 PARTICLES AND ANTIPARTICLES

As we have seen, the Dirac equation is the only equation for the wave function in accord with the requirements of quantum mechanics and relativity, and leads unambiguously to particles with spin $\frac{1}{2}$ and an extraordinarily accurate description of the properties of the electron, either free or bound in atoms. However, the Dirac equation has solutions with negative energy, which present a difficult quantum interpretation.

From general principles, it follows that a quantum state should evolve in time according to:

$$|t>= e^{-i\boldsymbol{H}t}|0> \tag{7.1}$$

from which, if $\psi$ is an eigenfunction of energy, it follows that:

$$\psi(\boldsymbol{x},t) = e^{-iEt}\psi(\boldsymbol{x},0)\ . \tag{7.2}$$

Moreover, to have a system which is stable under small perturbations, $E$ must be bounded from below when other quantum numbers vary, something which is evidently not true for the solutions of the Dirac equation with $E = -\omega(p) = -\sqrt{(\boldsymbol{p})^2 + m^2} \leq -m$.

The solution to the stability problem proposed by Dirac is simple and radical; all the negative energy states are occupied and the vacuum is in reality a *sea* of electrons which fill all the levels with $E \leq -m$. The exclusion principle prevents an electron with positive energy from ending in one of the forbidden states.

Exciting an electron from a state of negative energy to a state of positive energy creates a *hole* in the Dirac Sea, which behaves in every way like a particle of mass $m$, equal to the mass of the electron, and with spin $\frac{1}{2}$ and *positive electric charge*.

A particle consistent with this description, the *positron*, was discovered by


DOI: 10.1201/9781003436263-7

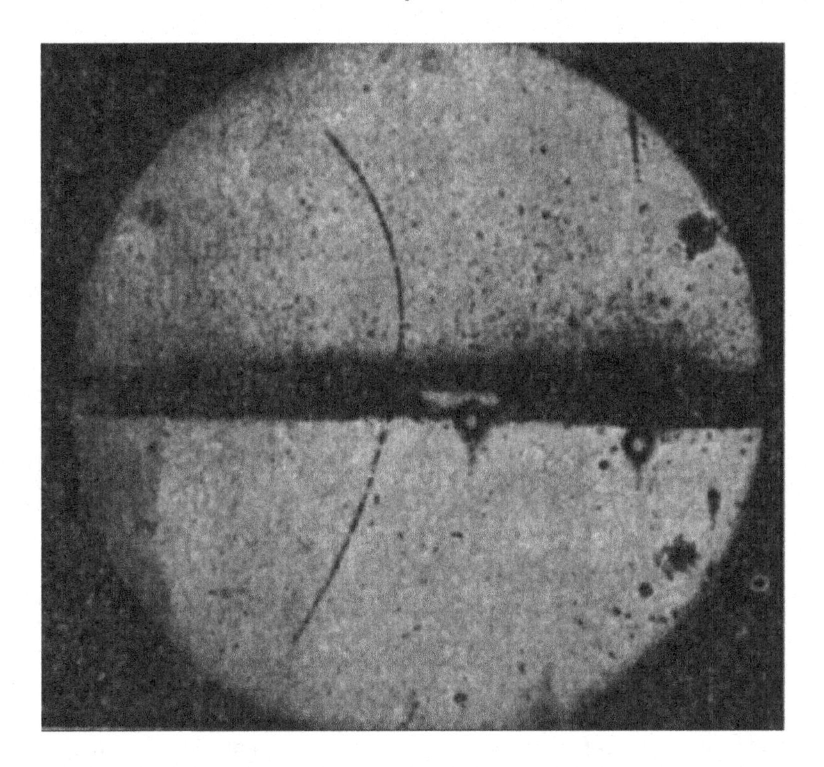

Figure 7.1 A cloud chamber photograph by Anderson (C. D. Anderson, Phys. Rev. **44** (1933), 406 [8]) with one of the first images of a positron. The positron travels from *bottom to top*, as shown by the fact that the track has a lower curvature in the upper part of the trajectory, owing to the loss of energy as the positron passes through the layer of lead, whose side view is visible in the middle of the chamber. From this information, it can be deduced that the particle has a *positive* charge, while the mass is consistent with the mass of an electron.

Anderson in 1932, by observing the products of cosmic ray interactions in a cloud chamber, Fig. 7.1.

However, the solution proposed by Dirac is not satisfactory from several points of view. Among them is the fact the the solution would not work for spin zero particles, which are bosons. One can answer, with Dirac, that for these particles the *$x$-coordinate is not observable* and therefore a spin zero particle does not have a wave function; *for such particles, there is still a representation of the momentum which is sufficient for practical matters* [9].

It can be argued that, in reality, on the basis of the uncertainty principle, the position of a relativistic particle is not an observable, regardless of the value of its spin.

Imagine the measurement of the position of an electron with a microscope. Obviously, we must use light of a sufficiently low wavelength, $\lambda$, to obtain adequate precision. Consequently, the electron experiences a random recoil of the order of the photon momentum, $k = 1/\lambda$, such that the uncertainty in the position and the momentum satisfy the uncertainty relation:

$$\Delta x \sim \lambda = \frac{1}{\Delta p} . \tag{7.3}$$

When the photon energy exceeds the value $\omega = k \sim 2m$ a new phenomenon occurs: the creation of an electron-positron pair. In the language of the Dirac sea, an electron in a negative energy state absorbs the photon and passes to a positive energy state leaving behind a hole. Now we have *two* electrons in the state and it is no longer possible to speak of the *electron position.*

In order for the locations of the two electrons to be confused, they must be separated by a distance of the order of the photon wavelength. Therefore the effect of non-locality begins when we arrive at a precision in the coordinate measurement of the electron of the order of:

$$\Delta x \sim \lambda = \frac{1}{m} \simeq 4.0 \cdot 10^{-11} \text{ cm.} \tag{7.4}$$

The length defined by (7.4) is known as the *Compton wavelength*, $\lambda_C$. For the proton, $\lambda_C \sim 0.2$ fermi $=0.2 \cdot 10^{-13}$ cm.

The same result is found if we try to construct ever smaller wave packets from the solutions to the Dirac equation. Using the positive energy solutions, which obviously do not form a complete set, we cannot obtain dimensions for the packet below $\lambda_C$.

The conclusion is that the *representation of $x$*, and therefore the wave function, simply does not exist for relativistic particles. Even for the electron, however, *there is still a representation of the momentum which is sufficient for practical purposes.* The way in which these purposes are satisfied is provided by *second quantisation.*

According to the second quantisation, more modernly known by the name of *relativistic quantum field theory*, the object which satisfies the Dirac equation is a *quantised field*, mathematically described by a linear operator which is a function of the location in space-time, $\psi(\mathbf{x}, t)$.

The field is a dynamic quantum variable in the Heisenberg representation. As an operator, the field acts on the space of physical states which, also in the Heisenberg representation, are constant in time.

The difficulty connected with solutions with negative energy (now, more properly, negative frequency) is solved because the time dependence of an operator in the Heisenberg representation is an exponential *whose sign is undefined.* The sign can be positive or negative, according to the energy difference of the states between which the operator induces transitions.

In the second quantisation, the quantities referred to in (6.89), (6.90) and (6.91) are also linear operators, rather than average values represented by complex numbers. As we will see in the following section, the first of these quantities must be identified with the charge associated with the particles, which must be opposite for the particles destroyed by $\psi^{(+)}$ to those created by $\psi^{(-)}$, as happens for the electron and positron.

It is reasonable to identify (6.90) and (6.91) with the Hamiltonian and the total momentum of the field. The operator structure of the creation and destruction operators and the further physical characteristics of the associated particles are obtained from the requirement that the Hamiltonian has a lower bounded spectrum (stability). The stability condition leads unambiguously to Fermi–Dirac statistics for the particles created or destroyed by the components of the Dirac field.

## 7.2 SECOND QUANTISATION: HOW IT WORKS

We separately consider in (6.88) the solutions with positive and negative frequencies, which we write generally as:

$$\psi^{(+)}(x) = Xe^{-i(px)};\ \ \text{(positive frequency)} \tag{7.5}$$
$$\psi^{(-)}(x) = Ye^{+i(px)};\ \ \text{(negative frequency)} \tag{7.6}$$
$$p^\mu = (E(p), \mathbf{p});\ \ E > m;\ \ p^2 = m^2. \tag{7.7}$$

The field in $x$ can be obtained from the field at the origin by a space-time translation according to the equation (see Appendix A):

$$\psi(x) = e^{i\boldsymbol{P}_\mu x^\mu}\psi(0)e^{-i\boldsymbol{P}_\mu x^\mu} \tag{7.8}$$

where $\boldsymbol{P}^\mu$ is the energy-momentum operator. We now take the matrix element of this equation between the 4-momentum eigenstates:

$$< E', \boldsymbol{P}'|\psi(x)|E, \boldsymbol{P} >= e^{-i(P-P')_\mu x^\mu} < E', \boldsymbol{P}'|\psi(0)|E, \boldsymbol{P} > . \tag{7.9}$$

If we compare (7.9) with (7.5) we see that we must have:

$$P' = P - p;\ \ \text{(positive frequency)} \tag{7.10}$$
$$P' = P + p;\ \ \text{(negative frequency)} \tag{7.11}$$

or:

- the field components with positive or negative frequencies respectively destroy or create a particle with mass $m$.

On the basis of these considerations, we rewrite the expansion (6.88) substituting the amplitudes of the normal modes of oscillation by, respectively, destruction and creation operators for electrons and positrons, $a$, $a^\dagger$ and $b$, $b^\dagger$:

$$\psi(x) = \sum_{\mathbf{p},r}\sqrt{\frac{m}{E(p)V}}[a_r(\mathbf{p})u_r(\mathbf{p})e^{-i(px)} + b_r(\mathbf{p})^\dagger v_r(\mathbf{p})e^{i(px)}] . \tag{7.12}$$

The algebraic structure of these operators is determined by considering the expressions in (6.90) and (6.91) which now we interpret as the energy and momentum of the field. Carrying out the multiplication of the operators without changing the order in which they appear in (6.90) and (6.91) we obtain:

$$\boldsymbol{H} = \int d^3x\, \psi^\dagger(\mathbf{x},t)[-i\boldsymbol{\alpha}\cdot\boldsymbol{\nabla} + \beta m]\psi(\mathbf{x},t) =$$
$$= \sum_{\boldsymbol{p},r} E(p)[a_r(\boldsymbol{p})^\dagger a_r(\boldsymbol{p}) - b_r(\boldsymbol{p})b_r(\boldsymbol{p})^\dagger] \tag{7.13}$$
$$\boldsymbol{P} = \int d^3x\, \psi^\dagger(\mathbf{x},t)(-i\nabla)\psi(\mathbf{x},t) =$$
$$= \sum_{\boldsymbol{p},r} \boldsymbol{p}[a_r(\boldsymbol{p})^\dagger a_r(\boldsymbol{p}) - b_r(\boldsymbol{p})b_r(\boldsymbol{p})^\dagger]\,. \tag{7.14}$$

We consider the second term in the Hamiltonian. The operator $bb^\dagger$ is semi-positive definite[1]. This prevents the assignment of commutation relations to the creation and destruction operators equal to those of the quantum harmonic oscillator. In this case, for fixed $\boldsymbol{p}$, we have:

$$-bb^\dagger = -b^\dagger b - 1 = -N(\boldsymbol{p}) - 1 \tag{7.15}$$

where $N(\boldsymbol{p})$ is the occupation number of the mode $\boldsymbol{p}$. The right-hand side can take arbitrarily negative values and the resulting Hamiltonian will be unbounded from below.

To obtain a consistent theory, the operator $b^\dagger b$ must be *limited*, as happens for the Fermi oscillator[2] which satisfies *anticommutation* rules. In this case we obtain:

$$-bb^\dagger = +b^\dagger b - 1 = +N(\boldsymbol{p}) - 1 \tag{7.16}$$

which has eigenvalues $0, -1$ ($-1$ for the vacuum). Using equation (7.16) for all the values of $\boldsymbol{p}$ and $r$, we obtain:

$$H = \sum_{pr} E_p \left(N_r(\boldsymbol{p}) + \bar{N}_r(\boldsymbol{p})\right)\,, \tag{7.17}$$

where $N_r(\boldsymbol{p}) = a_r^\dagger(\boldsymbol{p})a_r(\boldsymbol{p})$ and $\bar{N}_r(\boldsymbol{p}) = b_r^\dagger(\boldsymbol{p})b_r(\boldsymbol{p})$ are the number operators for the particles and antiparticles. In going from (7.13) to (7.17) we have omitted a constant term which corresponds to the zero point energy of the oscillators. Physically, this is equivalent to choosing the value $\langle 0|H|0\rangle$ as the zero of the energy scale.

---

[1]Actually, $< A|bb^\dagger|A >= \sum_n < A|b|n >< n|b^\dagger|A >= \sum_n |< A|b|n >|^2 \geq 0$, whatever the state $|A >$.

[2]The Fermi oscillator is defined by the anticommutation rule $\{b,b\} = \{b^\dagger,b^\dagger\} = 0$; $\{b,b^\dagger\} = 1$. From the first it follows that $b^2 = 0$ and therefore $N^2 = b^\dagger b b^\dagger b = b^\dagger b = N$, or $N(N-1) = 0$. The eigenvalues of $N$ are therefore $0, 1$.

The preceding equations are generalised by the following anticommutation rules:

$$\{a_r(\mathbf{p}), a^\dagger{}_{r'}(\mathbf{p}')\} = \{b_r(\mathbf{p}), b^\dagger{}_{r'}(\mathbf{p}')\} = \delta_{rr'}\delta_{\mathbf{pp}'} \tag{7.18}$$

$$\{a_r(\mathbf{p}), a_{r'}(\mathbf{p}')\} = \{b_r(\mathbf{p}), b_{r'}(\mathbf{p}')\} = \{a_r(\mathbf{p}), b_{r'}(\mathbf{p}')\} = \{a_r^\dagger(\mathbf{p}), b^\dagger{}_{r'}(\mathbf{p}')\} = 0. \tag{7.19}$$

Now we consider the normalisation factor in (6.89), which we recognise to be the integral of the time component of the 4-current $J^\mu = \bar{\psi}\gamma^\mu\psi$. Proceeding in a similar way to before and omitting a constant (infinity!) which represents the charge of the vacuum, we obtain:

$$\int d^3x\ J^0(x) = \sum_{\mathbf{p}r} \left(N_r(\mathbf{p}) - \bar{N}_r(\mathbf{p})\right). \tag{7.20}$$

We can therefore see that the particles created by $b^\dagger$, while they have mechanical properties, mass and spin, equal to those created by $a^\dagger$, have opposite charge.

## 7.3 CANONICAL QUANTISATION OF THE DIRAC FIELD

We will follow, as far as possible, a similar path to the one of the preceding section in constructing a quantised field which is subject to the Dirac equation.

The Lagrangian density in the absence of interactions is written (with $\bar{\psi} = \psi^\dagger\gamma_0$):

$$\mathcal{L} = \bar{\psi}(i\partial_\mu\gamma^\mu - m)\psi\ . \tag{7.21}$$

From here, we find;

$$\frac{\partial\mathcal{L}}{\partial\partial_\mu\psi} = i\bar{\psi}\gamma^\mu; \qquad \frac{\partial\mathcal{L}}{\partial\psi} = -m\bar{\psi}$$
$$\frac{\partial\mathcal{L}}{\partial\partial_\mu\bar{\psi}} = 0; \qquad \frac{\partial\mathcal{L}}{\partial\bar{\psi}} = (i\partial_\mu\gamma^\mu - m)\psi\ . \tag{7.22}$$

The Euler–Lagrange equations are therefore:

$$(i\partial_\mu\gamma^\mu - m)\psi = 0; \qquad i\partial_\mu(\bar{\psi}\gamma^\mu) + m\bar{\psi} = 0\ . \tag{7.23}$$

Using the relation $\gamma^0(\gamma^\mu)^\dagger\gamma^0 = \gamma^\mu$, it can be seen that the second equation is the adjoint of the first. In other words, if we consider $\psi$ and $\bar{\psi}$ as independent variables, the two equations tell us that $\bar{\psi}\gamma^0$ is the hermitian conjugate of $\psi$.

The Lagrangian (7.21) is invariant under translations and Lorentz transformations. The latter, for infinitesimal transformations can be written:

$$x'^\mu = \Lambda^\mu_\nu x^\nu = x^\mu + \frac{1}{4}\epsilon_{\alpha\beta}(g^{\alpha\mu}\delta^\beta_\nu - g^{\beta\mu}\delta^\alpha_\nu)x^\nu$$
$$\psi'(x') = S(\Lambda)\psi(x) = \psi(x) - i\frac{1}{4}\epsilon_{\alpha\beta}\sigma^{\alpha\beta}\psi(x) \tag{7.24}$$

with:

$$\sigma^{\alpha\beta} = \frac{i}{2}[\gamma^\alpha, \gamma^\beta] \tag{7.25}$$

Moreover, the Lagrangian (7.21) possesses a global symmetry, associated with the transformation of the fields by a *constant phase*

$$\psi(x) \to e^{i\alpha}\psi(x); \quad \bar{\psi} \to e^{-i\alpha}\bar{\psi}(x) \tag{7.26}$$

From (7.22), (7.24) and (7.26) we immediately find:

- the canonical energy-momentum tensor:

$$T^{\mu,\nu} = i\bar{\psi}\gamma^\mu\partial^\nu\psi - g^{\mu\nu}\mathcal{L} \,, \tag{7.27}$$

- the energy and the momentum of the field:

$$E = \int d^3x \psi^\dagger(-i\boldsymbol{\alpha}\cdot\boldsymbol{\nabla} + m\beta)\psi$$
$$\boldsymbol{P} = \int d^3x \psi^\dagger(-i\boldsymbol{\nabla})\psi \,, \tag{7.28}$$

- the symmetric energy-momentum tensor:

$$\theta^{\mu\nu} = \frac{1}{4}[\bar{\psi}\gamma^\mu\partial^\nu\psi - (\partial^\nu\bar{\psi})\gamma^\mu\psi + (\mu \to \nu)] \,, \tag{7.29}$$

- the conserved current connected to the transformation (7.26):

$$j^\mu(x) = \bar{\psi}(x)\gamma^\mu\psi(x); \qquad \partial_\mu j^\mu(x) = 0. \tag{7.30}$$

Comment. To construct the symmetric energy-momentum tensor according to the Belinfante–Rosenfeld procedure, we begin from the spin part of the angular momentum tensor. On the basis of (7.24) we can write

$$\Sigma^{\mu,\alpha\beta} = \frac{1}{2}\bar{\psi}(\gamma^\mu\sigma^{\alpha\beta})\psi = \frac{i}{4}\bar{\psi}(\gamma^\mu[\gamma^\alpha, \gamma^\beta])\psi \,. \tag{7.31}$$

Defining:

$$S_1^{\alpha\beta} = \partial_\mu\Sigma^{\mu,\alpha\beta}; \qquad S_2^{\alpha\beta} = \partial_\mu\Sigma^{\alpha,\mu\beta} = S_3^{\beta\alpha} \tag{7.32}$$

we have:

$$\theta^{\alpha\beta} = T^{\alpha,\beta} + \frac{1}{2}(S_1 - S_2 - S_3)^{\alpha\beta} \tag{7.33}$$

(the tensor $\frac{1}{2}S_1$ cancels the antisymmetric part of $T^{\alpha,\beta}$ while the subtraction of the symmetric tensor $S_2 + S_3$ balances the 4-divergence of $S_1$). Explicitly, we have

$$\frac{1}{2}(S_1 - S_2 - S_3) = \frac{i}{8}(\partial_\mu\bar{\psi})\mathbf{X}^\mu\psi + \bar{\psi}\mathbf{X}^\mu(\partial_\mu\psi);$$
$$\mathbf{X}^{\mu,\alpha\beta} = \frac{1}{4}[\gamma^\mu[\gamma^\alpha, \gamma^\beta] - \gamma^\alpha[\gamma^\mu, \gamma^\beta] - \gamma^\beta[\gamma^\mu, \gamma^\alpha]) \,. \tag{7.34}$$

Using the anticommutation rule for the $\gamma$ matrices, we can make the operators $\partial_\mu \gamma^\mu$ operate on the fields, where they give $\pm im$. It is easy to see that the terms proportional to $m$ cancel and we are left with the results of the anticommutations, from which:

$$\frac{1}{2}(S_1 - S_2 - S_3)^{\alpha\beta} = \frac{i}{4}\bar{\psi}(\partial^\alpha \gamma^\beta - 3\partial^\beta \gamma^\alpha)\psi - \frac{i}{4}[(\partial^\alpha \bar{\psi})\gamma^\beta \psi + (\partial^\beta \bar{\psi})\gamma^\alpha \psi] \,. \quad (7.35)$$

Summing this result with (7.27) leads to (7.29).

Hamiltonian Formalism. The conjugate momentum of the field $\psi$ is found from (7.22):

$$\Pi(x) = i\psi^\dagger(x) \quad (7.36)$$

while the conjugate momentum of $\bar{\psi}$ vanishes; in the Hamiltonian scheme there is only one pair of conjugate variables, which are $\psi$ and $i\psi^\dagger$. The Hamiltonian density agrees with the first result from (7.28), which has already been expressed in terms of conjugate variables.

The canonical quantisation foresees that the conjugate variables satisfy commutation rules similar to those of non-relativistic quantum mechanics (3.26). However, the requirement to have a lower bounded energy requires *anticommutation* rules for the electron and positron creation and destruction operators, as we saw in the preceding section. This corresponds to translating (3.26) into *equal-time anticommutation rules*:

$$\{\psi_\alpha(\boldsymbol{x}, t), \Pi_\beta(\boldsymbol{y}, t)\} = i\delta_{\alpha\beta}\delta^{(3)}(\boldsymbol{x} - \boldsymbol{y})$$

or

$$\{\psi_\alpha(\boldsymbol{x}, t), \bar{\psi}_\beta(\boldsymbol{y}, t)\} = \gamma^0_{\alpha\beta}\delta^{(3)}(\boldsymbol{x} - \boldsymbol{y}) \,. \quad (7.37)$$

Equations (7.37) determine the commutators of the dynamic variables with the Hamiltonian and thus the equations of motion. We use the identity:

$$[a, b \cdot c] = \{a, b\}c - b\{a, c\} \quad (7.38)$$

to find:

$$i\frac{\partial \psi}{\partial t} = [\psi(x), \boldsymbol{H}] = (-i\,\boldsymbol{\alpha} \cdot \boldsymbol{\nabla} + \beta m)\psi \,. \quad (7.39)$$

*The Heisenberg equations of motion reproduce the Dirac equation*, as they should. We can also calculate the commutator of the field with the conserved charge, with the result:

$$[\psi(x), \boldsymbol{Q}] = +\psi(x) \,. \quad (7.40)$$

The Space of States. Proceeding in a similar way to what was done for the Klein–Gordon equation, we can derive from (6.88) the $a_s(\boldsymbol{p})$ and $b_s(\boldsymbol{p})$

operators, taking advantage of the orthonormality conditions for the $u$ and $v$ spinors. We find:

$$a_s(\boldsymbol{p}) = \int d^3x \sqrt{\frac{m}{EV}} e^{i(px)} (\bar{u}_s(\boldsymbol{p})\gamma^0\psi(x));$$
$$b_s(\boldsymbol{p}) = \int d^3x \sqrt{\frac{m}{EV}} e^{i(px)} (\bar{\psi}(x)\gamma^0 v_s(\boldsymbol{p})) . \tag{7.41}$$

Starting from the canonical anticommutators, the anticommutators of these operators are calculated, which as expected agree with those already found earlier:

$$\{a_s(\boldsymbol{p}), a^\dagger_{s'}(\boldsymbol{p}')\} = \{b_s(\boldsymbol{p}), b^\dagger_{s'}(\boldsymbol{p}')\} = \delta_{s,s'}\ \delta_{\boldsymbol{p},\boldsymbol{p}'} \tag{7.42}$$

and all the other anticommutators equal to zero.

Using the anticommutation rules, we can express the conserved quantities in a way formally identical to (4.52):

$$\boldsymbol{H} = \int d^3x\ \theta^{00} = \Sigma_{\boldsymbol{k}}\ \omega(k)\ [a^\dagger(\boldsymbol{k})a(\boldsymbol{k}) + b^\dagger(\boldsymbol{k})b(\boldsymbol{k})]\ +\ \text{constant}$$
$$\boldsymbol{P}^i = \int d^3x\ \theta^{0i} = \Sigma_{\boldsymbol{k}}\ k^i\ [a^\dagger(\boldsymbol{k})a(\boldsymbol{k}) + b^\dagger(\boldsymbol{k})b(\boldsymbol{k})]$$
$$\boldsymbol{Q} = \int d^3x\ J^0 = \Sigma_{\boldsymbol{k}}\ [a^\dagger(\boldsymbol{k})a(\boldsymbol{k}) - b^\dagger(\boldsymbol{k})b(\boldsymbol{k})] . \tag{7.43}$$

We see that the Hilbert space of the theory is that of a set of Fermi oscillators. Explicitly, it consists of:

- the state $|0>$, the *vacuum*, which vanishes following the application of the destruction operators:

$$|0>\ \text{such that} : a_s(\boldsymbol{p})|0>= b_r(\boldsymbol{q})|0>= 0,\ \text{for each}\ s, r, \boldsymbol{p}, \boldsymbol{q}; \tag{7.44}$$

- the states with given occupation numbers, which are obtained by applying the creation operators $a^\dagger$ and $b^\dagger$ to the vacuum:

$$|n_1, n_2, \ldots; m_1, m_2, \cdots >=$$
$$= [a^\dagger_{s_1}(\boldsymbol{p}_1)]^{n_1}[a^\dagger_{s_2}(\boldsymbol{p}_2)]^{n_2} \ldots [b^\dagger_{r_1}(\boldsymbol{q}_1)]^{m_1}[b^\dagger_{r_2}(\boldsymbol{q}_2)]^{m_2} \ldots |0> . \tag{7.45}$$

As can be seen from the expression for the momentum, the $a$ and $b$ operators destroy relativistic particles of mass $m$ and spin $\frac{1}{2}$. The quantum states are those of a perfect gas made up of fermions of two different types, with equal mechanical properties.

Further information on the nature of these particles is provided by the conserved charge. We consider an element of the commutator matrix (7.40). We obtain[3]:

$$< q'|[\psi, \boldsymbol{Q}]|q >= (q - q') < q'|\psi|q >= + < q'|\psi|q >$$
$$\text{or}: \; q' = q - 1. \tag{7.46}$$

In both cases, the action of the field reduces the conserved charge by one unit, therefore:

- the particle destroyed by $\psi^{(+)}$ has opposite charge to that created by $\psi^{(-)}$.

In conclusion, the duplication of the sign of $p^0 = \pm\omega(p)$, which is inevitable in a relativistic theory, is reflected in the characterisation of the positive and negative frequency components as operators for the destruction and creation of particles. In the presence of a conserved charge with non-zero eigenvalues, the particles created by $\psi^{(-)}$ are a sort of mirror image of those destroyed by $\psi^{(+)}$, in the sense that they have equal mechanical properties (mass and spin) and opposite charge.

The combination of special relativity with quantum mechanics requires the existence of *antimatter.*

Comment. The antiparticle of the proton, the negatively charged antiproton, was discovered in 1955 with the Bevatron particle accelerator in Berkeley, by Segré, Chamberlain, Wiegand and Ypsilantis. Experimentally, the antiproton mass agrees with the mass of the proton to within a very small error, of order one part in 100 million [15]. The neutron also has an antiparticle, the antineutron. This is due to the existence of a conserved charge independent of electric charge. This new conserved charge, the *baryon number*, is necessary to take account of the extreme stability of matter.[4] The baryon number is associated with the current:

$$B^\mu = \bar{\psi}_P \gamma^\mu \psi_P + \bar{\psi}_N \gamma^\mu \psi_N \tag{7.47}$$

where $\psi_{P,N}$ denote the Dirac fields associated with the proton and neutron. The value $+1$ is conventionally assigned to the baryon number of the proton and neutron (and thus $-1$ to the antiproton and antineutron), while the lighter particles, electrons and neutrinos, have zero baryon numbers. Many other particles are now known which decay into protons or neutrons and therefore also have non-zero baryon number.

---

[3]The same result is obtained for the Klein–Gordon field starting from (4.46).

[4]Electric charge conservation alone would allow the decay of the proton into e positron and a neutral meson: $P \to e^+\pi^0$. The present non-observation of this decay leads to the limit pf the proton's lifetime: $\tau_P > 1.6\ 10^{34}$ years, see [16].

The simultaneous conservation of electric charge and baryon number gives rise to the stability of the hydrogen atom (which has $Q = 0$ but $B = 1$), which could otherwise convert into purely electromagnetic radiation.

## 7.4 THE REPRESENTATION OF THE LORENTZ GROUP

We will now show what was anticipated in Section 6.1.2, that using the $S(\Lambda)$ matrices we can construct a set of unitary operators that represent the Lorentz transformations in the space of occupation numbers (following the method of Wigner). It is sufficient to limit the discussion to states with only one particle. We therefore consider the states:

$$|p, r >= a_r(\boldsymbol{p})^\dagger|0> \ . \tag{7.48}$$

We must define the action of the $L_+^\uparrow$ transformations on these states, by means of unitary operators $U(\Lambda)$ which satisfy the composition rules of the group. The Lorentz transformations act on 4-vectors on the mass hyperboloid defined by $p_\mu p^\mu = p^2 = m^2$.

We choose a specific 4-vector, $p_0$. The $L_+^\uparrow$ transformations convert $p_0$ into 4-vectors in the region of the mass hyperboloid with $p_0 > +m$. Conversely, every $p^\mu$ is characterised by the Lorentz transformation which converts $p_0$ into $p$. For particles with mass $m \neq 0$, we choose

$$p_0^\mu = (m, \boldsymbol{0}) \ . \tag{7.49}$$

For each $p$ we can define a transformation we call a *boost*, which takes $p_0$ into $p$. We write concisely:

$$L(p)p_0 = p \tag{7.50}$$

and define the action of $U(L(p))$ on the states $|p_0, r>$ in the following way:

$$|p, r >= U(L(p))|p_0, r> \ . \tag{7.51}$$

(Note that $U$ is defined so as not to change the spin indices.)

Now we consider a general transformation:

$$p'^\mu = \Lambda^\mu{}_\nu p^\nu. \tag{7.52}$$

The crucial observation is that;

$$p' = L(\Lambda p)p_0 = \Lambda p = \Lambda L(p)p_0 \ . \tag{7.53}$$

Therefore, the transformation $L^{-1}(\Lambda p)\Lambda L(p)$ changes $p_0$ into itself and must belong to the subgroup of Lorentz transformations which *leaves $p_0$ invariant*. Wigner called this subgroup the *little group* of the representation. In our case, given the form of $p_0$ in (7.49), these transformations are *rotations* *in three-dimensional space*, which we denote as $R$. The explicit form of the transformations follows from the unitary transformations $U(\Lambda)$. We set

$$U(\Lambda) = U(L(\Lambda p))U(R)U(L^{-1}(p)); \tag{7.54}$$

from which:

$$\begin{aligned} U(\Lambda)|p,r> &= U(L(\Lambda p))U(R)U(L^{-1}(p))|p,r>= \\ &= U(L(\Lambda p))U(R)|p_0,r>= \\ &= U(L(\Lambda p))\sum_s S(R)_{r,s}(R)|p_0,s>= \\ &= \sum_s S(R)_{r,s}|\Lambda p,s>; \end{aligned} \tag{7.55}$$

$$S(R)_{r,s} = \bar{u}_s(\mathbf{0})S(R)u_r(\mathbf{0}) = \chi^\dagger{}_s S(R)\chi_r \tag{7.56}$$

where $S(R)$ are the matrices defined in (6.39).

The mapping $\Lambda \to R$ confirms the composition rule for $L_+^\uparrow$. To be exact, if $\Lambda = \Lambda_1\Lambda_2$:

$$U(\Lambda_1)U(\Lambda_2) = \{U(L(\Lambda_1 q))U(R_1)U(L^{-1}(q))\}\{U(L(\Lambda_2 p))U(R_2)U(L^{-1}(p))$$

We set $q = \Lambda_2 p$. Because $U(L^{-1}(q))U(L(q)) = 1$, we obtain:

$$U(\Lambda_1)U(\Lambda_2) = U(L(\Lambda_1\Lambda_2(p))U(R_1)U(R_2)U(L^{-1}(p)),$$

therefore,

$$\begin{aligned} &\text{if} : \Lambda_1 \to R_1;\ \Lambda_2 \to R_2; \\ &\text{then} : \Lambda_1\Lambda_2 \to R_1R_2 \end{aligned}$$

and equation (7.55) defines a representation of the $L_+^\uparrow$ group.

Corresponding to every unitary representation of the little group, the equation (7.55) therefore produces a unitary representation of the Lorentz group on the eigenstates of the momentum of the particle. For Dirac particles, the non-unitarity nature of $S(\Lambda)$ already noted in (6.1.2) does not change the result. Effectively, according to the method of Wigner, the representation of the states uses only the component of $S$ which corresponds to the group of spatial rotations, for which $S$ is unitary:

$$\begin{aligned} &S(R) = S = 1 - \frac{i}{4}\,\omega_{ij}\sigma^{ij} \qquad (i,j = 1,2,3) \\ &(\sigma^{ij})^\dagger = \sigma^{ij},\ \to\ \ S(R)^\dagger S(R) = 1\ . \end{aligned}$$

In conclusion, relativistic particles with non-zero mass are characterised by two quantum numbers

- the value of the mass, $m \neq 0$,

- the (integer or half-integer) value of the spin, which characterises the representation of the little group of $p_0$.

For particles of zero mass, like the photon, $p_0$ can be chosen as:

$$p_0 = (\omega, \omega \boldsymbol{n}_3) \tag{7.57}$$

where $\boldsymbol{n}_3$ is the unit vector along the $z$-axis. The little group is the one-dimensional group of rotations around the $z$-axis.

We leave to the reader the task of constructing the corresponding representations, to show the result already given in Section 2.2.

Normal Products. The prescription for obtaining multilinear expressions of the fermion fields with zero vacuum value must be modified appropriately to take account of the anticommutation rules.

Once the fields are separated into positive and negative frequency parts, the correct prescription is to write the operators with negative frequency to the right of those with positive frequency, with positive or negative sign according to whether the number of exchanges necessary to arrive at this configuration from the starting point, is even or odd. For example:

$$\begin{aligned} N(\psi(x)\bar{\psi}(y)) &=: \psi(x)\bar{\psi}(y) :=: \\ &:=: (\psi^{(+)}(x) + \psi^{(-)}(x))(\bar{\psi}^{(+)}(y) + \bar{\psi}^{(-)}(y)) := \\ &= \psi^{(+)}(x)\bar{\psi}^{(+)}(y) - \bar{\psi}^{(-)}(y)\psi^{(+)}(x) + \psi^{(-)}(x)\bar{\psi}^{(+)}(y) + \psi^{(-)}(x)\bar{\psi}^{(-)}(y). \end{aligned} \tag{7.58}$$

## 7.5 MICROCAUSALITY

As we saw in Section 1.3, given an event $x$, space-time divides into distinct regions regarding the causal connection of different events $y$ with event $x$. The region located outside the two light cones originating from $x$ represents the *absolute present* of $x$. These events are characterised by the fact that the interval $y - x$ is *spacelike*, or $(y - x)^2 < 0$. For brevity, we write $y \sim x$.

Measurements carried out on two observables localised at $x$ and $y$ respectively cannot influence one another when $x \sim y$, because this implies propagation of signals faster than the limit set by the speed of light in a vacuum. From quantum mechanical principles it follows that, under these conditions, the corresponding operators commute among themselves:

$$[O_1(x), O_2(y)] = 0, \text{ if } x \sim y \; . \tag{7.59}$$

The relation (7.59) is known as the *microcausality condition*. The hypothesis that it should hold for any value of the $x$–$y$ separation is a very stringent condition, which could be violated on microscopic length scales. However, the experimental consequences of microcausality have been confirmed down to the smallest distances so far tested, of the order of $10^{-15}$ cm.

Using relativistic invariance, we can extend the canonical quantisation rules (4.44) and (7.37) from the region $\boldsymbol{y} \neq \boldsymbol{x}$, $y^0 = x^0$ to the whole of the region of the present of $x$. For a general field $\chi$, we find

$$[\chi_a(x), \chi_b^\dagger(y)]_\pm = [\chi_a(x), \chi_b(y)]_\pm = 0 \text{ for } x \sim y \tag{7.60}$$

where the $\pm$ sign denotes the anticommutator or commutator of the fields, according to whether $\chi$ is a Dirac or Klein–Gordon field (the indices $a, b$ denote possible components of the field, either spinors or related to internal symmetries).

In the case of boson fields, (7.60) tells us that the components of the field commute for spatial separations and are therefore potential observables, such as the components of electric and magnetic fields[5].

In contrast, fermion fields *do not commute* among themselves for spacelike separations. The only possible interpretation of this result is that fermion fields cannot be observables.

The non-observability of the fermion field is confirmed by its transformation properties under rotation. We restrict equation (7.24) to the case of rotations around the $z$-axis. In this case, the only non-zero components among the infinitesimal parameters are $\epsilon_{12} = -\epsilon_{21} = \epsilon$ and we find:

$$\psi'(x') = (1 - i\frac{\epsilon}{2}\sigma^{12})\psi(x); \quad \sigma^{12} = \Sigma^3 = \begin{pmatrix} \sigma_3 & 0 \\ 0 & \sigma_3 \end{pmatrix}. \tag{7.61}$$

For finite rotations, the previous equation becomes an exponential, so:

$$\psi'(x') = e^{-i\frac{\theta}{2}\Sigma_3}\psi(x). \tag{7.62}$$

Given that $\Sigma_3$ has eigenvalues $\pm 1$, the equation simply expresses the fact that the field is associated with a spin $\frac{1}{2}$ particle. However, for $\theta = 2\pi$, when $x' = x$, we find:

$$\psi' = -\psi. \tag{7.63}$$

This result is obviously absurd for an observable quantity, which should return to itself after a $2\pi$ rotation.

However, if $\psi$ cannot be an observable, the physical quantities constructed from $\psi$, such as the energy density or charge density, must be. On the basis of equation (7.63), we can conclude that *homogeneous functions of even powers* of the field are good candidates to be observables.

The same conclusion is reached by starting from the canonical anticommutation rules. We consider as an example the commutators of two charge densities:

$$\left[j^0(x), j^0(y)\right] \quad \text{for} \quad x \sim y. \tag{7.64}$$

[5] A complex scalar field is not observable, not being hermitian, but it's real and imaginary parts are in principle observables.

Repeatedly applying the identify (7.38), and similarly for the commutators, we find:

$$[\psi^\dagger(x)\psi(x), j^0(y)] = \psi^\dagger(x)[\psi(x), j^0(y)] + [\psi^\dagger(x), j^0(y)]\psi(x) =$$
$$= \psi^\dagger(x)(-\psi^\dagger(y)\{\psi(x), \psi(y)\} + \{\psi(x), \psi^\dagger(y)\}\psi(y)) + \ldots = 0. \quad (7.65)$$

because all the anticommutators vanish for $x \sim y$.

## 7.6 THE RELATION BETWEEN SPIN AND STATISTICS

We have seen in a comprehensive way that the quanta of the Dirac field, particles of spin $\frac{1}{2}$, must obey Fermi–Dirac (F–D) statistics. It is easy to be convinced that the quanta of the Klein–Gordon field and of the electromagnetic field, spin 0 and 1 respectively, should obey Bose–Einstein (B–E) statistics. We consider the classical expression for the Hamiltonian of a real Klein–Gordon field in terms of the amplitudes of the normal modes of oscillation, cf. equation (4.52):

$$\boldsymbol{H} = \Sigma_{\boldsymbol{k}}\, \omega(k)\, \frac{1}{2}[a^*(\boldsymbol{k})a(\boldsymbol{k}) + a(\boldsymbol{k})a^*(\boldsymbol{k})]. \quad (7.66)$$

If we wish to quantise according to the rules of the Dirac field:

$$a \to a;\; a^* \to a^\dagger$$
$$\{a, a^\dagger\} = 1 \quad (7.67)$$

we find the absurd result that the Hamiltonian is a multiple of the identity operator so that every observable has to be a constant of the motion. The same holds for the electromagnetic field and the photons, cf. equation (5.112).

For particles of spin $j \leq 1$ we can therefore state the spin-statistics theorem:

- *identical particles with integer spin obey Bose–Einstein statistics, while those with half-integer spin conform to Fermi–Dirac statistics.*

We can generalise the theorem to assemblies of arbitrary spin composed from spin $\frac{1}{2}$ particles. In fact, the particles which we know with spin greater than 1 are all of this type:

- atomic nuclei, with mass number $A$ and atomic number $Z$, composed of $Z$ protons and $A - Z$ neutrons,
- atoms, composed of $Z$ electrons and a nucleus with mass number $A$; in all $Z + A$ spin $\frac{1}{2}$ particles,
- subnuclear particles, classified as baryons, made of three quarks which are fundamental spin $\frac{1}{2}$ particles, and mesons, composed of a quark-antiquark pair.

In general, since spin $\frac{1}{2}$ is the fundamental representation of the rotation group, we can construct any value of angular momentum $j$ by combining an appropriate number $N$ of spin $\frac{1}{2}$, with

- N=even, $j = \frac{1}{2}N =$ integer;
- N=odd, $j = \frac{1}{2}N =$ half-integer.

The effective field which creates and destroys these assemblies (for example the iron nucleus) is obtained starting from $\psi^N$, if $\psi$ is the field of the constituent.

After rotations by 360°, each field $\psi$ gains a factor $-1$, or:

$$\Psi = \psi^N \to (-1)^N \Psi = e^{i2\pi j}\Psi \quad \text{(rotation through } 2\pi\text{)} \tag{7.68}$$

while under the exchange of constituents of two identical assemblies, the state has a phase factor:

$$|N> \to (-1)^N |N> = |N> e^{i2\pi j}. \tag{7.69}$$

Therefore the assemblies take a sign appropriate for either B–E or F–D statistics, according to whether $N$ is even or odd, or whether $j$ is integer or half-integer, in agreement with the theorem.

In a relativistic theory, a state with a definite $j$ value can in general be a superposition of one state with $N$ constituents plus an indefinite number of fermion–antifermion pairs. This corresponds to adding to $\Psi$ components of the type $\psi^N(\psi\psi^\dagger)^m$, with $m$ variable. However, because $\psi\psi^\dagger$ corresponds to integer angular momentum and the pairs are equivalent to $2m$ fermions, the result does not change. The same holds if we introduce the expression for $\Psi$ the derivatives of $\psi$ or $\psi^\dagger$. This corresponds to introducing orbital angular momenta which, having integer eigenvalues, do not change the integer or half-integer property of $j$. Finally, the same result arises if, as in nuclei or atoms, the constituents are groups of different kinds of fermions; each fermion contributes a factor $-1$ to (7.68) and the same factor to (7.69).

Comment 1. The first determination of the statistics of several nuclei was due to Franco Rasetti, with an experiment carried out in 1928 (for a description of Rasetti's experiment and the discussion which followed, see [17]). The experiment concerned the nuclei ${}^{14}_{7}Na$ and ${}^{16}_{8}O$ (the upper and lower indices represent the values of A and Z) and both were shown to obey B–E statistics. At that time, it was thought that a nucleus of mass $A$ and charge $Z$ was composed of $A$ protons, the only heavy particle then known, and by $A - Z$ electrons, the *nuclear electrons*, confined within the nucleus, a total of $2A - Z$ fermions. This number is even or odd according to $Z$, while the nuclei considered by Rasetti have $Z$ both even and odd, while obeying the same statistics. The paradox was resolved by the discovery of the neutron which, as we explained, causes the number of fermions present in a nucleus to be $A$, in agreement with the result of Rasetti and the spin-statistics theorem.

Rasetti's result was noted by Pauli, who cited it in the letter in which he proposed the existence of the neutrino. Pauli thought the electron and the neutrino emitted in $\beta$ decays of nuclei existed permanently in the initial nucleus. Assuming a spin $\frac{1}{2}$ neutrino and one neutrino for each nuclear electron, the number of fermions present became $A + 2(A - Z) = 3A - 2Z$ which depends only on the evenness of $A$, in agreement with Rasetti. After the discovery of the neutron, Fermi proposed that the $e\bar{\nu}p$ system (which has half-integer angular momentum) was not permanently present in the nucleus but was simply the product of the neutron decay, as an effect of the weak interaction. The $e\bar{\nu}$ pair is created from the vacuum by the creation operators included in the interaction Hamiltonian, in a similar way to the photon produced by the deexcitation of an atom, cf. Chapter 9.

Comment 2. A similar paradox occurred at the origin of quark theory, in which baryons are thought to be bound states of three quarks. The solution to the problem introduced a new quark quantum number, *colour*, cf. Sects. 9.3, 18.1 and Ref. [13].

## 7.7 PROBLEMS FOR CHAPTER 7

### Sect. 7.3

**1.** Starting from the equal time anticommutators

$$\{\psi_\alpha(\boldsymbol{x},t), \bar{\psi}_\beta(\boldsymbol{y},t)\} = \gamma^0_{\alpha\beta}\delta^{(3)}(\boldsymbol{x}-\boldsymbol{y}) \tag{7.70}$$

prove that:

$$\{a_s(\boldsymbol{p}), a^\dagger_{s'}(\boldsymbol{p}')\} = \{b_s(\boldsymbol{p}), b^\dagger_{s'}(\boldsymbol{p}')\} = \delta_{s,s'}\ \delta_{\boldsymbol{p},\boldsymbol{p}'}$$

i.e. that the quanta obey the Fermi–Dirac statistics.

**2.** Consider the explicitly hermitian Lagrangian density:

$$\mathcal{L} = \frac{1}{2}(\mathcal{L}_D + \mathcal{L}_D^\dagger)$$

where $\mathcal{L}_D$ is the Dirac Lagrangian density

$$\mathcal{L}_D = \bar{\psi}(i\gamma_\mu\partial^\mu - m)\psi\ .$$

Use its explicit form to obtain:

- the equations of motions for the fields $\psi$ and $\bar{\psi}$;
- the expressions of the variables conjugate to the fields $\psi$ and $\bar{\psi}$;

– the energy, written in terms of the energy-momentum tensor $\theta^{\mu\nu}$.

**3.** With $\psi$ a spinor field, consider the Lagrangian density

$$\mathcal{L} = i\bar{\psi}\gamma^{\mu}\left(\partial_{\mu}\psi\right) - \frac{1}{2}i\left(\partial_{\mu}\bar{\psi}\right)\gamma^{\mu}\psi - M\bar{\psi}\psi \ .$$

– Derive the equations of motion for $\psi$ and $\bar{\psi}$.
– Determine the spin of the quanta described by the field $\psi$.
– Determine the mass of the quanta.

**4.** From the Heisenberg equation of motion and the Dirac Hamiltonian:

$$i\frac{\partial\psi}{\partial t} = [\psi(x), \boldsymbol{H}]$$
$$\boldsymbol{H} = \bar{\psi}(i\boldsymbol{\alpha}\cdot\boldsymbol{\nabla} + \beta m)\psi$$

and using the equal time anticommutators (7.70), derive the Dirac equation for $\psi$ (hint: use the identity $[a, b\cdot c] = \{a,b\}c - b\{a,c\}$).

**5.** Consider the plane wave representation:

$$\psi(x) = \sum_{\boldsymbol{p},r}\sqrt{\frac{m}{E(p)V}}\left[a_r(\boldsymbol{p})u_r(\boldsymbol{p})e^{-i(px)} + b_r(\boldsymbol{p})^{\dagger}v_r(\boldsymbol{p})e^{i(px)}\right]$$

and the commutation rule

$$[\psi(x), \boldsymbol{Q}] = q\psi(x) \ .$$

Compute the matrix elements of $\boldsymbol{Q}$ between vacuum and the one-electron state $|e, \boldsymbol{p}, r >$ and between $< \bar{e}, \boldsymbol{p}, r|$ and vacuum, where $e$ and $\bar{e}$ denote the particles created by $a^{\dagger}$ and $b^{\dagger}$, with electric charges $q_e$ and $q_{\bar{e}}$, respectively. Show that:

$$q_e = -q_{\bar{e}} = q \ .$$

## Sect. 7.5

**1.** Consider the bilinear densities

$$J_A = \bar{\psi}(x)\Gamma_A\psi(x)$$

where $\Gamma_A$, $A = 1, \cdots 16$ is one of the Dirac matrices introduced in Sect. 6.1.2.

Prove the formula for the equal time commutator:

$$[J_A(\boldsymbol{x}, 0), J_B(\boldsymbol{y}, 0)] = \delta^{(3)}(\boldsymbol{x}-\boldsymbol{y})\bar{\psi}(\boldsymbol{y}, 0)\left[\gamma^0\Gamma_A, \gamma^0\Gamma_B\right]\psi(\boldsymbol{y}, 0) \ .$$

This shows, in particular, that the time component of the conserved current $J_{\mu}$ has vanishing equal time commutators with all the current's components, a fact at the basis of the so-called Ward identities, see the problems in Chapter 14.

CHAPTER 8

# FREE FIELD PROPAGATORS

In this section we set $\hbar = c = 1$.

## 8.1 THE TIME-ORDERED PRODUCT

The *time-ordered* product of two scalar fields is defined as:

$$T\{\phi(x)\phi^\dagger(y)\} = \begin{cases} \phi(x)\phi^\dagger(y) & x^0 - y^0 > 0 \\ \phi^\dagger(y)\phi(x) & x^0 - y^0 < 0 \end{cases} . \tag{8.1}$$

The vacuum expectation value:

$$iD_F(x,y) =< 0|\, T\{\phi(x)\phi^\dagger(y)\}\, |0> \tag{8.2}$$

gives the quantum amplitude for the simplest observable process involving a quantised field:

(i) creation of a particle by a source located at $y$ and corresponding absorption at $x$, if $x^0 > y^0$, or

(ii) creation of an antiparticle by a source located at $x$ and corresponding absorption at $y$, if $y^0 > x^0$.

The function (8.2) is known as the *propagator* of the corresponding field and its exact form depends on the mass and the spin of the quanta associated with the field.

In view of the anticommutation relations satisfied by the Dirac field, the time-ordered product of fermion fields must be antisymmetric:

$$T\{\psi(x)\bar{\psi}(y)\} = \begin{cases} \psi(x)\bar{\psi}(y) & x^0 - y^0 > 0 \\ \bar{\psi}(y)\psi(x) & x^0 - y^0 < 0 \end{cases} , \tag{8.3}$$

and the fermion propagator is defined by:

$$i(S_F)_{\alpha\beta}(x,y) =< 0|\, T\{\psi_\alpha(x)\bar{\psi}_\beta(y)\}\, |0> . \tag{8.4}$$


DOI: 10.1201/9781003436263-8

In a field theory invariant under translations, the propagators are functions only of the difference $x$–$y$.

This can be seen by inserting into (8.2) the products $U^\dagger U = 1$, where $U$ is the operator which translates by $-y$ (the argument is repeated identically in the fermion case (8.4)). The vacuum being invariant on the application of $U$, we obtain:

$$\begin{aligned} <0|\, T\{\phi(x)\phi^\dagger(y)\}\, |0> &=<0|\, T\{\phi(x-y)\phi^\dagger(0)\}\, |0>= \\ &= iD_F(x-y,0) = iD_F(x-y). \end{aligned} \tag{8.5}$$

When the particle associated with the field has a charge, as happens if the field is complex, the charge flows in both cases from $y$ towards $x$. The propagator can be represented by a line directed from $y$ to $x$.

## 8.2 PROPAGATORS OF THE SCALAR FIELD

To calculate $iD_F(x)$ explicitly from (8.2), we proceed as follows. We write the fields in terms of the positive and negative frequency components, and use the fact that the former gives zero when applied to the vacuum. We obtain:

$$\begin{aligned} &<0|\, T\{\phi(x)\phi^\dagger(0)\}\, |0> = \\ &= \begin{cases} <0|\phi^{(+)}(x)(\phi^\dagger)^{(-)}(0)\, |0>=<0|[\phi^{(+)}(x),(\phi^\dagger)^{(-)}(0)]|0>;\ x^0>0 \\ <0|(\phi^\dagger)^{(+)}(0)\phi^{(-)}(x)\, |0>= -<0|[\phi^{(-)}(x),(\phi^\dagger)^{(+)}(0)]|0>;\ x^0<0. \end{cases} \end{aligned}$$

We define the two functions:

$$i\Delta^{(+)}(x) =<0|[\phi^{(+)}(x),(\phi^\dagger)^{(-)}(0)]|0>; \tag{8.6}$$

$$i\Delta^{(-)}(x) =<0|[\phi^{(-)}(x),(\phi^\dagger)^{(+)}(0)]|0>. \tag{8.7}$$

We can quickly convince ourselves that these two functions are exactly the solutions of the homogeneous Klein–Gordon equation which we introduced, with the same name, in Section 4.2, equations (4.26) and (4.27). For example, starting from the expansion (4.47) and taking account of the canonical commutators, we find:

$$\begin{aligned} i\Delta^{(+)}(x) &=<0|[\phi^{(+)}(x),(\phi^\dagger)^{(-)}(0)]|0>= \sum_{\mathbf{k}} \frac{1}{2\omega(k)V} e^{-i(kx)} = \\ &= \frac{1}{(2\pi)^3}\int \frac{d^3k}{2\omega(k)} e^{-i(kx)} \end{aligned}$$

which agrees with (4.26).

For the propagator, we therefore find:

$$iD_F(x) = \theta(x^0)i\Delta^{(+)}(x) - \theta(-x^0)i\Delta^{(-)}(x) \tag{8.8}$$

which agrees with the solution of the inhomogeneous equation with the Feynman boundary conditions referred to in equation (4.33) with the same name:

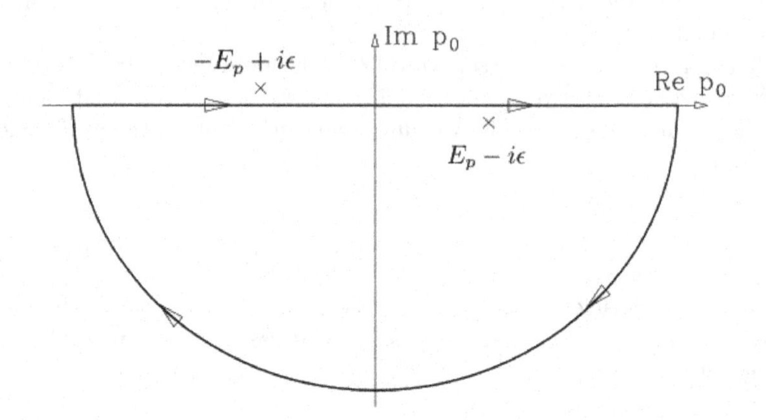


Figure 8.1 Following the $i\epsilon$ prescription of Feynman, the poles of the Green's function move in the complex plane as shown in the figure. The path of integration to obtain $iD_F$, equation (8.10), is now the real axis. The figure shows the way we can close the path to obtaining $iD_F = i\Delta^{(+)}$ for $x^0 > 0$. For $x^0 < 0$ the path should be closed in the upper half-plane.

- *The Feynman boundary conditions are the correct ones to create the quantum propagator.*

In summary, if the scalar field satisfies the equation:

$$(-\Box - \mu^2)\phi = 0 \tag{8.9}$$

the Feynman propagator is given, in terms of its Fourier transform, by

$$iD_F(x) =< 0|\ T\{\phi(x)\phi^\dagger(0)\}\ |0 >= \int \frac{d^4k}{(2\pi)^4} \frac{i}{k^2 - \mu^2 + i\epsilon} e^{-i(kx)}. \tag{8.10}$$

As described in Section 4.2 the integration in the complex plane of $k^0$ is carried out along the real axis and the prescription for $i\epsilon$ in (8.10) moves the poles of the integrand as shown in Fig. 8.1.

Unlike the commutators of the fields, which satisfy the microcausality condition, the propagator does not vanish for spatial separations. To be exact:

$$\begin{aligned} iD_F(\mathbf{x}, 0) &= \lim_{t\to 0^+} i\Delta^{(+)}(\mathbf{x}, t) = \lim_{t\to 0^-} i\Delta^{(-)}(\mathbf{x}, t) = \\ &= \frac{1}{(2\pi)^3} \int d^3k\, \frac{1}{2\sqrt{\mathbf{k}^2 + \mu^2}} e^{i\mathbf{k}\cdot\mathbf{x}} \neq 0. \end{aligned} \tag{8.11}$$

For large values of $r = |\mathbf{x}|$ the propagator decreases exponentially (for a detailed study of the functions $\Delta^{(\pm)}(x)$ and $D_F(x)$, see Bogoliubov–Shirkov [2]):

$$iD_F(r, 0) \simeq \frac{1}{8\pi^2} \sqrt{\frac{2}{\pi}} \frac{\mu^{1/2}}{r^{3/2}} e^{-\mu r} \tag{8.12}$$

consistent with the arguments of Section 7.1, by which a relativistic particle can be localised at most to within its Compton wavelength, $\lambda_C = 1/\mu$ in natural units. According to (8.11), we obtain a function localised only in the limit $\mu \to \infty$, as is reasonable to expect:

$$lim_{\mu\to\infty} iD_F(\mathbf{x}, 0) = \frac{1}{2\mu}\delta^{(3)}(\mathbf{x}). \tag{8.13}$$

In the limit, we find again the non-relativistic situation in which a pointlike source at the origin produces a particle in an eigenstate of the coordinate with $\mathbf{x} = 0$.

In the limit of infinite mass, the propagator is completely localised in space-time. In this limit, the singularity in the integration of $k^0$ in (8.10) becomes infinite, and we obtain[1]:

$$lim_{\mu\to\infty} D_F(x) = \frac{-i}{\mu^2}\,\delta^{(4)}(x). \tag{8.14}$$

Comment. It is useful to show by direct calculation that the time-ordered product of scalar fields satisfies the Klein–Gordon equation with pointlike sources. We consider the expression in (8.1) which we write explicitly as:

$$T\{\phi(x)\phi^\dagger(0)\} = \theta(t)\phi(x)\phi^\dagger(0) + \theta(-t)\phi^\dagger(0)\phi(x) \tag{8.15}$$

From the relations:

$$\frac{\partial}{\partial t}\theta(t) = -\frac{\partial}{\partial t}\theta(-t) = \delta(t) \tag{8.16}$$

and from the commutation rules (4.43), we find:

$$\begin{aligned}\partial_t\, T\{\phi(x)\phi^\dagger(0)\} &= T\{\partial_t\phi(x)\phi^\dagger(0)\} + \delta(t)[\phi(\mathbf{x},0), \phi^\dagger(0)] = T\{\partial_t\phi(x)\phi^\dagger(0)\}\\ \partial_t^2\, T\{\phi(x)\phi^\dagger(0)\} &= T\{\partial_t^2\phi(x)\phi^\dagger(0)\} + \delta(t)[\partial_t\phi(\mathbf{x},0), \phi^\dagger(0)] =\\ &= T\{\partial_t^2\phi(x)\phi^\dagger(0)\} - i\delta^4(x)\end{aligned} \tag{8.17}$$

or, given that $\phi$ satisfies (8.9), we obtain:

$$(-\Box - \mu^2)T\{\phi(x)\phi^\dagger(0)\} = i\delta^{(4)}(x) \tag{8.18}$$

consistent with (8.10).

## 8.3 PROPAGATORS OF THE DIRAC FIELD

By analogy to the case of the scalar field, the Feynman propagator is defined as:

$$iS_F(x) = \langle 0|\, T\{\psi(x)\bar{\psi}(0)\}\, |0\rangle. \tag{8.19}$$

[1]The discrepancy between the two preceding equations is due to the fact that they are obtained with two different sequences of limits; (8.13) is obtained in the limit $t \to 0$ followed by $\mu \to \infty$, while in (8.14) $\mu \to \infty$ for a general $x$; the two limits do not commute.

We separate into positive and negative frequency parts in the fields and omit terms with operators which vanish when applied to the vacuum. We obtain:

$$\begin{aligned}\langle 0|\,\psi(x)\bar{\psi}(0)\,|0\rangle &= \langle 0|\,\psi^{(+)}(x)\bar{\psi}^{(-)}(0)\,|0\rangle = \\ &= \langle 0|\{\psi^{(+)}(x),\bar{\psi}^{(-)}(0)\}|0\rangle = iS^{(+)}(x)\end{aligned} \tag{8.20}$$

$$\begin{aligned}\langle 0|\,\bar{\psi}(0)\psi(x)\,|0\rangle &= \langle 0|\,\bar{\psi}^{(+)}(0)\psi^{(-)}(x)\,|0\rangle = \\ &= \langle 0|\{\psi^{(-)}(x),\bar{\psi}^{(+)}(0)\}|0\rangle = iS^{(-)}(x).\end{aligned} \tag{8.21}$$

Using the 4-spinor completeness relations, (6.84) and (6.85), we can rewrite equations (8.20) and (8.21) in the form

$$\begin{aligned}iS^{(+)}(x) &= \frac{1}{V}\sum_{\mathbf{p}s}\left(\frac{m}{E_p}\right)\frac{\not{p}+m}{2m}\mathrm{e}^{-ip(x')} = \\ &= (i\not{\partial}+m)\frac{1}{(2\pi)^3}\int\frac{d^3p}{2E(k)}e^{-i(px)} = \\ &= (i\not{\partial}+m)[i\Delta^{(+)}(x)].\end{aligned} \tag{8.22}$$

Similarly:

$$\begin{aligned}iS^{(-)}(x) &= \frac{1}{V}\sum_{\mathbf{p}s}\left(\frac{m}{E_p}\right)\frac{\not{p}-m}{2m}\mathrm{e}^{ip(x)} \\ &= -(i\not{\partial}+m)\frac{1}{(2\pi)^3}\int\frac{d^3p}{2E(k)}e^{+i(px)} = \\ &= (i\not{\partial}+m)[i\Delta^{(-)}(x)]\end{aligned} \tag{8.23}$$

Taking account of the minus sign in the time-ordered product, the Feynman propagator (8.19) is then (with the integration path in the plane $p^0$ of Fig. 8.1):

$$\begin{aligned}iS_F(x) &= \theta(x^0)iS^{(+)}(x) - \theta(-x^0)iS^{(-)}(x) = \\ &= (i\not{\partial}+m)[\frac{1}{(2\pi)^4}\int d^4p\frac{i}{p^2-m^2+i\epsilon}e^{-i(px)}] = \\ &= \frac{1}{(2\pi)^4}\int d^4p\frac{i(\not{p}+m)}{p^2-m^2+i\epsilon}e^{-i(px)}.\end{aligned} \tag{8.24}$$

Sometimes, using the relation:

$$(\not{p}+m)(\not{p}-m) = p^2 - m^2 \tag{8.25}$$

equation (8.24) is symbolically rewritten as:

$$iS_F(x) = \frac{1}{(2\pi)^4}\int d^4p\frac{i}{\not{p}-m+i\epsilon}e^{-i(px)}. \tag{8.26}$$

## 8.4 THE PHOTON PROPAGATOR

In the preceding section, we calculated the propagators of the scalar field and the Dirac field starting from their definition in terms of quantum fields. In principle, the calculation of the propagation function of the electromagnetic field, necessary for the applications we will discuss in Chapter 14, requires the quantisation of the electromagnetic field in covariant form, which we will not tackle in this volume. However, we can arrive at the same result using a shortcut.

As we saw, the quantum propagation function of the scalar field coincides with the Green's function of the classical Klein–Gordon equation, with Feynman boundary conditions. Extending by analogy this results to the case of the electromagnetic field, we can use the results of the analysis of the Green's function from Chapter 5.

From (5.23) it follows immediately that, in the Lorenz gauge, the Green's the function we are seeking has the form

$$iD_F^{\mu\nu}(q) = \frac{-ig^{\mu\nu}}{q^2 + i\epsilon} , \tag{8.27}$$

and therefore, in the space of coordinates:

$$iD_F^{\mu\nu}(x) = \int \frac{d^4q}{(2\pi)^4} \frac{-ig^{\mu\nu}}{q^2 + i\epsilon} \, e^{iqx} . \tag{8.28}$$

Equation (8.28) defines the photon propagator in the *Feynman gauge.* The completeness condition:

$$g^{\lambda\mu} = g^{\lambda\lambda} \epsilon_\lambda^\mu \epsilon_\lambda^\nu \tag{8.29}$$

(see Chapter 5), would seem to indicate that the function $D_F^{\mu\nu}(x)$ represents the propagation of virtual photons in four different polarisation states, described by the vectors $\epsilon_\lambda$. In addition to the two transverse polarisation states which describe real photons, there is one longitudinal polarisation state, described by a vector parallel to $\mathbf{q}$, and one state of polarisation in time, described by the timelike vector $\eta^\mu$; cf. equation (5.24).

However, by carrying out the decomposition (5.27), it can easily be shown that only photons in transverse states of polarisation contribute to the pole at $q^2 \to 0$, so only those states are observable at large distances from the interaction vertex.

The role of the other terms in (8.29) becomes clear if we note that in the calculation of the probability amplitudes of physical processes, the electromagnetic field propagator always appears saturated by the electromagnetic current $j_\mu$, in the form $j_\mu(-q) D_F^{\mu\nu}(q) j_\nu(q)$. From the current conservation condition, $q^\mu j_\mu = 0$, it follows that the terms proportional to $q^\mu$ or $q^\nu$ in $D_F^{\mu\nu}(x-y)$ gives zero contribution to the physical amplitudes.

As we saw in Chapter 5, the contributions to (8.29) from the longitudinal ($\lambda = 3$) and temporal ($\lambda = 0$) polarisation vectors can be combined, with the result

$$\epsilon_3^\mu \epsilon_3^\nu - \eta^\mu \eta^\nu = \frac{q^2}{|\boldsymbol{q}|} \eta^\mu \eta^\nu + \dots , \tag{8.30}$$

where the dots denote the terms proportional to $q^\mu$ or $q^\nu$, which do not contribute to the amplitudes. The term which remains in $D_F^{\mu\nu}(q)$ is proportional to $1/|\mathbf{q}|$ and therefore describes the instantaneous Coulomb interaction due to the charge distribution $j^0$.

- *The propagator (8.28) describes the propagation of two transverse photons and the Coulomb interaction between charges separated by a distance $x$.*

The possibility that the fields $A^\mu(x)$ could be subject to a gauge transformation implies that the propagator of the electromagnetic field is not uniquely determined. If we limit ourselves to covariant gauges, the propagators corresponding to different gauges differ by terms proportional to $q^\mu$ or $q^\nu$ and therefore give the same result when they are combined with the currents to obtain physical amplitudes. The expression for the propagator in a general covariant gauge is obtained in the following way:

We recall that the propagator in the Feynman gauge, $D_F^{\mu\nu}(x)$, is the inverse of the operator $g^{\mu\nu}\Box$. In momentum space:

$$g^{\mu\nu}\Box \to -g^{\mu\nu} q^2 = K^{\mu\nu}(q) \tag{8.31}$$

and $D_F^{\mu\nu}(q)$ is determined by the equation:

$$D^{\mu\lambda}(q) K_{\lambda\nu}(q) = \delta_\nu^\mu . \tag{8.32}$$

We can characterise the general gauge with the substitution:

$$g^{\mu\nu}\Box \to g^{\mu\nu}\Box + \left(\frac{1}{\xi} - 1\right) \partial^\mu \partial^\nu , \tag{8.33}$$

specified by a parameter $\xi$ that can take arbitrary values. $D_F^{\mu\nu}(q)$ is obtained by inverting the operator

$$K_{\mu\nu} = -g^{\mu\nu} q^2 - \left(\frac{1}{\xi} - 1\right) q^\mu q^\nu . \tag{8.34}$$

In general, the tensor $D_F^{\mu\nu}(q)$ will be of the form:

$$D_F^{\mu\nu}(q) = A(q^2) g^{\mu\nu} + B(q^2) q^\mu q^\nu , \tag{8.35}$$

and the result is easily obtained from (8.32):

$$D_F^{\mu\nu}(q) = \frac{1}{q^2} \left[ -g^{\mu\nu} + (1 - \xi) \frac{q^\mu q^\nu}{q^2} \right] . \tag{8.36}$$

The choice of $\xi = 1$ corresponds to the Feynman gauge, while the choice $\xi = 0$ is known as the *Landau gauge.*

In practical calculations, it is usual to keep $\xi$ general. The disappearance of any dependence on $\xi$ from the physical amplitudes provides a very effective check of the correctness of the calculation.

## 8.5 PROBLEMS FOR CHAPTER 8

### Sect. 8.3

**1.** Prove the relation:

$$(i\partial\!\!\!/ - m)T\{\psi(x)\bar{\psi}(0)\} = i\delta^{(4)}(x) \ .$$

**2.** Using the result of Problem **1** to Sect. 7.5, prove the relation:

$$\partial^{\mu}T\{J_{\mu}(x)\ J_{\nu}(y)\} = 0 \ .$$

### Sect. 8.4

**1.** Starting from the Lagrangian:

$$\mathcal{L} = -\frac{1}{4}F_{\mu\nu}F^{\mu\nu} - \frac{1}{2\xi}(\partial_{\mu}A^{\mu})^2, \quad F^{\mu\nu} = \partial^{\nu}A^{\mu} - \partial^{\mu}A^{\nu}$$

derive the photon propagator in the generic $\xi$-gauge, given in (8.36).

**2.** Verify that, in a generic $\xi$-gauge

$$\partial^{\mu}T\{A_{\mu}(x)A_{\nu}(0)\} = -\xi\ \partial_{\nu}D_F(x, m = 0)$$

where $D_F(x, m = 0)$ is the propagator of a massless scalar field.

CHAPTER 9

# INTERACTIONS

The theory of the free field describes an unchangeable world in which the energy and momentum of every particle of the system are conserved separately.

The variety of phenomena which we observe requires instead some form of interaction between the fields. In this case, as we saw in the classical limit, Section 5.3, particles can exchange energy and momentum giving rise to scattering processes or to the emission and absorption of light; the Sun can shine, the sky can be blue and our eyes can perceive the external world via photons absorbed by the retina.

In a relativistic theory, not only photons, but also particles associated with matter, like electrons, protons and others, can be created or annihilated. For a system isolated from the rest of the world, the elementary processes must respect the conservation laws imposed by the symmetry of the system: energy, momentum, angular momentum as well as possible internal conserved charges; for example, electric charge. This requires that the *interaction Lagrangian*, which we add to the free field Lagrangian should be *invariant under the transformations of special relativity*: the Poincaré group, comprised of translations in space-time and proper Lorentz transformations, $L_+^\uparrow$.

Invariance under the Poincaré group still permits a wide variety of forms for the interactions, for example between the field of the electron and the electromagnetic field. In principle, we should identify the *correct* interaction by an iterative comparison between prediction and experiment (*trial and error*). One starts from a type of interaction which explains at least part of the initial data and then the theory is extended to other processes, successively comparing predictions with the experimental data. When new data are found to be in contradiction with the theory, it is *falsified*, in Popper's terminology, and nothing remains but to modify the interaction to bring it into agreement with the complete set of old and new data[1].

[1]Following Popper, we can say that theories are never confirmed, in that every set of data inevitably has ranges of errors which make it compatible with very many, even an infinite number of, theories for the interaction. However, a theory can be eliminated when the data *falsify it* in favour of a theory in agreement with the overall data set. For example, classical


DOI: 10.1201/9781003436263-9

In this process of trial and error, heuristic *a priori* principles, such as the requirement to lead to the classical theory in the limit of large systems or the presence of certain further symmetries, provide a powerful aid to restrict the form of the interaction, and therefore the choice of discriminating experiments, naturally keeping in mind the necessity to give greater weight to the scrutiny by experimental facts compared to heuristic criteria.

A result of great significance from the physics of the last century has been the assignment of observed processes in the subatomic and subnuclear world to the action of three different categories of interaction: electromagnetic, weak and strong, as already pointed out in Section 5.3. The form of these interactions, in terms of fundamental fields, is greatly restricted by symmetry principles which are the extension of the gauge symmetry encountered in the theory of the electromagnetic field, Section 5.1, and of the phase transformations of Sections 4.3 and 7.3.

In what follows, we will derive the form of the electromagnetic interaction for fundamental spinor particles, such as the electron and the muon, the Fermi interaction is suitable for describing $\beta$ decay of the neutron and discuss qualitatively several aspects of nuclear interactions, above all to point out the difficulty of a description in terms of fields associated with the observed nuclear particles, nucleons and pions.

We delay to later volumes [13, 14] a deeper treatment of the interactions in terms of fundamental constituents (quark, leptons and gauge fields).

## 9.1 QUANTUM ELECTRODYNAMICS

The free theory which describes photons and electrons is obtained by combining the Maxwell Lagrangian (5.14) with the Dirac Lagrangian (7.21):

$$\mathcal{L}_0 = -\frac{1}{4}F_{\mu\nu}F^{\mu\nu} + \bar{\psi}(i\partial\!\!\!/ - m)\psi. \tag{9.1}$$

In classical theory, the interaction between field and particle is described by the minimal substitution $p^\mu \to p^\mu - qA^\mu$ (Section 5.4). On the other hand, in quantum mechanics the 4-momentum is represented by the operators:

$$p^0 = +i\frac{\partial}{\partial t};\ p^i = -i\frac{\partial}{\partial x^i} = -i\partial_i = +i\partial^i$$

and the minimal substitution takes the form:

$$i\partial^\mu \to i\partial^\mu - qA^\mu. \tag{9.2}$$

mechanics is replaced by quantum mechanics when we wish to include the phenomena of atomic physics. Scientific progress comes about through the elimination of hypotheses, rather than their verification.

To reobtain classical electrodynamics from the quantum theory in the limit $\hbar \to 0$ (correspondence principle) we must introduce the substitution (9.2) into (9.1). For the electron, which has charge $q = -e$, we obtain:

$$\begin{aligned}\mathcal{L} &= -\frac{1}{4}F_{\mu\nu}F^{\mu\nu} + \bar{\psi}(i\not{\partial} + e\not{A} - m)\psi = \\ &= \mathcal{L}_0 + eA_\mu\bar{\psi}\gamma^\mu\psi. \end{aligned} \tag{9.3}$$

The interaction Lagrangian requires the inclusion of the current $J^\mu_e = \bar{\psi}\gamma^\mu\psi$, the Noether current associated with the invariance of (9.3) for the global phase transformations of the Dirac field, (7.26).

Gauge Invariance. The Lagrangian we have obtained is invariant under a wider group of transformations, the transformations of the Dirac field by a phase which is *variable in space-time*, counterbalanced by a suitable gauge transformation of the field $A^\mu$.

We consider the simultaneous transformations (which we denote as *gauge transformations of the second kind*, or simply gauge transformations):

$$\begin{aligned}&\psi(x) \to e^{i\alpha(x)}\psi(x); \quad \bar{\psi}(x) \to e^{-i\alpha(x)}\bar{\psi}(x); \\ &A^\mu \to A^\mu + \frac{1}{e}\partial^\mu\alpha(x)\end{aligned} \tag{9.4}$$

with $\alpha(x)$ an arbitrary function of space-time. It can immediately be confirmed that (9.4) leaves (9.3) invariant.

We can reverse the argument, as observed by Pauli [20], and show that the Lagrangian (9.3) is the simplest solution to the problem of constructing a Lagrangian for the electron which is invariant under the transformations (9.4).

To show this we note that the combination (the *covariant derivative*):

$$D^\mu\psi = (\partial^\mu - ieA^\mu)\psi \tag{9.5}$$

is transformed by a simple phase change, exactly like $\psi$, if $\psi$ and $A^\mu$ are subjected to (9.4):

$$\begin{aligned}(D^\mu\psi)' &= \partial^\mu(e^{i\alpha}\psi) - ie(A^\mu + \frac{1}{e}\partial^\mu\alpha)e^{i\alpha}\psi = \\ &= e^{i\alpha}D^\mu\psi\end{aligned} \tag{9.6}$$

from which it follows that:

- if we carry out the substitution $\partial^\mu \to D^\mu$ in a Lagrangian invariant under global phase transformations, we obtain a new Lagrangian invariant under (9.4). The new Lagrangian contains an interaction term prescribed exactly by the symmetry, which agrees with the interaction in (9.3).

In particular, the Dirac Lagrangian thus modified:

$$\mathcal{L}_D(\psi, D^\mu\psi) = \bar{\psi}(i\not{D} - m)\psi \tag{9.7}$$

is gauge invariant. The *general solution* to the problem is obtained by adding to (9.7) terms which should be:

- gauge-invariant functions of $A^\mu$, namely functions only of $F^{\mu\nu}$; which is the case of the Maxwell Lagrangian $\mathcal{L}_{e.m.}$ in (5.14), which when added to (9.7) exactly reproduces (9.3);
- functions of $F^{\mu\nu}$ and of $\psi$ and $\bar{\psi}$, but not of their derivatives, which should be invariant under global transformations obtained from (9.4) with $\alpha$ = constant.

The minimal transformation now assumes a clearer meaning. It provides the simplest Lagrangian for the electron which is gauge invariant; the possibility of adding terms of the second type from those listed above is not used.

A *non-minimal* gauge-invariant term which we can add to the interaction is the so-called Pauli term:

$$\mathcal{L}_{non\ min.} = \frac{-e\kappa}{4m}\ \bar{\psi}\sigma_{\mu\nu}\psi\ F^{\mu\nu} \tag{9.8}$$

which is obviously gauge-invariant.

In the presence of the Pauli term, the equation of motion of the electron becomes:

$$(i\not{\partial} + e\not{A} - m)\psi = \frac{e\kappa}{4m}\psi. \tag{9.9}$$

The non-relativistic limit of this equation is easily found by repeating the arguments of Section 6.1.5. The result is the equation of motion for the two-dimensional spinor:

$$i\frac{\partial}{\partial t}\chi = \boldsymbol{H}\chi$$

$$\boldsymbol{H} = \frac{(\mathbf{p} + e\mathbf{A})^2}{2m} + \frac{e(1+\kappa)}{2m}\boldsymbol{\sigma}\cdot\boldsymbol{B}$$

which replaces (6.114). The Pauli term would change the gyromagnetic ratio of the electron to:

$$g_{tot} = 2(1+\kappa). \tag{9.10}$$

QED for the Charged Leptons. Equation (9.10) provides a strong argument for limiting the electromagnetic interaction of the electron purely to the minimal substitution, which already well describes the magnetic moment of the electron. The result is confirmed by the calculation of higher order corrections to the quantity $g - 2$, first carried out by Schwinger, which shows that these corrections reproduce with high precision the small experimentally observed deviations from the Dirac prediction of $g = 2$, cf. [14].

The same conclusion holds for the muon, $\mu$, which has spin $\frac{1}{2}$ and a mass about 200 times that of the electron. The muon appears in all respects a *heavy version* of the electron. The magnetic moment of the muon is known experimentally with great precision and the deviations from the Dirac prediction $g = 2$ are well described by higher-order corrections.

In 1976, another charged spin $\frac{1}{2}$ lepton similar to the electron and muon, the $\tau$ particle, was discovered.

The electron, $\mu$, $\tau$ and the corresponding neutrinos are classified as *leptons*, a family of particles not sensitive to nuclear forces. The most convincing hypothesis in agreement with our experimental knowledge is to describe the electromagnetic interactions of the three charged leptons with only the minimal substitution, and therefore with the Lagrangian:

$$\begin{aligned}
\mathcal{L}_{QED} &= -\frac{1}{4}F_{\mu\nu}F^{\mu\nu} + \bar{\psi}_e(i\not{D} - m_e)\psi_e + \bar{\psi}_\mu(i\not{D} - m_\mu)\psi_\mu + \bar{\psi}_\tau(i\not{D} - m_\tau)\psi_\tau \\
&= \mathcal{L}_{e.m.} + \mathcal{L}_{0e} + \mathcal{L}_{0\mu} + \mathcal{L}_{0\tau} - eA_\lambda J^\lambda_{lept}; \\
J^\lambda_{lept} &= -\{\bar{\psi}_e\gamma^\lambda\psi_e + \bar{\psi}_\mu\gamma^\lambda\psi_\mu + \bar{\psi}_\tau\gamma^\lambda\psi_\tau\} \qquad (9.11)
\end{aligned}$$

where we denote by $\mathcal{L}_0$ the free Lagrangian.

The fermion fields appear in the interaction Lagrangian with the same coupling constant, the electric charge of the electron, a property known as the *universality* of the electromagnetic interaction (see Table 9.1).

The theory described by (9.11) is known as *spinor QED* ($QED$ = quantum electrodynamics).

Nuclear Particles. The minimal substitution does not correctly reproduce the electromagnetic interactions of the proton and neutron. Both of these particles should be described by a Dirac field, because they have spin $\frac{1}{2}$, but their gyromagnetic ratios are not in agreement with the minimal values of $g = 2$ and $g = 0$ respectively. For the nucleon magnetic moment, we set $(N = p, n)$:

$$\mu_N = g_N \mu_p \mathbf{S}; \quad \mu_p = \frac{e}{2M_p}$$

Table 9.1 Electromagnetic properties of the charged leptons, from the Particle Data Group [?]. Numbers in parentheses denote the error on the last digit of each quantity.

| | $m$ (MeV) | $g$ |
|---|---|---|
| $e$ | 0.510998902(21) | 2.002319304374(4) |
| $\mu$ | 105.658357(5) | 2.0023318320(6) |
| $\tau$ | 1776.99(28) | 2.000(58) |

with $M_p$ the proton mass. Experimentally, we find [?]:

$$\frac{g_N}{2} = \begin{cases} 2.792847351 \pm 0.000000028 & \text{(proton)} \\ -1.9130427 \pm 0.0000005 & \text{(neutron).} \end{cases} \tag{9.12}$$

We must therefore add a Pauli term for each of the two nucleons, with:

$$\begin{cases} \kappa_p = +1.792847351 \pm 0.000000028 \\ \kappa_n = -1.9130427 \pm 0.0000005. \end{cases} \tag{9.13}$$

The corresponding Lagrangian of the overall system (leptons and nucleons) is written:

$$\begin{aligned} \mathcal{L} = \mathcal{L}_{QED} + \bar{\psi}_p(i\partial\!\!\!/ - M_p)\psi_p + \bar{\psi}_n(i\partial\!\!\!/ - M_n)\psi_n - eA_\mu\bar{\psi}_p\gamma^\mu\psi_p + \\ + \frac{e}{4M_p}\{\kappa_p\, \bar{\psi}_p\sigma_{\mu\nu}\psi_p + \kappa_n\, \bar{\psi}_n\sigma_{\mu\nu}\psi_n\}\, F^{\mu\nu}. \end{aligned} \tag{9.14}$$

We can rewrite the Pauli terms as additive corrections to the overall electromagnetic current of the nuclear particles. We write, in general:

$$\begin{aligned} \frac{e\kappa}{4m}\bar{\psi}\sigma_{\mu\nu}\psi\, F^{\mu\nu} &= \frac{e\kappa}{2m}\bar{\psi}\sigma_{\mu\nu}\psi(\partial^\nu A^\mu) = \\ &= -eA^\mu\{\frac{\kappa}{2m}(\partial^\nu(\bar{\psi}\sigma_{\mu\nu}\psi)\} + 4\text{–divergence.} \end{aligned} \tag{9.15}$$

Omitting the 4-divergence, which does not contribute to the principle of minimum action, we can rewrite the Lagrangian (9.14) as:

$$\begin{aligned} \mathcal{L} &= \mathcal{L}_{QED} + \bar{\psi}_p(i\partial\!\!\!/ - M_p)\psi_p + \bar{\psi}_n(i\partial\!\!\!/ - M_n)\psi_n \;-\; eA_\mu J^\mu_{nucl}; \\ J^\mu_{nucl} &= \bar{\psi}_p\gamma^\mu\psi_p + \frac{\kappa_p}{2m}\partial^\nu(\bar{\psi}_p\sigma_{\mu\nu}\psi_p) + \frac{\kappa_n}{2m}\partial^\nu(\bar{\psi}_n\sigma_{\mu\nu}\psi_n). \end{aligned} \tag{9.16}$$

The Pauli terms become additional terms of the Noether current of the proton (that of the neutron is obviously zero) which describe the deviations of the magnetic moment from the Dirac value. We note that the additional terms are the divergence of an antisymmetric tensor, therefore the ambiguity associated with the Noether current reoccurs (cf. Section 3.5); they are identically conserved and do not contribute to the conserved charge:

$$\partial_\mu J^\mu_{nucl} = 0; \int d^3x\, J^0_{nucl} = \int d^3x\, \bar{\psi}_p\gamma^0\psi_p.$$

Finally, we can write the Lagrangian for electromagnetic interactions of charged leptons and nuclear particles in the following way:

$$\mathcal{L}_{QED} = \mathcal{L}_{0,tot} - eA_\mu J^\mu_{tot} \tag{9.17}$$

where:

$$\begin{aligned} \mathcal{L}_{0,tot} &= -\frac{1}{4}F_{\mu\nu}F^{\mu\nu} + \sum_{i=e,\mu,\tau,p\,n} \bar{\psi}_i(i\partial\!\!\!/ - m_i)\psi_i; \\ J^\mu_{tot} &= J^\mu_{lept} + J^\mu_{nucl} \end{aligned} \tag{9.18}$$

with $J^{\mu}_{lept}$ and $J^{\mu}_{nucl}$ given by (9.11) and (9.16).

Today we know many particles in addition to the proton and neutron which are sensitive to strong interactions, collectively called *hadrons*, cf. Section 9.3. If we want to describe each of them with a quantised field, we must add their free Lagrangian to $\mathcal{L}_{0,tot}$, and include their contribution to the total electromagnetic current. The Lagrangian (9.17) keeps its form with the extension of $J^{\mu}_{nucl}$ to the overall hadron current.

## 9.2 THE FERMI INTERACTION FOR $\beta$ DECAY

The first quantitative theory of nuclear $\beta$ decay is due to Fermi [18]. Following the proposal of Pauli, Fermi assumed that the electron in $\beta$ decay is emitted together with an unobserved neutral particle, the neutrino, and that the process is caused by an interaction independent of both electromagnetic interactions and the forces which bind nuclei. The corresponding reaction is written:

$$N(A, Z) \to N(A, Z+1) + e + \bar{\nu} \tag{9.19}$$

where $A$ and $Z$ are the mass number and the charge of the nucleus. The coupling constant associated with this new interaction proved to be so small, on the scale of nuclear phenomena, to justify the name *weak interaction* given to the interaction identified by Fermi.

Fermi assumed that the interaction Lagrangian must be the product of two terms: an operator which induces the transition from the initial to final nucleus, and an operator which *creates* the pair of light particles. There is a deep analogy here with electromagnetic transitions in atoms and nuclei:

$$A^* \to A + \gamma. \tag{9.20}$$

The interaction Lagrangian of (9.20) is also the product of an operator which causes the transition from $A^*$ to $A$, and of an operator which creates the photon. For example, as we saw in the previous section, equation (9.16), for a nuclear transition we have:

$$\mathcal{L}_\gamma = -eJ^{\mu}_{nucl}A_\mu. \tag{9.21}$$

In the case of the reaction (9.19), Fermi assumed that it was induced by the basic process of neutron disintegration:

$$n \to p + e + \bar{\nu} \tag{9.22}$$

and wrote the simplest possible interaction Lagrangian as:

$$\mathcal{L}_F = -G_F\left[\bar{\psi}_p\gamma^\mu\psi_n\right]\left[\bar{\psi}_e\gamma_\mu\psi_\nu.\right] \tag{9.23}$$

We have assumed that all the particles can be represented by Dirac fields, and $G_F$ is the Fermi constant. Conventionally the particle emitted in neutron $\beta^-$ decay is an *antineutrino*.

In modern notation, the Lagrangian for the decay of the neutron is instead written as:

$$\mathcal{L}_F = -\frac{G_F}{\sqrt{2}}\left[\bar{\psi}_p\gamma^\mu(1+\frac{g_A}{g_V}\gamma_5)\psi_n\right]\left[\bar{\psi}_e\gamma_\mu(1-\gamma_5)\psi_\nu\right] \tag{9.24}$$

Equation (9.24) condenses a wealth of experimental and theoretical developments from studies of weak interactions, from the formulation of Fermi to today:

- $V$–$A$ theory: the hadronic and leptonic terms in (9.24) are combinations only of bilinears $V^\mu$ and $A^\mu$.
- Parity violation: the Lagrangian is a consistent superposition of polar and axial vectors.
- The neutrino field appears in the combination $(1-\gamma_5)\psi_\nu$. This ensures that, for zero mass, neutrinos and antineutrinos are emitted or absorbed only in negative or positive helicity states, respectively. Neutrinos (antineutrinos) with positive (negative) helicity are not coupled and may not actually exist (for the theory of two-component neutrinos, see Section 13.1).
- The nuclear particles do not appear in the $V$–$A$ combination, but with a normalising coefficient, $g_A/g_V$, to be determined by experiment.

**The Sign of $G_F$.** In his work on $\beta$ decays [18], Fermi wrote the Hamiltonian density with a sign consistent with the relation $\mathcal{H}_{int} = -\mathcal{L}_{int} = -\mathcal{L}_F$. The applications we will discuss in Chapter 15 depend only on $|\langle f|\mathcal{L}_F|i\rangle|^2$. The sign of $G_F$ is determined from the interference of the weak and electromagnetic contributions in the forward–backward asymmetry in, for example, $e^+e^- \to \mu^+\mu^-$, cf. [13], and in the $MSW$ effect on the propagation of neutrinos through the body of the Sun, Chapter 16. The sign in (9.23) and (9.24) with $G_F > 0$ is consistent with what is obtained if we derive the Fermi interaction from the second order of the theory of the intermediate vector boson, cf. Chapter 15.

## 9.3 STRONG INTERACTIONS

This is the name which describes the force which binds together protons and neutrons (*nucleons* in atomic nuclei, overcoming electrostatic repulsion of the protons. Nuclear forces have a range limited to dimensions of the order of nuclear radii:

$$\begin{aligned} V(r) &\simeq \frac{e^{-r/R}}{r} \\ R &\sim 1\cdot 10^{-13}\ \text{cm} = 1\ \text{fermi}. \end{aligned} \tag{9.25}$$

A first theory of nuclear interactions, proposed by Yukawa, described them as due to the exchange of a particle similar to the photon but endowed with mass, the $\pi$ meson. In this case, the effective range is determined by the mass of the intermediate particle:

$$R \simeq \frac{\hbar}{m_\pi c}. \tag{9.26}$$

Comparing equation (9.26) with (9.25) a value of $m_\pi \simeq$ 100–200 MeV is found[2], in agreement with the mass of the $\pi$ meson ($m_\pi \simeq$ 140 MeV) discovered in cosmic ray interactions by Lattes, Muirhead, Occhialini and Powell in 1947.

The pion–nucleon coupling is characterised by a dimensionless constant similar to the fine structure constant, but about 1000 times larger; nuclear forces overcome the electrostatic repulsion of protons inside nuclei. Strong nuclear forces are by a long way the most powerful forces in Nature, hence their name, but from atomic distances upwards they are irrelevant owing to the exponential decline with distance, equation (9.25).

## 9.4 HADRONS, LEPTONS AND FIELDS OF FORCE

Since the Second World War, numerous particles which, like nucleons and $\pi$ mesons, are affected by strong interactions have been identified in high-energy collisions of cosmic rays and particle accelerators. All these particles are known collectively as *hadrons*[3] and are divided into two large families:

- *mesons*: particles which do not possess any absolutely conserved quantum number except for electric charge; consequently mesons decay into $\pi$ mesons, nucleon-antinucleon pairs, electromagnetic radiation and, ultimately, electrons and neutrinos,

- *baryons*: particles which possess a non-zero baryon number (cf. Section 7.3) and therefore decay into states with a nucleon plus mesons, nucleon-antinucleon pairs, electromagnetic radiation and, ultimately, electrons and neutrinos.

The proliferation of hadrons has made particularly pressing the problem of identifying the fundamental degrees of freedom to describe them in the framework of a quantum field theory. According to the present picture, the fundamental degrees of freedom are associated with several types of spin $\frac{1}{2}$

[2]The numerical result is obtained recalling that in natural units 1 fm$^{-1}$=197 MeV.

[3]The term was introduced in 1962 by L.Okun from the Greek $\alpha\delta\rho\acute{o}\sigma$= thick, strong, in opposition to the particles which are sensitive to Electromagnetic and Weak interactions only, that Okun called *leptons*, from $\lambda\epsilon\pi\tau\acute{o}\sigma$= small, weak.

field, the *quarks*, confined inside hadrons according to the scheme in which mesons are $q\bar{q}$ states and baryons are $qqq$ states.

The hadrons are accompanied by the family of *leptons*, particles which do not feel the strong interaction, divided into *charged leptons*, whose electrodynamic interactions were described in Section 9.1, and neutrinos, subject only to the weak interaction. As already mentioned, three charged leptons are known today: $e$, $\mu$ and $\tau$ and their associated neutrinos.

The fundamental forces acting on quarks and leptons are attributed to the exchange of particles of spin 1, all similar to the photon, and of a spin 0 particle, the *Higgs boson*, coupled to the mass of the different particles. In more detail (see [13] for an extended discussion):

- QED, Section 9.1, describes electromagnetic forces caused by the exchange of photons,
- the weak interactions are associated with the exchange of massive, electrically charged vector bosons ($W^{\pm}$) responsible for the interactions identified by Fermi, Section 9.2, and a neutral boson ($Z^0$) responsible for *neutral current process*,
- the primary strong interactions between quarks are due to the action of fields similar to the photon, *gluons*, from the word 'glue' because of their property of binding the quarks together to form hadrons,
- the field associated with the Higgs boson, the particle observed by the ATLAS and CMS Collaborations at CERN in 2012 with a mass of approximately 125 GeV, has a fundamental role in breaking the symmetry which connects the photon to the fields of the weak interaction, generating the mechanism by which the $W^{\pm}$ and $Z^0$ particles, as well as quarks and leptons, gain their masses.

Particle Names. Apart from the term hadron, recently coined from the Greek (*adrós* = strong), the terms lepton, meson and baryon were born in a historical period when the only known particles were the electron and the neutrino (leptons, from *leptós* = light), the Yukawa meson with intermediate mass (from *mesos* = in between) and the nucleons (baryons, from *barús* = heavy). Today we know of leptons and mesons which are much heavier than nucleons and the names really reflect the interactions and the quantum numbers of the particles. The name quark was given to the fundamental constituents of hadrons by Gell–Mann, from a passage in *Finnegans Wake* by James Joyce (... *Three quarks for Muster Mark!*), probably alluding to the fact that three quarks are needed to describe the nucleon.

## 9.5 PROBLEM FOR CHAPTER 9

### Sect. 9.1

1. Scalar electrodynamics. Consider the Lagrangian density describing a charged scalar field interacting with the electromagnetic field

$$\mathcal{L} = (D_\mu \phi)^\dagger (D^\mu \phi) - m^2 \phi^\dagger \phi - \frac{1}{4} F_{\mu\nu} F^{\mu\nu}$$

where

$$D_\mu \phi = (\partial_\mu - ieA_\mu)\phi \quad , \quad F_{\mu\nu} = \partial_\nu A_\mu - \partial_\mu A_\nu \ .$$

- Determine the equation of motions.
- Indicate the differences of the interactions of scalar vs. spinor QED.
- Obtain the Nöther current $j^\mu$ associated with the invariance of the fields $\phi$ and $\phi^\dagger$ under phase transformations;
- Verify that $\partial_\mu j^\mu = 0$.

CHAPTER 10

# TIME EVOLUTION OF QUANTUM SYSTEMS

In general, the expectation values of observable quantities depend on time. In quantum mechanics, these values are given by the expression:

$$< X >_t = < A(t)|X(t)|A(t) >. \tag{10.1}$$

There is an intrinsic ambiguity in determining the time dependence on the various elements (bra, ket and operator) which comprise the right-hand side of equation (10.1) because we can transfer this dependence from one element to another, while keeping unchanged the expectation value $< X >_t$, which is all we can measure on the system. The ambiguity gives rise to different descriptions of the motion, connected to unitary time-dependent transformations; thus, equivalent to each other. In the following sections, we describe the Schrödinger and Heisenberg representations. Subsequently, we will introduce a third description of the motion or interactions: the Dirac representation, also called the *interaction representation*, which is particularly useful in the case of weakly interacting systems.

## 10.1 THE SCHRÖDINGER REPRESENTATION

In this representation, the dynamic variables (position, momentum, etc) are associated with fixed operators. The time variation of the expectation value (10.1) is due to the time variation of the ket which represents the physical state at time $t$. Given the ket $|A>$ at time $t_0$ (the initial state), the principle of superposition requires that $|A(t)>$ is obtained from $|A>$ via application of a linear operator, $U(t,t_0)$, independent of $|A>$:

$$|A(t)> = U(t,t_0)|A> . \tag{10.2}$$

Moreover, if $|A>$ is normalised, so that its components $c_n$, on the basis of a given observable $\boldsymbol{O}$, are the probability amplitudes of possible results

DOI: 10.1201/9781003436263-10

of a measurement of $O$, it is natural to require that $|A(t)>$ should also be normalised, so that:

$$\sum_n |c_n|^2 = \sum_n |c_n(t)|^2 = 1. \tag{10.3}$$

Equation (10.3) corresponds to the conservation of the probability of all possible results. Given this condition, the operator $U(t, t_0)$ must be unitary:

$$U(t, t_0)^\dagger U(t, t_0) = 1. \tag{10.4}$$

We can transform (10.2) into a differential equation:

$$\frac{d}{dt}|A(t)> = \frac{dU(t, t_0)}{dt} U(t, t_0)^\dagger . |A(t)> = -iH|A(t)> \tag{10.5}$$

$H$ is the generator of infinitesimal time translations. Because $U$ is unitary, $H$ is hermitian:

$$\begin{aligned} H &= i\frac{dU(t, t_0)}{dt} U(t, t_0)^\dagger = \\ &= i\frac{d}{dt}[U(t, t_0)U(t, t_0)^\dagger] - iU(t, t_0)\frac{dU(t, t_0)^\dagger}{dt} = H^\dagger . \end{aligned} \tag{10.6}$$

Equation (10.5) is the Schrödinger equation. It is a first order differential the equation in $t$, in agreement with the hypothesis that, at time $t_0$, the ket $|A>$ gives a complete description of the state of the system and that therefore the time evolution should be determined by a single initial condition.

In the classical limit, $H$ becomes the Hamilton function of the system, and for this reason is called the Hamiltonian operator, or simply Hamiltonian, of the system.

In the case in which $H$ is independent of time, we can integrate the equation (10.5) and write directly the solution of the Schrödinger equation which reproduces the state $|A>$ at the time $t_0$:

$$|A(t)> = e^{-iH(t-t_0)}|A> . \tag{10.7}$$

If we expand $|A>$ in the basis of the eigenvectors of $H$:

$$|A(t)> = \sum_n c_n(t)|h_n> \tag{10.8}$$

we obtain from (10.5):

$$c_n(t) = e^{-iE_n(t-t_0)} c_n(t_0). \tag{10.9}$$

Energy is conserved; equation (10.5) shows that if $|A>$ is an eigenstate of $H$, so is $|A(t)>$, with the same eigenvalue. For a general state, this conserves the expected value of $H$:

$$< A(t)|H|A(t) > = < A|H|A > = constant. \tag{10.10}$$

An interesting aspect of these results is invariance with respect to time translations, or the fact that $|A(t)>$ depends only on $t - t_0$. If we carry out an experiment preparing the system in the state $|A>$ at 9:30 am today and we make a measurement at 10 am, the result is the same as if we had carried out the same operations between 5 pm and 5:30 pm yesterday. This follows directly from the time of independence of $H$.

Reasoning in the reverse direction, if we expect *a priori* that a given system should be invariant for translations in time, its Hamiltonian will be independent of $t$ and therefore conserved in time; *conservation of energy is a direct consequence of invariance under time translation.*

Everything we know makes us believe that systems sufficiently isolated from the rest of the universe are independent of time. An isolated system, therefore, must respect the conservation of energy.

## 10.2 THE HEISENBERG REPRESENTATION

As an alternative to the Schrödinger representation, we can associate the state to a fixed vector, and attribute the time dependence of the expectation values (10.1) to the change of the operator which represents the observable. Formally, the Heisenberg representation is obtained by applying to the ket $|A(t)>_S$ of the Schrödinger representation of the unitary transformation which returns it to the value which it had at a fixed time $t_0$:

$$|A>_H = e^{+iH(t-t_0)}|A(t)>_S \, . \tag{10.11}$$

At time $t_0$, the two representations coincide.

The dependence of observable quantities on $t$ is fixed by the requirement that their expectation values (10.1) should be the same in the two representations at all $t$, and that the ket $|A>_H$ should be constant. From the equation:

$${}_S<A(t)|X_S|A(t)>_S = {}_H<A|X_H(t)|A>_H \tag{10.12}$$

we obtain, using (10.11):

$$X_H(t) = e^{+iH(t-t_0)} X_S e^{-iH(t-t_0)}. \tag{10.13}$$

Differentiating with respect to time, we obtain:

$$i\frac{dX_H(t)}{dt} = [X_H(t), H]. \tag{10.14}$$

To visualise the Heisenberg representation, we return to the analogy with the motion of a classical system.

We can describe the state of the system at time $t$ by giving the instantaneous position of the system in phase space, $(p_t, q_t)$; this corresponds to the Schrödinger representation. Instead, the Heisenberg representation corresponds to describe the state of motion by giving the initial conditions,

$(p_{t_0}, q_{t_0})$, at an arbitrary but fixed time, $t_0$. The initial conditions completely determine the trajectory in phase space and naturally, like the Heisenberg state, they do not change with time.

There is an unusual aspect of the Heisenberg representation which is implicit in what has been said but is worth underlining.

The Heisenberg state is independent of time $t$. However, the vector which represents it depends implicitly on the value of time $t_0$, which was chosen to fix the initial conditions (in other words, the time at which the Heisenberg representation coincides with the Schrödinger representation). The choice of $t_0$ is arbitrary, but we must fix $t_0$ in the same way for all states of motion, if we wish to compare them with different states.

The vectors which represent the state of motion for a given choice of $t_0$ differ from those relative to an alternative choice by a unitary transformation:

$$|A; t_0 >_H = e^{+iH(t_0 - t_0')} |A; t_0' >_H . \qquad (10.15)$$

Equation (10.15) leaves invariant the expectation values (10.1). However, as $t_0$ changes, the vectors which represent the same state of motion can assume a considerably different appearance.

## 10.3 THE INTERACTION REPRESENTATION

In many physically interesting cases, the Hamiltonian is the sum of two terms:

$$H = H_0 + V_0 \qquad (10.16)$$

in which $H_0$ is exactly diagonalisable and $V_0$ can be considered a "small" modification to $H_0$. In these cases, we can try to approximate a solution to $H$ starting from the solutions for $H_0$, with an expansion in powers of $V_0$ limited to a finite number of terms. This is the quantum version of perturbation theory, widely used in classical mechanics.

The most relevant example is quantum electrodynamics (QED). $H_0$ is the Hamiltonian which describes free electrons and photons while $V_0$ describes the interaction of the electron with the electromagnetic field. The strength of the interaction is determined by the electric charge of the electron, and is expressed in terms of a dimensionless quantity (the fine structure constant):

$$\alpha = (\frac{e^2}{4\pi}) = \frac{1}{137} \qquad (10.17)$$

much less than unity.

To obtain the perturbative expansion in a systematic way, it is convenient to describe the motion of the system in the *interaction representation*, introduced by Dirac.

The state at time $t$ in the interaction representation is obtained from the state in the Schrödinger representation with the unitary transformation:

$$|A(t) >_I = e^{+iH_0 t} |A(t) >_S . \qquad (10.18)$$

The expectation values of the observables must be the same in the two representations:

$$_I < A(t)|O_I(t)|A(t)>_I = {}_S < A(t)|O_S|A(t)>_S \tag{10.19}$$

and therefore:

$$O_I(t) = e^{+iH_0t}O_Se^{-iH_0t}. \tag{10.20}$$

Differentiating with respect to $t$, we obtain the equations of motion of the states and the observables:

$$i\frac{\partial}{\partial t}|A(t)>_I = V_{0I}(t)|A(t)>_I \tag{10.21}$$

$$i\frac{dO_I(t)}{dt} = [H_0, O_I(t)] \tag{10.22}$$

where $V_{0I}(t)$ is the interaction Hamiltonian in the interaction representation:

$$V_{0I}(t) = e^{+iH_0t}V_0e^{-iH_0t}. \tag{10.23}$$

The observables change in time with the free Hamiltonian, while the time variation of the states is due only to the interaction. It should be noted that $V_{0I}$ depends explicitly on time since $V_0$ and $H_0$ generally do not commute.

Equation (10.21) defines a translation operator in time between $t_0$ and $t$:

$$|t>_I = U_I(t, t_0)|t_0>_I \tag{10.24}$$

where we have denoted with $|t>_I$ the state at time $t$ which becomes $|t_0>_I$ at time $t_0$.

$U_I(t, t_0)$ is a linear operator. Also, it is subject to the relation:

$$U_I(t, t_0) = U_I(t, t_1)U_I(t_1, t_0) \qquad (t > t_1 > t_0). \tag{10.25}$$

$U_I$ is unitary, as a consequence of the fact that the interaction Hamiltonian, $V_{0I}$, is hermitian:

$$U(t, t_0)^\dagger U(t, t_0) = U(t, t_0)U(t, t_0)^\dagger = I. \tag{10.26}$$

Clearly, we can also solve the equation of motion with a *final* condition, i.e. determining the state $|t>_I$ which reduces to a given state $|t_1>$ at a time $t_1 > t$. The corresponding operator, $\bar{U}_I(t, t_1)$, is defined by:

$$|t>_I = \bar{U}_I(t, t_1)|t_1>_I \qquad (t_1 > t) \tag{10.27}$$

and it is not difficult to see that it is the hermitian conjugate of $U_I(t_1, t)$:

$$\bar{U}_I(t, t_1) = U_I(t_1, t)^\dagger. \tag{10.28}$$

### 10.3.1 Theory of Time-dependent Perturbations

We can express $U_I$ as a power series in $V_{0I}$. To do this, we integrate (10.21) between $t_0$ and $t$, in this way obtaining the integral equation:

$$|t>_1 = |t_0>_I - i\int_{t_0}^{t} dt' V_{0I}(t')|t'>_I \, . \tag{10.29}$$

Obviously, $|t>_I$ differs from $|t_0>_I$ in terms that are at least of order $V_{0I}$. We obtain a solution to first order in $V_{0I}$ by simply substituting $|t'>_I$ with $|t_0>_I$:

$$|t>_I = |t_0>_I - i\int_{t_0}^{t} dt' V_{0I}(t')|t_0>_I + O(V_0^2). \tag{10.30}$$

The procedure can be repeated. For example, by substituting (10.30) back into (10.29), we obtain the solution to second order in $V_{0I}$:

$$\begin{aligned} |t>_1 = {} & |t_0>_I - i\int_{t_0}^{t} dt' V_{0I}(t')|t_0>_I + \\ & +(-i)^2 \int_{t_0}^{t} dt' V_{0I}(t') \int_{t_0}^{t'} dt'' V_{0I}(t'')|t_0>_I + O(V_0^3) \end{aligned} \tag{10.31}$$

and so on.

Continuing the process indefinitely, we find the series:

$$\begin{aligned} |t>_I = {} & |t_0>_I + \sum_{n=1}^{+\infty}(-i)^n \int_{t_0}^{t} dt_1 V_{0I}(t_1)\times \\ & \int_{t_0}^{t_1} dt_2 V_{0I}(t_2)\ldots \int_{t_0}^{t_{n-1}} dt_n V_{0I}(t_n)|t_0>_I = \\ = {} & [1+\sum_{n=1}^{+\infty}(-i)^n \int_{t_0}^{t} dt_1 V_{0I}(t_1) \int_{t_0}^{t_1} dt_2 V_{0I}(t_2)\ldots \int_{t_0}^{t_{n-1}} dt_n V_{0I}(t_n)]|t_0>_I \\ = {} & U_I(t,t_0)|t_0>_I \end{aligned} \tag{10.32}$$

which gives a formal solution to the equation of motion with the initial condition $|t>_I = |t_0>_I$, as is easily confirmed by substituting the result (10.32) into equation (10.21).

### 10.3.2 Time-ordered Products

It is helpful to express the terms of the perturbative series (10.32) as the result of an integration independent of all the variables $t_1, \ldots t_n$, in the fixed interval between $t_0$ and $t$. To do this, we introduce the time-ordered product (T-product) of the operators $V_{0I}(t_1), V_{0I}(t_2)\ldots, V_{0I}(t_n)$ defined, for fixed values of the variables $t_1, t_2, \ldots t_n$, as the product of operators in the order which corresponds to decreasing values of time, reading from left to right:

$$T(V_{0I}(t_1), V_{0I}(t_2)\ldots, V_{0I}(t_n)) = V_{0I}(t_{a1})V_{0I}(t_{a2})\ldots V_{0I}(t_{an}) \quad (10.33)$$

where $t_{a_1}, t_{a_2}\ldots t_{a_n}$ is the permutation of $t_1, t_2, \ldots, t_n$ for which:

$$t_{a_1} > t_{a_2} > \ldots > t_{a_n}. \quad (10.34)$$

In the case of two operators, explicitly:

$$T(V_{0I}(I_1), V_{0I}(t_2)) = \theta(t_1 - t_2)V_{0I}(t_1V_{0I}(t_2 + \theta(t_2 - t_1)V_{0I}(t_2)V_{0I}(t_1) \quad (10.35)$$

with $\theta(x) = 0, 1$ according to whether $x < 0$ or $x > 0$.

We now consider the integral:

$$\int_{t_0}^{t} T(V_{0I}(t_1), V_{0I}(t_2)\ldots, V_{0I}(t_n))dt_1dt_2\ldots dt_n \quad (10.36)$$

with all the variables between $t_0$ and $t$.

The region of integration is factorised into $n!$ regions, corresponding to the possible time ordering of the variables. The contribution of the particular region in which the ordering given by (10.33) holds is given by:

$$\int_{t_0}^{t} dt_{a1}V_{0I}(t_{a1})\int_{t_0}^{t_{a1}} dt_{a2}dV_{0I}(t_{a2})\ldots\int_{t0}^{t_{an-1}} dt_{an}V_{0I}(t_{an}) \quad (10.37)$$

and therefore it coincides exactly with the integral which appears in (10.32), apart from a trivial change in the name of the variables. The instruction of the time ordering in (10.36) is such that the integrals extended to the $n!$ regions are all equal to (10.37), so that, finally, we can rewrite $U_I(t, t_0)$ in the symmetric form:

$$U_I(t, t_0) = 1 + \sum_{n=1}(-i)^n\frac{1}{n!}\int_{t0}^{t} dt_1dt_2\ldots dt_nT(V_{0I}(t_1)V_{0I}(t_2)\ldots V_{0I}(t_n)). \quad (10.38)$$

## 10.4 SYMMETRIES AND CONSTANTS OF THE MOTION

Observables which commute with the Hamiltonian are constants of the motion; their expectation values are independent of $t$ (cf. equation (10.14)). As in classical mechanics (cf. Noether's theorem), in quantum mechanics there is a direct relation between the symmetries of a physical system and the constants of the motion.

Qualitatively, we speak of symmetry every time we can carry out a transformation on apparatus which prepares the system (in different states $|A>, |B>$, etc) in such a way as to leave invariant the relations between them. To be concrete, if we denote by $|RA>, |RB>$, etc. the states which are obtained from $|A>, |B>$, etc. with the transformation $R$, this means that:

$$| < RB|RA > |^2 = | < B|A > |^2 \tag{10.39}$$

for all the states $|A>$ and $|B>$.

As Wigner showed, equation (10.39) gives rise to two alternatives, which are:

$$< RB|RA > = < B|A > \tag{10.40}$$

or:

$$< RB|RA > = < B|A > *. \tag{10.41}$$

If the condition (10.40) holds, the transformation is represented by a linear and unitary operator, $U(R)$:

$$\begin{aligned} &|RA> = U(R)|A> \\ &U(R)U(R)^\dagger = 1. \end{aligned} \tag{10.42}$$

Condition (10.41) instead requires that $U(R)$ should be anti-linear and anti-unitary [21]:

$$\begin{aligned} &U(R)(\alpha|A> + \beta|B>) = \alpha^* U(R)|A> + \beta^* U(R)|B> \\ &U(R)U(R)^\dagger = 1. \end{aligned} \tag{10.43}$$

The second alternative applies to the case of time reversal. All transformations which leave the direction of time unchanged should be represented by unitary operators, and this is the case we will consider in this section. In particular, operators that represent those groups of transformations which are continuously connected to the identity transformation (which can always be represented by the operator 1) should be unitary, as for example are spatial translations and rotations, discussed in Section A.4.1, or translations in time, Section 10.1.

Given an observable $X$, we define the transformed observable to be one which has the same expectation value on the state $|RA>$, as $X$ has on $|A>$, so:

$$< RA|X_R|RA > = < A|X|A > . \tag{10.44}$$

From (10.42) we therefore find:

$$X_R = U(R) X U(R)^\dagger. \tag{10.45}$$

The most important case is that of the Hamiltonian. If $H$ is left invariant by the transformation:

$$H_R = U(R)HU(R)^\dagger = H \tag{10.46}$$

then:

$$[U(R), H] = 0 \tag{10.47}$$

and *all the operators* $U(R)$ *are constants of the motion.*

We compare two experiments which start from two states which are transforms of each other, $|A>$ and $|RA>$, at time $t = 0$. In the two cases, at time $t$, we have:

$$\begin{aligned} &|A(t)>= e^{-iHt}|A> \\ &|RA(t)>= e^{-iHt}|RA>= e^{-iHt}U(R)|A>= \\ &\qquad = U(R)e^{-iHt}|A>= U(R)|A(t)>. \end{aligned} \tag{10.48}$$

In this case, we speak of an exact, or conserved, symmetry and the previous formulas show that the relation by which one state is transformed into the other is conserved in time.

Given a group of continuous transformations, we consider those infinitesimally close to the identity. For these, we can set:

$$U(R) = U(\lambda) \equiv 1 - i\lambda T \tag{10.49}$$

where $\lambda$ denotes the transformation parameter. The infinitesimal generator, $T$, must be hermitian given that $U(R)$ is unitary:

$$T^\dagger = T. \tag{10.50}$$

On the basis of (10.48), the generator of an exact symmetry commutes with $H$:

- *a continuous group of exact symmetries implies the existence of conserved observables*, as many as the number of independent parameters which characterise the group.

The conservation of momentum and angular momentum are the most conspicuous examples of these results, which are otherwise completely general.

The observables of quantum field theories, for example, the energy density, are local quantities, operators which we can think of as determined by measurements in a small region of space-time.

We denote one of these operators as $\Pi(\boldsymbol{x}, t)$. The vector $\boldsymbol{x}$ is a numerical variable (and not the position operator) since it is connected to the position of the macroscopic apparatus which measures $\Pi(\boldsymbol{x}, t)$.

$\Pi(\boldsymbol{x}, t)$ is obtained from a space-time transformation, starting from $\Pi(\mathbf{0}, 0) = \Pi(0)$. For the time development, given that $\Pi(\boldsymbol{x}, t)$ is an operator in the Heisenberg representation, we obtain from (10.12):

$$\Pi(\boldsymbol{x}, t) = e^{+iHt}\Pi(\boldsymbol{x}, 0)e^{-iHt}. \tag{10.51}$$

Using equations (10.45) and (A.39), we find for the spatial variables:

$$\Pi(\boldsymbol{x}, t) = e^{+iHt}e^{-i\boldsymbol{P}\cdot\boldsymbol{x}}\Pi(\mathbf{0}, 0)e^{+i\boldsymbol{P}\cdot\boldsymbol{x}}e^{-iHt}. \tag{10.52}$$

In the exponents, the following relativistically invariant combination appears:

$$Ht - \boldsymbol{P}\cdot\boldsymbol{x} = P^{\mu}x^{\nu}g_{\mu\nu} = P_{\mu}x^{\mu}. \tag{10.53}$$

Consequently the matrix element between energy and momentum eigenstates of a local operator has a characteristic space-time dependence:

$$\begin{aligned} &< P', E'|\Pi(\boldsymbol{x}, t)|P, E >= \\ &=< P', E'|e^{+i(Ht-\boldsymbol{P}\cdot\boldsymbol{x})}\Pi(\mathbf{0}, 0)e^{-i(Ht-\boldsymbol{P}'\cdot\boldsymbol{x})}|P, E >= \\ &= e^{-i[(E-E')t-(\boldsymbol{P}i-\boldsymbol{P}')\cdot\mathbf{x}]} < P', E'|\Pi(\mathbf{0}, 0)|P, E > . \end{aligned} \tag{10.54}$$

CHAPTER 11

# RELATIVISTIC PERTURBATION THEORY

In this section, we derive the perturbative expansion of the scattering matrix ($S$-matrix) for a relativistic field theory.

The starting point is the expansion of the transfer matrix, $U_I(t_2, t_1)$, the operator which transforms the state at time $t_1$ into that at time $t_2$, in the interaction representation (IR) introduced in the previous chapter.

The fundamental hypothesis to derive the scattering matrix is to assume that after a long time, the interaction Hamiltonian in the IR *tends to zero*:

$$lim_{t\to\pm\infty}\boldsymbol{H}_I(t) \to 0. \tag{11.1}$$

Intuitively, the condition arises from the fact that when the time tends towards infinity, particles separate indefinitely and their mutual interactions tend to zero.

However, condition (11.1) is actually not trivial. For example, it is not satisfied in the presence of long range forces, such as the electrostatic interaction between electric charges. This case requires particular care to define the probability of observable processes (cf. for example [11]).

Also in the case of short-range forces, to satisfy condition (11.1) we must carefully define the interaction Hamiltonian, so as to subtract the interaction of each particle with the *field generated by the particle itself*. The procedure of subtraction of the effects of self-interaction will be discussed in detail in the third volume of this series, titled *Introduction to Gauge Theories* [14].

Assuming the validity of (11.1), the state vector tends to a constant vector in the limit $t \to +\infty$ and the same thing happens, with a different vector, if we let $t \to -\infty$:

$$lim_{t\to\pm\infty}|t> = |\pm\infty> = constant. \tag{11.2}$$

DOI: 10.1201/9781003436263-11

For finite times, the state vector is given by the application of the linear operator $U(t_0, t)$ to the initial state $|t_0>$. From equations (11.2) it follows equally that $|+\infty>$ depends linearly on $|-\infty>$. We can therefore define a linear operator independent of time, the scattering matrix, or *S-matrix*, which transforms one to the other:

$$|+\infty>= S|-\infty> . \tag{11.3}$$

Comparing with the definition of $U_I(t_2, t_1)$, we see that:

$$S = lim_{t_2 \to +\infty, t_1 \to -\infty} U_I(t_2, t_1) \tag{11.4}$$

or, using the perturbative expansion discussed in Chapter 10:

$$S = \sum_n \frac{(-i)^n}{n!} \int_{-\infty}^{+\infty} dt_1 \int_{-\infty}^{+\infty} dt_2 \dots \int_{-\infty}^{+\infty} dt_n \; T(\boldsymbol{H}_I(t_1)\boldsymbol{H}_I(t_2)\dots\boldsymbol{H}_I(t_n)). \tag{11.5}$$

In a scattering process, the state at time $-\infty$ represents the *initial state*, $|i>$, while the state at time $+\infty$ is a linear superposition of the states which correspond to *all possible results* of the interaction. If we denote one of these states with $|f>$, the probability amplitude to observe $|f>$ at the end of the process and the corresponding probability are given by:

$$Amplitude(i \to f) =< f| \, S \, |i>;$$
$$Probability(i \to f) = |A(i \to f)|^2.$$

Unitarity. The state $|f\rangle$ is one of the possible results of the scattering process. Summing over all the states $|f_n\rangle$ of a complete basis, the total probability must be unity:

$$\sum_n P(i \to f_n) = \sum_n \langle i| \, S \, |f_n\rangle\langle f_n| \, S^\dagger \, |i\rangle = \langle i| \, SS^\dagger |i\rangle = 1.$$

Being the same for every $|i\rangle$, this result implies that the $S$-matrix must be *unitary*:

$$SS^\dagger = S^\dagger S = 1. \tag{11.6}$$

Relativistic invariance. The principle of special relativity takes a particularly simple form for the $S$-matrix. For simplicity, we set our system in a finite volume, $V$. In this case, the states belong to a discrete spectrum and are normalised as $|<n|m>|^2 = \delta_{n,m}$. The probability of observing a certain result from given initial conditions and after an interval of infinite time *must be the same in any inertial frame of reference.* Therefore, if we denote with $|Ui>, |Uf>$ the initial and final states transformed to another IF, we must have:

$$|<f|S|i>|^2 = |<Uf|S|Ui>|^2 = |<f|U^\dagger S U|i>|^2.$$

For transformations continuously connected to the identity transformation, therefore:

$$< f|U^\dagger SU|i >=< f|S|i >$$

or, given that the initial and final states are arbitrary:

$$U^\dagger SU = S. \tag{11.7}$$

The $S$-matrix is *invariant under Lorentz transformations.*

## 11.1 THE DYSON FORMULA

In a field theory, the interaction Hamiltonian is in general the spatial integral of a Hamiltonian density:

$$\boldsymbol{H}_I(t) = \int d^3x\, \mathcal{H}_I(x).$$

Furthermore, we restrict ourselves to the case in which the Lagrangian density is the sum of a free Lagrangian and an interaction Lagrangian *which does not contain the field derivatives,* as happens in spinor QED. In this case, we can obtain the Hamiltonian density of interactions as follows:

$$\begin{aligned}\mathcal{H}_{tot} &= \frac{\partial \mathcal{L}}{\partial \partial_0 \phi}\partial_0\phi - \mathcal{L} = \frac{\partial \mathcal{L}_0}{\partial \partial_0 \phi}\partial_0\phi - (\mathcal{L}_0 + \mathcal{L}_I) = \\ &= \mathcal{H}_0 + \mathcal{H}_I.\end{aligned}$$

By comparison, we obtain:

$$\mathcal{H}_I = -\mathcal{L}_I. \tag{11.8}$$

Consequently, we obtain the formula for the $S$-matrix due to Dyson:

$$S = \sum_n \frac{(i)^n}{n!} \int \int \ldots \int d^4x_1 d^4x_2 \ldots d^4x_n\, T\left[\mathcal{L}_I(x_1)\mathcal{L}_I(x_2)\ldots\mathcal{L}_I(x_n)\right]. \tag{11.9}$$

The Dyson formula allows us to directly verify the relativistic invariance of the $S$-matrix.

We note first that the volume of integration and the interaction Lagrangian density are Lorentz invariants. We must examine only the invariance of the time-ordered product. For simplicity, we consider the product of two operators:

$$T\left[\mathcal{L}_I(x_1)\mathcal{L}_I(x_2)\right] = \theta(x_1^0 - x_2^0)\mathcal{L}_I(x_1)\mathcal{L}_I(x_2) + \theta(x_2^0 - x_1^0)\mathcal{L}_I(x_2)\mathcal{L}_I(x_1). \tag{11.10}$$

If the separation between two events is time-like, the order is maintained in every IF. Conversely, for space-like separations, the temporal order of two events can be different in a different IF. However, the microcausality condition, Section 7.5, comes to our aid. Since under these conditions the two interaction Lagrangians must commute, the order in which the two operators appear is immaterial and the T-product is in fact a relativistic invariant.

Comment. The fields which appear in (11.9) are operators in the interaction representation, therefore *they satisfy the free equations of motion* and are in fact, the operators given by expansions of the type referred to in (4.47) and (7.12), with the canonical quantisation rules. If we add the convention of defining $\mathcal{L}_I$ as normal products of the fields, these rules completely determine the products which appear in the Dyson formula *as long as we maintain* $x_1 \neq x_2 \neq \ldots \neq x_n$. When one or more points coincide, the products of $\mathcal{L}_I$ exhibit an unusual behaviour (for this aspect, cf. in particular [2]). The overcoming of the difficulty requires a redefinition of $\mathcal{L}_I$ and the parameters which appear in it. The possibility to *renormalise* the theory in interactions will be studied in [14]. We can disclose in advance that the renormalisation procedure can be completed only for a restricted class of interactions: *renormalisable interactions*. QED generated by the minimal substitution is in this category. Conversely, the electrodynamic interaction with the addition of Pauli terms, for example, the Lagrangian (9.16), is not renormalisable and must be considered a phenomenological theory, valid only in a restricted range of energy.

## 11.2 CONSERVATION LAWS

Consider the element of the $S$-matrix between states of defined 4-momentum:

$$< P_{fin}, \beta| \, S \, |P_{in}, \alpha > \tag{11.11}$$

($\alpha$ and $\beta$ are further quantum numbers necessary to specify the states). We consider the $n$th term of the $S$-matrix and apply a translation of $-x_1$. If the interaction Lagrangian is invariant under translations, i.e. depends only on the coordinates through the fields[1], then:

$$\mathcal{L}_I(x_1) = e^{i\mathbf{P}_\mu x_1^\mu} \mathcal{L}_I(0) e^{-i\mathbf{P}_\mu x_1^\mu} \tag{11.12}$$

$$\begin{aligned} &< P_{fin}, \beta|T\left[\mathcal{L}_I(x_1)\mathcal{L}_I(x_2)\ldots\mathcal{L}_I(x_n)\right]|P_{in}, \alpha> = \\ &\quad = < P_{fin}, \beta|U^\dagger(-x_1)U(-x_1)T\times \\ &\qquad\qquad \left[\mathcal{L}_I(x_1)\mathcal{L}_I(x_2)\ldots\mathcal{L}_I(x_n)\right]U^\dagger(-x_1)U(-x_1)|P_{in}, \alpha> \\ &\quad = e^{-i[(P_{in}-P_{fin})x_1]}\times \\ &\qquad\qquad < P_{fin}, \beta|T\left[\mathcal{L}_I(0)\mathcal{L}_I(x_2-x_1)\ldots\mathcal{L}_I(x_n-x_1)\right]|P_{in}, \alpha> . \end{aligned} \tag{11.13}$$

In carrying out the integral in (11.9) we can translate the integration

[1]For example, this excludes the presence of a classical field, a complex numerical function of the coordinates, which would remain unaltered under the unitary transformations in (11.12).

variables ($x_2 \to x_2 - x_1 = \zeta_2, x_3 \to x_3 - x_1 = \zeta_3, \ldots$) and obtain:

$$< P_{fin}, \beta|\, S^{(n)} |P_{in}, \alpha > = \int d^4x_1 e^{-i[(P_{fin}-P_{in})x_1]} \cdot \frac{(i)^n}{n!} \cdot$$
$$\cdot < P_{fin}, \beta| \int \ldots \int d^4\zeta_2 \ldots d^4\zeta_n\, T\left[\mathcal{L}_I(0)\mathcal{L}_I(\zeta_2)\ldots\mathcal{L}_I(\zeta_n)\right] |P_{in}, \alpha > . \tag{11.14}$$

Therefore, in conclusion:

$$< P_{fin}, \beta|\, S\, |P_{in}, \alpha >= (2\pi)^4\delta^{(4)}(P_{fin} - P_{in}) \cdot F(P_{fin}, \beta, P_{in}, \alpha) \tag{11.15}$$

where $F$ is a well-behaved function of its variables.

Equation (11.15) expresses the conservation of 4-momentum in a translationally invariant theory, according to (11.12).

We can treat all the conservation laws which derive from the invariance of the interaction Lagrangian (and thus the $S$-matrix) in a similar manner. This is the case, for example, for the law of conservation of angular momentum, when $\mathcal{L}_I$ is invariant for rotations. If:

$$e^{+i\alpha\boldsymbol{Q}} S e^{-i\alpha\boldsymbol{Q}} = S; \quad \text{or}$$
$$[\boldsymbol{Q}, S] = 0.$$

we also have, for the eigenstates of $\boldsymbol{Q}$:

$$0 =< q'|\, [\boldsymbol{Q}, S]\, |q >= (q' - q) < q'|\, S\, |q >$$

or the $S$-matrix element must vanish, unless the conservation law

$$q' = q \tag{11.16}$$

is satisfied.

## 11.3 COLLISION CROSS SECTION AND LIFETIME

Scattering processes start from an initial state, $|i\rangle$, in which a particle beam is directed onto a fixed target or, in so-called *colliders*, against another beam of particles. At large distances from the interaction region, a system of detectors produces a signal when a possible final state, $|f\rangle$, is identified.

Normally a single quantum state is not selected. In general, the detectors are not able to discriminate between the states of a certain set (for example, particles with momentum around an average value $\bar{\boldsymbol{p}}$ in a given volume $\Delta\boldsymbol{p}$). In this case, we must *sum the probabilities over the final states* which are not separated by the experimental apparatus. Similarly, the initial state can be an incoherent mixture of a certain set of states; for example, an unpolarised beam contains equal proportions of different spin states of the incident particle. In this case, we must *average the probability over the initial states.*

For simplicity, we consider a fixed target experiment. We enclose the system in a large box with volume $V$ and surface area $S$ transverse to the beam and direct a single incident particle at a time onto a target, also considered to consist of a single particle. In a steady state (replacing the target and projectile after each collision) we have a constant flux of events in the detectors, characterised by the *probability per unit of time*, $W$. It is easy to convince oneself that $W$ must be equal to the flux of particles incident on the target multiplied by the interaction probability of a projectile with the target particle, $P$:

$$W = \Phi \cdot P;$$
$$\Phi = \rho \cdot S \cdot v; \; \rho = \frac{1}{V} \tag{11.17}$$

where $\rho$ is the density of incident particles and $v$ their velocity. The probability $P$ should be inversely proportional to the surface area $S$, if the beam is uniformly distributed transversely:

$$P = \frac{\Delta\sigma}{S}. \tag{11.18}$$

The quantity $\Delta\sigma$ represents the interaction probability and has dimensions of an area. $\Delta\sigma$ is called the *collision cross-section*; it all goes *as if* the target would offer a transverse area $\Delta\sigma$ to a beam of incident particles randomly distributed over the area $S$.

Clearly, the cross-section depends on the initial conditions (for example the energy and type of incident particle we choose) and the specific final conditions fixed by our detectors. $\Delta\sigma$ is all that can be measured for a given scattering process.

Summarising, we find:

$$\Delta\sigma = \frac{W}{\rho \cdot v}. \tag{11.19}$$

If we denote with $T$ the time over which the passage from the initial to final state takes place (as in the case of $V$, at the end we will take the limit $T \to \infty$), we have:

$$W = \frac{|\langle f| \, S \, |i\rangle|^2}{T} \tag{11.20}$$

and therefore:

$$\Delta\sigma = \frac{V|\langle f| \, S \, |i\rangle|^2}{T \cdot v}. \tag{11.21}$$

Consider the process:

$$a \, + \, b \, \to \, 1 + 2 + \ldots + n_f \tag{11.22}$$

with initial and final states:

$$|i\rangle = a_a^\dagger(\boldsymbol{p}_a) a_b^\dagger(\boldsymbol{p}_b)|0\rangle$$
$$|f\rangle = a_1^\dagger(\boldsymbol{p}_1) a_2^\dagger(\boldsymbol{p}_2) \ldots a_{n_f}^\dagger(\boldsymbol{p}_{n_f})|0\rangle. \tag{11.23}$$

In a situation invariant for space-time translations, the $S$-matrix element has the form (11.15) and in calculating the probability $W$ we must *square* the Dirac $\delta$-function, an operation normally not allowed. In this case, for $V$ and $T$ finite, the $\delta$-function can be substituted by:

$$\int_\Gamma d^4x\, e^{-i[(P_f-P_i)x]} \tag{11.24}$$

where $\Gamma$ is the finite region of space-time of the 4-volume $V \cdot T$. It is easy to show[2] that, in the limit of large $V$ and $T$:

$$|\int_\Gamma d^4x\, e^{-i[(P_f-P_i)x]}|^2 \to V \cdot T \cdot (2\pi)^4\delta^{(4)}(P_f - P_i). \tag{11.25}$$

From this, we find:

$$\Delta\sigma = V^2(2\pi)^4\delta^{(4)}(P_f - P_i)\frac{1}{v}|F(P_f, \ldots, P_i \ldots)|^2 \tag{11.26}$$

where $F$ is the well-behaved function defined in (11.15) and the dots indicate the other variables needed to fully describe the process.

To any finite order of perturbation theory, the $S$-matrix, (11.9), is a polynomial of the fields. Among the various terms, only those which contain fields capable of annihilating the particles of the initial state and creating those in the final state gives a non-zero matrix element. Using the appropriate commutation or anticommutation rules[3], we can arrange these fields to operate

[2] For simplicity, we consider the one-dimensional case

$$\int_{-T/2}^{+T/2} dt\, e^{i\omega t} = \frac{2}{\omega} sin(\frac{\omega T}{2}).$$

If we integrate with a trial function, $f(\omega)$ and take the limit $T \to \infty$, we find:

$$\int_{-\infty}^{+\infty} d\omega \frac{2}{\omega} sin(\frac{\omega T}{2}) f(\omega) = 2\int_{-\infty}^{+\infty} dx \frac{sin(x)}{x} f(\frac{2x}{T}) \to 2f(0)\int_{-\infty}^{+\infty} dx \frac{sin(x)}{x} = 2\pi f(0)$$

or, as expected:

$$\frac{2}{\omega} sin(\frac{\omega T}{2}) \to 2\pi\delta(\omega).$$

Now, we consider the square and integrate it with a trial function:

$$\int_{-\infty}^{+\infty} d\omega \frac{4}{\omega^2} sin^2(\frac{\omega T}{2})\, f(\omega) =$$

$$= 2T\int_{-\infty}^{+\infty} dx \frac{sin^2 x}{x^2} f(\frac{x}{T}) \to 2Tf(0))\int_{-\infty}^{+\infty} dx \frac{sin^2 x}{x} = T \cdot 2\pi f(0)$$

or:

$$|\int_{-T/2}^{+T/2} dt\, e^{i\omega t}|^2 \to T \cdot 2\pi\delta(\omega).$$

[3] As discussed in Section 11.1, the field operators which appear in the Dyson formula are free fields, whose commutators or anticommutators are calculable complex numbers.

on the initial and final states respectively:

$$
\begin{aligned}
&\langle 1,2,\ldots|\, S\, |a,b\rangle = \\
&= \int d^4x \int d^4x' \ldots \langle 1,2,\ldots|\, \chi_1^\dagger(x_1)\chi_2^\dagger(x_2)\ldots[\ldots]\,\chi_a(x_a)\chi_b(x_b)\,|a,b\rangle = \\
&= \langle 1,2,\ldots|\, (\chi_1^\dagger)^{(-)}(x_1)(\chi_2^\dagger)^{(-)}(x_2)\ldots[\ldots]\,\chi_a^{(+)}(x_a)\chi_b^{(+)}(x_b)\,|a,b\rangle.
\end{aligned}
\tag{11.27}
$$

The points $x_1, x_2, \ldots$ are a permutation of $x, x', \ldots$, not necessarily different from each other. In parentheses, there is a polynomial of other fields, provided that the product of the fields shown explicitly with those implied forms a Lorentz-invariant combination.

In the expansion of the fields $\chi$ in plane waves, only terms with the destruction or creation operators appropriate for the initial and final states are counted, therefore the matrix element contains the corresponding normalisation coefficients as factors:

$$
N_i = \begin{cases} \sqrt{\frac{m}{E_i V}} & \text{fermions} \\ \\ \sqrt{\frac{1}{2\omega_i V}} & \text{bosons} \end{cases}, \tag{11.28}
$$

each multiplied by a Lorentz-invariant exponential, $e^{i(px)}$, and by a solution of the wave equation (for example a Dirac spinor) which has the transformation properties of the original field. This last factor, multiplied by the fields in parentheses in (11.27), form a Lorentz-invariant combination.

Introducing the sum over the final states corresponding to the detector resolution, we find, finally:

$$
\Delta\sigma = V^2 \left[ \sum_f (2\pi)^4 \delta^{(4)}(P_f - P_i) \left(\Pi_{i=1,2,\ldots} N_i^2\right) |M_{i\to f}|^2 \right] \frac{(N_a N_b)^2}{v}. \tag{11.29}
$$

$M_{i\to f}$ is called the *Feynman amplitude* of the process and is, from its derivation, *relativistically invariant.* The factor $V^2$ which arises from (14.200) is cancelled by the factors $1/V$ in the normalisation of the initial states. The factors $1/V$ in the normalisation of the final states are balanced by corresponding factors in the phase space of each final particle:

$$
\sum_f \Pi_{i=1,2,\ldots} N_i^2\ (\ldots) = \frac{1}{(2\pi)^{3n_f}} \int \frac{d^3p_1}{2E_1}\frac{d^3p_2}{2E_2} \ldots \left(\Pi_{i=1,2,\ldots} n_i^2\right)(\ldots) \tag{11.30}
$$

where we have set $2E_i N_i^2 = n_i^2 = 2m_i$ or 1 for fermions or bosons, respectively. It should be noted that the momentum integration volume associated with each particle is invariant for $L_+^\uparrow$ transformations[4]. as seen from the equality:

$$\frac{d^3p}{2E(p)} = d^4p\, \delta(p^2 - m^2)\theta(p^0) \tag{11.31}$$

Taking into account all these observations, we arrive at the expression ($E_b = m_b$ for a fixed target):

$$d\sigma = \frac{(n_a n_b)^2}{v \cdot (4E_a m_b)}(2\pi)^4 \delta^{(4)}(P_f - P_i)\left[\Pi_i \frac{d^3p_i}{(2\pi)^3\, 2E_i} n_i^2\right] |M_{i\to f}|^2. \tag{11.32}$$

We can see that the flux factor is also Lorentz invariant, by rewriting:

$$v \cdot (E_a m_b) = |\boldsymbol{p}_a| m_b = \sqrt{(E_a m_b)^2 - m_a^2 m_b^2} = \sqrt{(p_a \cdot p_b)^2 - p_a^2 p_b^2}, \tag{11.33}$$

therefore, by direct inspection of (11.32), we reach an important conclusion:

- the collision cross-section is a relativistic invariant.

Equation (11.33) permits us to calculate the cross-section in reference systems in which both the initial particles are in motion, as is the case in modern colliders. In this case, a particularly useful reference frame is the centre of the mass system ($\boldsymbol{p}_a + \boldsymbol{p}_b = 0$) which, for collisions between symmetric beams, coincides with the laboratory frame of reference. Assuming also, for simplicity, equal masses for particles $a$ and $b$, we find:

$$\sqrt{(p_a \cdot p_b)^2 - p_a^2 p_b^2} = 2|\boldsymbol{p}|E = 2vE^2 = |v_a - v_b|\; E^2. \tag{11.34}$$

The flux is determined by the relative velocity between the two particles (which can also be larger than $c$). The differential cross-section in the general case of particles with velocity $v_a$ and $v_b$ is given by:

$$d\sigma = \frac{(n_a n_b)^2}{|v_a - v_b| \cdot (4E_a E_b)}(2\pi)^4 \delta^{(4)}(P_f - P_i)\left[\Pi_i \frac{d^3p_i}{(2\pi)^3\, 2E_i} n_i^2\right] |M_{i\to f}|^2. \tag{11.35}$$

Lifetime. The interaction may cause decays of particles which would be stable in the free theory. The decay probability per unit time of a state $|f\rangle$ is given by:

$$\Gamma(f) = \frac{|\langle f|\, S\, |a\rangle|^2}{T}. \tag{11.36}$$

Summing over all final states the *total decay probability* is found:

$$\Gamma = \sum_f \Gamma(f). \tag{11.37}$$

[4]Which leave the sign of $p^0$ invariant, $p^\mu$ being timelike or lightlike.

For a given final state $f$, the ratio:

$$B(f) = \frac{\Gamma(f)}{\Gamma} \tag{11.38}$$

represents the probability to find the state $f$ among the decay products and is called the *branching ratio* or *branching fraction.* Clearly:

$$\sum_f B(f) = 1. \tag{11.39}$$

For a system of $N$ particles, the change in the number present at time $t$ is given by:

$$\frac{dN(t)}{dt} = -\Gamma \cdot N(t)$$

from which:

$$N(t) = N(0)e^{-\Gamma t}.$$

The *lifetime* of the unstable particle is therefore:

$$\tau = \frac{1}{\Gamma}. \tag{11.40}$$

$\Gamma$ is known as the *total width* of the particle[5]. We consider the case of an interaction invariant for translations in space-time. Proceeding as before, we find, in the rest system of the unstable particle:

$$d\Gamma(a \to 1 + 2 + \cdots + n_f) =$$
$$= \frac{1}{(2\pi)^{3n_f}} \frac{n_a^2}{2m_a} \Pi_i \frac{d^3p_i}{2E_i} n_i^2 (2\pi)^4 \delta^{(4)}(P_f - P_a) |M(a \to 1 + 2 + \cdots + n_f)|^2. \tag{11.41}$$

## 11.4 PROBLEMS FOR CHAPTER 11

### Sect. 11.3

**1.** Prove the relation:

$$\frac{d^3p}{2E(p)} = d^4p\, \delta(p^2 - m^2)\theta(p^0)\ .$$

[5]The name originates from the fact that an unstable state necessarily has an uncertainty in the energy given the Heisenberg principle $\Delta E \Delta t \simeq \hbar$. If we take $\Delta t = \tau$ we find $\Delta E = \hbar/\tau = \Gamma$; therefore $\Gamma$ represents the width of the distribution of energies of the unstable state.

2. The Fabri–Dalitz plot. Consider a three-particle decay and assume the initial particle has spin zero (e.g. $K^0 \to \pi^+\pi^-\pi^0$) or it is not polarised, so we can assume isotropy in its rest system. In this frame, the momenta of the three final particles add to zero, which means that they are in a plane, which we take to be the $x - y$ plane, with the first particle (e.g. the $\pi^+$) along the $x$ axis. The momentum of particle 3 is fixed to be $p_3 = -p_1 - p_2$ so the only variables left are $|p_1|$, $|p_2|$ and the angle between the two vectors, $\theta_{12}$.The angle can be eliminated in favour of the energy of particle 3, from the relation

$$|p_3|^2 = |p_1|^2 + |p_2|^2 + 2\cos\theta_{12}\ |p_1| \cdot |p_2|$$

and we are left with three variables, the energies, restricted by the energy conservation relation:

$$E_1 + E_2 + E_3 = M = \text{const.}$$

Each set of allowed energies can be represented as a point inside a fixed equilateral triangle, the energies being the distances of the point from the three sides of the triangle (this property is known as *Viviani's theorem*).

The beauty of this construction and its usefulness is that one can prove that *the density of points from phase space only*; that is, for a constant decay matrix element squared, *is uniform*. Hence if we measure many decays and report a point for each decay in the *Fabri–Daliz plot* any density variation of the points reflects a true property of the matrix element, e.g. the presence of a resonance or possible spin correlations. Thus, the Fabri–Dalitz plot puts into evidence the dynamics of the decay.

The differential rate in the rest system of the decay is

$$d\Gamma = \text{Const} \ \times \ \delta^{(4)}(P - \sum_{i=1,2,3} p_i) \left( \prod_{i=1,2,3} \frac{d^3 p_i}{2E_i} \right) \ |\mathcal{M}|^2 \ .$$

Prove that, after elimination of irrelevant variables, one can write

$$d\Gamma = \text{Const}' \ \times \ \delta(M - E_1 - E_2 - E_3) \left( \prod_{i=1,2,3} dE_i \right) \ |\mathcal{M}|^2 \ .$$

3. Prove that the boundary of the allowed region of the Fabri–Dalitz plot corresponds to collinear configurations, namely $\theta_{12} = 0, \pi$.

4. Determine the boundary of the Fabri–Dalitz plot for decay in three equal mass particles (e.g. $K^0 \to \pi^+\pi^-\pi^0$).

CHAPTER 12

# THE DISCRETE SYMMETRIES: P, C, T

In this chapter, we discuss discrete symmetries in quantum field theories, i.e. the transformations:

- inversion of the spatial coordinate axes at a given time:

$$\mathbf{x} \to -\mathbf{x}, \; t \to t \tag{12.1}$$

  or *parity*, which we denote with $\mathcal{P}$,

- substitution of every particle by its antiparticle, and vice versa, or *charge conjugation*, $\mathcal{C}$,

- inversion of time, or *time reversal*

$$t \to -t, \; \mathbf{x} \to \mathbf{x} \tag{12.2}$$

  which we denote as $\mathcal{T}$.

The first two transformations are represented in the Hilbert space of states by unitary operators while time inversion must act as an anti-unitary operator. In this chapter, we will refer principally to the QED Lagrangian which, as we will see, is invariant under all these transformations, which are therefore *exact symmetries* of QED. At the end of the chapter, we will consider the Fermi Lagrangian, the prototype description of the weak interactions, in which the symmetries are individually violated, beginning with parity, and the only exact symmetry is the product of all three, the $TCP$ transformation.

## 12.1 PARITY

The action of parity on the field operators follows from (12.1): the transformed components of the field in $\mathbf{x}$ must be a superposition of the field


DOI: 10.1201/9781003436263-12

components in $-\mathbf{x}$. The application of the parity operation twice returns us to the starting point, therefore we impose the condition:

$$h\mathcal{P}\mathcal{P} = 1. \tag{12.3}$$

For the electromagnetic field, we know that in the classical limit, the spatial components of the vector potential are polar vectors, while the scalar potential is just that, a scalar. Therefore (the indices repeated three times are *not* summed):

$$\mathcal{P}A^\mu(\mathbf{x},t)\mathcal{P} = g^{\mu\mu}A^\mu(-\mathbf{x},t). \tag{12.4}$$

From here we find:

$$\mathcal{P}\partial^\lambda A^\mu(\mathbf{x},t)\mathcal{P} = g^{\mu\mu}g^{\lambda\lambda}\partial^\lambda A^\mu(-\mathbf{x},t) \tag{12.5}$$

from which follows the rule for the transformation of Maxwell's tensor.

$$\mathcal{P}F^{\mu\nu}(\mathbf{x},t)\mathcal{P} = g^{\mu\mu}g^{\nu\nu}F^{\mu\nu}(-\mathbf{x},t). \tag{12.6}$$

In the case of the spinor, the general rule is written:

$$\mathcal{P}\psi(\mathbf{x},t)_\alpha\mathcal{P} = [P\psi(-\mathbf{x},t)]_\alpha = P_{\alpha\beta}\psi(-\mathbf{x},t)_\beta \tag{12.7}$$

where $P$ is a matrix in the space of Dirac spinors to be defined so that observables constructed with the fields transform correctly under parity.

To respect the classical limit, the current must transform like the vector potential:

$$\mathcal{P}\bar{\psi}(\mathbf{x},t)\gamma^\mu\psi(\mathbf{x},t)\mathcal{P} = g^{\mu\mu}\bar{\psi}(-\mathbf{x},t)\gamma^\mu\psi(-\mathbf{x},t). \tag{12.8}$$

Substitution of (12.7) in the above equation yields

$$\mathcal{P}\bar{\psi}(\mathbf{x},t)\gamma^\mu\psi(\mathbf{x},t)\mathcal{P} = g^{\mu\mu}\bar{\psi}(-\mathbf{x},t)P^\dagger\gamma^0\gamma^\mu P\psi(-\mathbf{x},t). \tag{12.9}$$

- For $\mu = 0$, we find:

$$P^\dagger P = 1. \tag{12.10}$$

- For $\mu = i$ (i=1, 2, 3), we find:

$$P^\dagger\gamma^0\gamma^i P = P^\dagger\alpha^i P = -\alpha^i. \tag{12.11}$$

Finally, $P$ must anticommute with all the $\alpha^i$ and this implies:

$$P = c\gamma^0. \tag{12.12}$$

If we require that $P^2 = 1$, it follows that $|c|^2=1$. Without loss of generality, we can fix $c = 1$ and thus:

$$P = \gamma^0; \quad P = P^\dagger = P^{-1}. \tag{12.13}$$

As we saw in Section 7.5, observables are constructed from bilinear combinations of the Dirac fields, which in their turn are organised into Dirac *bilinear covariants*. Equations (12.7) and (12.13) define the parity transformation properties of each bilinear.

**Bilinear Covariants.** We leave to the reader the task of generalising (12.9), showing that the bilinears $S, P, V, A, T$ transform under parity as scalars, pseudoscalars, polar vectors, axial vectors and tensors.

$$\begin{aligned}
&\text{S}: \ \mathcal{P}\bar{\psi}(\mathbf{x},t)\psi(\mathbf{x},t)\mathcal{P} = \bar{\psi}(-\mathbf{x},t)\psi(-\mathbf{x},t)\\
&\text{P}: \ \mathcal{P}\bar{\psi}(\mathbf{x},t)i\gamma_5\psi(\mathbf{x},t)\mathcal{P} = \ -\ \bar{\psi}(-\mathbf{x},t)i\gamma_5\psi(-\mathbf{x},t)\\
&\text{V}: \ \mathcal{P}\bar{\psi}(\mathbf{x},t)\gamma^\mu\psi(\mathbf{x},t)\mathcal{P} = g^{\mu\mu}\ \bar{\psi}(-\mathbf{x},t)\gamma^\mu\psi(-\mathbf{x},t)\\
&\text{A}: \ \mathcal{P}\bar{\psi}(\mathbf{x},t)\gamma^\mu\gamma_5\psi(\mathbf{x},t)\mathcal{P} = \ -g^{\mu\mu}\ \bar{\psi}(-\mathbf{x},t)\gamma^\mu\gamma_5\psi(-\mathbf{x},t)\\
&\text{T}: \ \mathcal{P}\bar{\psi}(\mathbf{x},t)\sigma^{\mu\nu}\psi(\mathbf{x},t)\mathcal{P} = \ g^{\mu\mu}g^{\nu\nu}\ \bar{\psi}(-\mathbf{x},t)\sigma^{\mu\nu}\psi(-\mathbf{x},t).
\end{aligned} \quad (12.14)$$

A corollary of the relations we have demonstrated is that the QED action is parity invariant. Actually, in

$$\mathcal{L}_{QED} = -\frac{1}{4}F_{\mu\nu}(x)F^{\mu\nu}(x) + \bar{\psi}(x)(i\partial\!\!\!/ - m)\psi(x) + eA_\mu(x)\bar{\psi}\gamma^\mu\psi(x) \quad (12.15)$$

only scalar quantities or vector products appear. Therefore:

$$\begin{aligned}
\mathcal{P}S\mathcal{P} = \mathcal{P}\int d^4x\mathcal{L}_{QED}(\mathbf{x},t)\mathcal{P} &=\\
&= \int d^4x\mathcal{L}_{QED}(-\mathbf{x},t) = \int d^4x\mathcal{L}_{QED}(\mathbf{x},t) = S. \quad (12.16)
\end{aligned}$$

The invariance remains valid if we add to the action non-minimal terms which describe a possible anomalous magnetic moment of the form (cf. Section 9.1):

$$\mathcal{L}_{non\ min.} = \frac{\kappa}{2m}\bar{\psi}\sigma_{\mu\nu}\psi\ F^{\mu\nu}. \quad (12.17)$$

## 12.2 CHARGE CONJUGATION

The field $\psi$ destroys an electron and creates a positron, while $\psi^\dagger$ creates an electron and destroys a positron. As we have observed several times, the statement that the positron is the antiparticle of the electron is purely a matter of convention and the roles of the electron and positron in QED are completely symmetric. In other words, we may equally choose to formulate QED in terms of a new field which destroys the positron and creates the electron. If we call the new field $\psi_c$, from what has just been said *the components of* $\psi_c$ *must be linear combinations of the components of* $\psi^\dagger$.

In equation form (we do not include the coordinates which are the same on the left and right-hand sides):

$$[\psi_c]_\alpha = \mathcal{C}\psi_\alpha\mathcal{C} = \left[C\psi^\dagger\right]_\alpha = C_{\alpha\beta}\psi^\dagger_\beta. \quad (12.18)$$

where, as before, $C$ is a matrix to be determined so that the transformation of observables by $\mathcal{C}$ corresponds to the replacement of each electron by a positron and vice versa. Taking the hermitian conjugate of (12.18), we also find:

$$[\psi_c]^\dagger_\alpha = \mathcal{C}\psi^\dagger_\alpha\mathcal{C} = \left[\psi C^\dagger\right]_\alpha. \tag{12.19}$$

As in the case of parity, repeating $\mathcal{C}$ twice gives the identity transformation. Therefore we can require that $C^2 = 1$, and thus:

$$\psi = (\psi_c)^\dagger C^\dagger; \quad \psi^\dagger = C\psi_c \tag{12.20}$$

In order that the Dirac Lagrangian should be invariant under $\mathcal{C}$, we must have, in the first place:

$$\bar{\psi}\psi = \bar{\psi}_c\psi_c. \tag{12.21}$$

We use the Pauli representation, in which $\gamma^0$ is real and symmetric, for the $\gamma$ matrices. The field products in the formulae should all be understood as normal products, within which the fermion fields anticommute. Therefore we have:

$$\bar{\psi}\psi = -\ \psi(\gamma^0)^T\psi^\dagger = -\ (\psi_c)^\dagger C^\dagger\gamma^0 C\psi_c \tag{12.22}$$

from which:

$$C^\dagger\gamma^0 C = -\gamma^0. \tag{12.23}$$

Now we require that terms with derivatives in the Dirac action, $S_{kin}$, should maintain the same form when expressed in the conjugate fields:

$$\begin{aligned} S_{kin} &= \int d^4x\ i\bar{\psi}\partial\!\!\!/\psi = -\int d^4x\ i(\partial_\mu\psi^\dagger)\gamma^0\gamma^\mu\psi \\ &= +\int d^4x\ i\psi(\gamma^0\gamma^\mu)^T(\partial_\mu\psi^\dagger) \\ &= +\int d^4x\ i\bar{\psi}_c\gamma^0 C^\dagger(\gamma^\mu)^T\gamma^0 C(\partial_\mu\psi_c). \end{aligned} \tag{12.24}$$

Therefore we must have;

$$\gamma^0 C^\dagger(\gamma^\mu)^T\gamma^0 C = \gamma^\mu. \tag{12.25}$$

If we set $\mu = 0$, the previous equation becomes:

$$C^\dagger C = 1 \ \ \rightarrow \ \ C^\dagger = C = C^{-1} \tag{12.26}$$

and from equation (12.23) we see that $C$ must anticommute with $\gamma^0$. Furthermore, if we use the general relation:

$$\gamma^0(\gamma^\mu)^\dagger\gamma^0 = \gamma^\mu \tag{12.27}$$

equation (12.25) becomes:

$$C\gamma^{\mu}C = -(\gamma^{\mu})^{*}. \tag{12.28}$$

In the Pauli representation, $\gamma^{0,1,3}$ are real while $\gamma^2$ is imaginary. Equation (12.28) says that $C$ anticommutes with $\gamma^{0,1,3}$ and commutes with $\gamma^2$, therefore $C$ must be proportional to $\gamma^2$. If we require that its square should be equal to one, we find, to within an unimportant sign:

$$C = i\gamma^2. \tag{12.29}$$

$C$ is therefore a *real, symmetric* matrix. With this definition of $C$, we can determine the behaviour of the Dirac bilinears under $\mathcal{C}$. In the general case, we write:

$$\mathcal{O}(\Gamma) = \bar{\psi}\Gamma\psi = -\psi(\Gamma)^{T}\gamma^{0}\psi^{\dagger}. \tag{12.30}$$

Using (12.18) and (12.19), we have:

$$\mathcal{C}\Gamma\mathcal{C} = -\ \bar{\psi}_c\gamma^0 C(\Gamma)^T\gamma^0 C\psi_c = \bar{\psi}_c C\mathbf{\Gamma}^* C\psi_c \tag{12.31}$$

where we have assumed that the bilinears are defined to satisfy the hermiticity relation:

$$\gamma^0\Gamma\gamma^0 = \Gamma^{\dagger} \tag{12.32}$$

(This is the reason for the factor which appears in the pseudoscalar density).

Finally, we find that, under charge conjugation:

$$\mathcal{O}(\Gamma)\ \rightarrow\ \mathcal{O}(C\Gamma^{*}C). \tag{12.33}$$

From (12.33), it can easily be shown that each of the Dirac bilinears takes a characteristic sign under $\mathcal{C}$, which we denote as $\eta_C$, where

$$\begin{aligned} \eta_C &= +1, \quad \text{for} \quad S, P, A \\ \eta_C &= -1, \quad \text{for} \quad V, T. \end{aligned} \tag{12.34}$$

The QED Lagrangian, (12.15), including its non-minimal extensions, is invariant under $C$.

## 12.3 TIME REVERSAL

In classical mechanics, we can change the signs of the velocities of all the particles in the system at time $t = 0$. At subsequent times, the new system retraces all the configurations through which the original system passed. We denote a point in phase space of a classical Hamiltonian system with $A(q, p)$, and the transformation is known as *time reversal* with

$$A = (q, p) \rightarrow A_T = (q, -p). \tag{12.35}$$

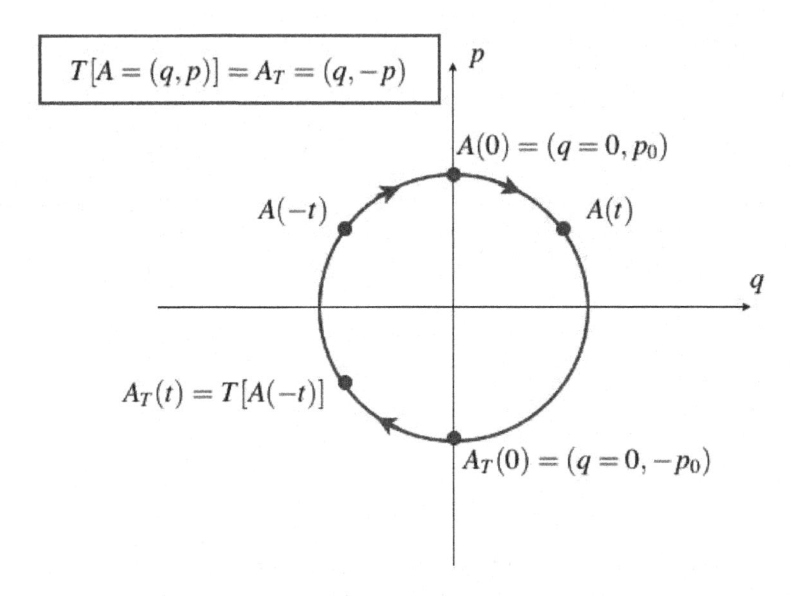


Figure 12.1 Time reversal of the motion of a classical harmonic oscillator.

If we compare the behaviour of the system which starts from $A$ at the time $t = 0$ with one which starts from $A_T$ we can express the previous statements in the following way:

$$\begin{aligned} & A(t) && \text{(state at time } t \text{ with } A(0) = A) \\ & A_T(t) && \text{(state at time } t \text{ with } A_T(0) = A_T) \\ & A_T(t) = T\left[A(-t)\right] && \text{(time reversal in classical mechanics)} \end{aligned} \tag{12.36}$$

Equation (12.36) is shown graphically in Fig. 12.1, for the case of the motion of a harmonic oscillator.

What happens in quantum mechanics? The time reversal transformation must be carried out on the quantum states by an operator, which we denote as $\mathcal{T}$, such that (working in the Schrödinger representation):

$$\begin{aligned} & |A; t = 0\rangle \rightarrow |A_T; t = 0\rangle = \mathcal{T}|A; t = 0\rangle \\ & |A_T; t\rangle = \mathcal{T}|A; -t\rangle \quad \text{(time reversal in quantum mechanics)}. \end{aligned} \tag{12.37}$$

However, the $\mathcal{T}$ operator must have very special properties, as we can see from the following argument. Suppose that the state $A$ is an energy eigenstate with eigenvalue $E$. We expect that $A_T$ should also have the same energy (as happens in classical mechanics). In that case:

$$|E; t\rangle = e^{-iEt}|E; t = 0\rangle; \quad |E_T; t\rangle = e^{-iEt}|E_T; t = 0\rangle. \tag{12.38}$$

However, it must be true that

$$|E_T;t\rangle = e^{-iEt}|E_T;t=0\rangle = \\ = \mathcal{T}[e^{-i(-t)E}|E;t=0\rangle] = \mathcal{T}(e^{+itE}|E;t=0\rangle) \quad (12.39)$$

which is impossible to realise if $\mathcal{T}$ is a linear operator, because in that case the phase of the exponentials on the two sides will have opposite signs.

The solution proposed by Wigner is to use an *antilinear* operator:

$$\mathcal{T}\left(\alpha|A\rangle + \beta|B\rangle\right) = \alpha^*\mathcal{T}|A\rangle + \beta^*\mathcal{T}|B\rangle. \quad (12.40)$$

In that case we have:

$$\mathcal{T}(e^{+iEt}|E;t=0\rangle) = e^{-itE}\mathcal{T}|E;t=0\rangle = e^{-itE}|E_T;t=0\rangle_T \quad (12.41)$$

as required.

Symmetries and Unitary or Antiunitary Operators. The condition set by Wigner to have a symmetry, is that the operator which represents it should leave the relations between quantum states unchanged. In their turn, these relationships are represented by *squared moduli of scalar products*, which specify the probability of results of quantum measurements. If:

$$|A_T\rangle = \mathcal{T}|A\rangle;\ \ |B_T\rangle = \mathcal{T}|B\rangle;\ \ldots \quad (12.42)$$

we must have:

$$|\langle A_T|B_T\rangle|^2 = |\langle A|B\rangle|^2 \quad (12.43)$$

which gives two possibilities for the operator which carries out the transformation $|A\rangle \to |A_T\rangle$:

$$\langle A_T|B_T\rangle = \langle A|B\rangle\ :\quad \text{unitary operator} \quad (12.44)$$

$$\langle A_T|B_T\rangle = \langle A|B\rangle^*:\quad \text{antiunitary operator.} \quad (12.45)$$

For transformations dependent on one or more parameters, which are connected to the identity transformation (proper Lorentz, translations), the first condition should hold, by continuity. From the argument we gave earlier, the second is the one which should apply to all symmetries which involve time reversal.

Action of $\mathcal{T}$ on Fields. We require that:

$$\mathcal{T}^\dagger = \mathcal{T}^{-1} = \mathcal{T}. \quad (12.46)$$

The action of the $\mathcal{T}$ operator on the vector potential is fixed by the classical limit. If $\mathcal{T}$ leaves the positions of charges unchanged and reverses their velocities, it must be true that:

$$\mathcal{T}A^0(\boldsymbol{x},t)\mathcal{T} = +A^0(\boldsymbol{x},-t);\;\; \mathcal{T}A^i(\boldsymbol{x},t)\mathcal{T} = -A^i(\boldsymbol{x},-t)$$

or

$$\mathcal{T}A^\mu(\boldsymbol{x},t)\mathcal{T} = g^{\mu\mu}A^\mu(\boldsymbol{x},-t). \tag{12.47}$$

The $\mathcal{T}$ operator must transform $\psi_\alpha(\boldsymbol{x},t)$ into a linear combination of the $\psi_\beta(\boldsymbol{x},-t)$ fields:

$$\mathcal{T}\psi_\alpha(\boldsymbol{x},t)\mathcal{T} = T_{\alpha\beta}\psi_\beta(\boldsymbol{x},-t) = [T\psi(\boldsymbol{x},-t)]_\alpha \tag{12.48}$$

from which, then:

$$\mathcal{T}\psi(\boldsymbol{x},t)^\dagger\mathcal{T} = [\psi(\boldsymbol{x},-t)T]^\dagger. \tag{12.49}$$

The form of the $T$ matrix is defined by requiring that the current transforms like $A^\mu$ in (12.47). Taking into account the antilinearity of $\mathcal{T}$, we have:

$$\begin{aligned}\mathcal{T}\bar{\psi}(\boldsymbol{x},t)\gamma^\mu\psi(\boldsymbol{x},t)\mathcal{T} &= \mathcal{T}\psi(\boldsymbol{x},t)^\dagger\mathcal{T}\mathcal{T}\left[\gamma^0\gamma^\mu\right]\psi(\boldsymbol{x},t)\mathcal{T} = \\ &= \mathcal{T}\psi(\boldsymbol{x},t)^\dagger\mathcal{T}\left[\gamma^0\gamma^\mu\right]^*\mathcal{T}\psi(\boldsymbol{x},t)\mathcal{T} = \psi(\boldsymbol{x},-t)^\dagger T^\dagger\left[\gamma^0\gamma^\mu\right]^* T\psi(\boldsymbol{x},-t)\end{aligned} \tag{12.50}$$

from which we find:

$$\begin{aligned}\mu = 0:&\;\; T^\dagger T = 1 \\ \mu = i:&\;\; T^\dagger(\alpha^i)^* T = -\alpha^i\end{aligned} \tag{12.51}$$

where we have introduced the Dirac matrices, $\alpha^i = \gamma^0\gamma^i$. If we require, as usual, that $T^2 = 1$, the previous conditions give:

$$\begin{aligned}&T^\dagger T = T^{-1} = T \\ \mu = i:&\;\; T(\alpha^i)^* T = -\alpha^i.\end{aligned} \tag{12.52}$$

In the Pauli representation, $\alpha^{1,3}$ are real and $\alpha^2$ imaginary, therefore $T$ must anticommute with $\alpha^{1,3}$ and commute with $\alpha^2$. Thus, from (12.52) we find:

$$T = i\gamma^1\gamma^3 = \sigma_2 \tag{12.53}$$

The transformation with matrix $\sigma_2$ leaves unchanged $\gamma^0$, which is real, and changes the signs of $\gamma^i$, therefore:

$$T(\gamma^\mu)^* T = g^{\mu\mu}\gamma^\mu \tag{12.54}$$

From this, the transformation rules of the Dirac bilinears, which are formed from products of the gamma matrices, can immediately be found. We find:

$$\begin{aligned}&\mathcal{O}(\Gamma)((\boldsymbol{x},t) = \bar{\psi}(\boldsymbol{x},t)\Gamma\psi(\boldsymbol{x},t) \\ &\mathcal{T}\mathcal{O}(\Gamma)(\boldsymbol{x},t)\mathcal{T} = \mathcal{O}(T\Gamma^* T)(\boldsymbol{x},-t) = \eta_T\mathcal{O}(\Gamma)(\boldsymbol{x},-t)\end{aligned} \tag{12.55}$$

where:

$$\begin{aligned}
\eta_T &= +1 && \text{(S, P)} \\
\eta_T &= g^{\mu\mu} && \text{(V, A)} \\
\eta_T &= -g^{\mu\mu}g^{\nu\nu} && \text{(T).}
\end{aligned} \tag{12.56}$$

## 12.4 TRANSFORMATION OF THE STATES

We consider the transformation properties of the electron and positron states. We restrict ourselves to the case of a free particle fields which describe asymptotic "in" and "out" states. For the reader's convenience, we repeat the positive and negative energy solutions of the Dirac equation from Section 6.1.4. The spin is quantised in a fixed direction which we choose to be along the $z$-axis.

$$\begin{aligned}
u_s(p) &= \sqrt{\frac{E(p)+m}{2m}} \begin{pmatrix} \chi_s \\ \frac{\sigma\cdot\mathbf{p}}{E(p)+m}\chi_s \end{pmatrix}; \quad \sigma_3\chi_s = s\chi_s \;\; (s=\pm 1) \\
v_s(p) &= \sqrt{\frac{E(p)+m}{2m}} \begin{pmatrix} \frac{\sigma\cdot\mathbf{p}}{E(p)+m}\xi_s \\ \xi_s \end{pmatrix}; \quad \sigma_3\xi_s = s\xi_s \;\; (s=\mp 1).
\end{aligned} \tag{12.57}$$

From the properties of the Pauli matrices we find:

$$\begin{aligned}
\sigma_2(\sigma^i)^*\sigma_2 &= -\sigma^i \qquad (i=1,\ 2,\ 3) \\
\sigma_3(\sigma_2\xi_s^*) &= -\sigma_2(\sigma_3\xi_s)^* = \mp(\sigma_2\xi_s^*)
\end{aligned}$$

or

$$(\sigma_2\xi_s^*) = \xi_{-s} = \chi_s$$

and also

$$(\sigma_2\chi_s^*) = \chi_{-s} = \xi_s. \tag{12.58}$$

Parity. From (12.7) and (12.13) we find:

$$\mathcal{P}\psi(\boldsymbol{x},t)\mathcal{P} = \gamma^0\psi(-\boldsymbol{x},t). \tag{12.59}$$

- The left-hand side, expanded from the solutions to the Dirac equation, is:

$$\int \frac{d^3p}{(2\pi)^{3/2}} \sqrt{\frac{m}{E(p)}} \left[\mathcal{P}a_s(p)\mathcal{P}e^{-ipx}u_s(p) + \mathcal{P}b_s(p)^\dagger\mathcal{P}e^{+ipx}v_s(p)\right]. \tag{12.60}$$

- The right-hand side is:

$$\int \frac{d^3p}{(2\pi)^{3/2}} \sqrt{\frac{m}{E(p)}} \left[a_s(p)e^{-ipx_P}(\gamma^0 u_s)(p) + b_s(p)^\dagger e^{+ipx_P}(\gamma^0 v_s)(p)\right] \tag{12.61}$$

where

$$x_P = (-\boldsymbol{x}, t) \quad px_P = Et + \boldsymbol{x} \cdot \mathbf{p} = p_P x. \tag{12.62}$$

Changing the integration variable, $\mathbf{p} \to -\mathbf{p}$, the right-hand side becomes:

$$\int \frac{d^3p}{(2\pi)^{3/2}} \sqrt{\frac{m}{E(p)}} \left[a_s(p_P)e^{-ipx}(\gamma^0 u_s)(p_P) + b_s(p_P)^\dagger e^{+ipx}(\gamma^0 v_s)(p_P)\right]. \tag{12.63}$$

From (12.57) we obtain:

$$\gamma^0 u_s(p_P) = +u_s(p); \quad \gamma^0 v_s(p_P) = -v_s(p). \tag{12.64}$$

Finally, comparing the terms in the expansion on each side of (12.59), we find:

$$\mathcal{P} a_s(p) \mathcal{P} = +a_s(p_P); \quad \mathcal{P} b_s(p) \mathcal{P} = -b_s(p_P). \tag{12.65}$$

The absolute sign in these relations could be changed with a different definition of the sign of $P$. The important fact is the relative sign of the electron and its antiparticle:

*The electron and positron have opposite parity.*

Charge Conjugation. From (12.18) we find:

$$\mathcal{C}\psi(x)\mathcal{C} = i\gamma^2[\psi(x)^\dagger]. \tag{12.66}$$

We proceed as before:

- Right-hand side:

$$\int \frac{d^3p}{(2\pi)^{3/2}} \sqrt{\frac{m}{E(p)}} \times$$
$$\left[b_s(p)e^{-ipx}(i\gamma^2)_{\alpha\beta}[(v_s)_\beta(p)]^* + a_s(p)^\dagger e^{+ipx}(i\gamma^2)_{\alpha\beta}[(u_s)_\beta(p)]^*\right].$$

- Using (12.57) and (12.58) we find:

$$i\gamma_2 v_s(p)^* = \sqrt{\frac{E(p)+m}{2m}} \begin{pmatrix} 0 & i\sigma_2 \\ -i\sigma_2 & 0 \end{pmatrix} \begin{pmatrix} \frac{\sigma\cdot\mathbf{p}}{E(p)+m}\xi_s \\ \xi_s \end{pmatrix}^*$$

$$= \sqrt{\frac{E(p)+m}{2m}} \begin{pmatrix} i\sigma_2 \xi_s^* \\ -i\sigma_2 \frac{(\sigma\cdot\mathbf{p})^*}{E(p)+m}\xi_s^* \end{pmatrix} =$$

$$= i\sqrt{\frac{E(p)+m}{2m}} \begin{pmatrix} \chi_s \\ \frac{\sigma\cdot\mathbf{p}}{E(p)+m}\chi_s \end{pmatrix} = iu_s(p) \tag{12.67}$$

and, similarly:

$$i\gamma_2 u_s(p)^* = -iv_s(p). \tag{12.68}$$

Comparing the two sides, we obtain:

$$\mathcal{C}a_s(p)\mathcal{C} = ib_s(p); \quad \mathcal{C}b_s(p)\mathcal{C} = -ia_s(p). \tag{12.69}$$

Time Reversal. We start from (12.48):

$$\mathcal{T}\psi(\mathbf{x},t)\mathcal{T} = [\sigma_2\psi(\mathbf{x},-t)]. \tag{12.70}$$

- The left-hand side, recalling the antilinearity of $\mathcal{T}$, is:

$$\int \frac{d^3p}{(2\pi)^{3/2}} \sqrt{\frac{m}{E(p)}} \left[\mathcal{T}a_s(p)\mathcal{T}e^{+ipx}u_s(p)^* + \mathcal{T}b_s(p)^\dagger\mathcal{T}e^{-ipx}v_s(p)^*\right]. \tag{12.71}$$

- The right-hand side is:

$$\int \frac{d^3p}{(2\pi)^{3/2}} \sqrt{\frac{m}{E(p)}} \left[a_s(p)e^{-ipx_T}[\sigma_2 u_s(p)]^\dagger + b_s(p)^\dagger e^{+ipx_T}[\sigma_2 v_s(p)]^\dagger\right] \tag{12.72}$$

where:

$$x_T = (\mathbf{x},-t); \quad px_T = -Et - \mathbf{p}\cdot\boldsymbol{x} = p_T x. \tag{12.73}$$

If we change the integration variable, (12.72) becomes:

$$\int \frac{d^3p}{(2\pi)^{3/2}} \sqrt{\frac{m}{E(p)}} \left[a_s(p_T)e^{+ipx}[\sigma_2 u_s(p_T)]^\dagger + b_s(p_T)^\dagger e^{-ipx}[\sigma_2 v_s(p_T)]^\dagger\right]. \tag{12.74}$$

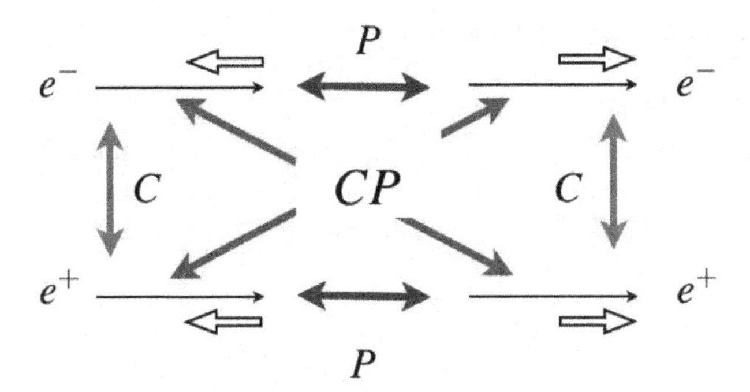


Figure 12.2 The Action of parity, $\mathcal{P}$, and charge conjugation, $\mathcal{C}$, on the electron and positron states with different helicity. The $\mathcal{T}$ transformation does not change the helicity since it changes both momentum and spin.

- We use (12.57) again, to obtain:

$$\begin{aligned}\{[\sigma_2 u_s(p_T)]^\dagger\} &= \{u_s(p_T)^\dagger \sigma_2\} \\ &= \sqrt{\frac{E(p)+m}{2m}} \left( \chi_s^\dagger, \quad \chi_s^\dagger \tfrac{-\sigma\cdot\mathbf{p}}{E(p)+m} \right) \begin{pmatrix} \sigma^2 & 0 \\ 0 & \sigma^2 \end{pmatrix} = \\ &= \sqrt{\frac{E(p)+m}{2m}} \left( \chi_s^\dagger \sigma^2, \quad \chi_s^\dagger \tfrac{-\sigma\cdot\mathbf{p}}{E(p)+m} \sigma^2 \right).\end{aligned} \tag{12.75}$$

- Using (12.58) we find:

$$\begin{aligned}(\chi_s^\dagger \sigma^2)_\alpha &= (\sigma^2 \chi_s^*)_\alpha = (\chi_{-s})_\alpha; \\ (\chi_s^\dagger \frac{-\sigma\cdot\mathbf{p}}{E(p)+m} \sigma^2)_\alpha &= (\sigma^2 \tfrac{-\sigma\cdot\mathbf{p}}{E(p)+m} \chi_s^*)_\alpha \\ &= (\tfrac{+\sigma\cdot\mathbf{p}}{E(p)+m} \chi_{-s})_\alpha\end{aligned} \tag{12.76}$$

or:

$$\{[\sigma_2 u_s(p_T)]^\dagger\}_\alpha = [u_{-s}(p)]^*_\alpha; \tag{12.77}$$

and similarly:

$$\{[\sigma_2 v_s(p_T)]^\dagger\}_\alpha = [v_{-s}(p)]^*_\alpha. \tag{12.78}$$

We can now compare the two sides, and find:

$$\mathcal{T} a_s(\mathbf{p}) \mathcal{T} = a_{-s}(-\mathbf{p}); \quad \mathcal{T} b_s(\mathbf{p}) \mathcal{T} = b_{-s}(-\mathbf{p}). \tag{12.79}$$

Fig. 12.2 shows the action of the three symmetries on the electron and positron states.

## 12.5 SOME APPLICATIONS

### 12.5.1 Furry's Theorem

This theorem, a consequence of the invariance of QED under $\mathcal{C}$, concerns Green's functions which *involve only* $A^\mu$ and states that:

*the Green's functions with an odd number of external photon lines are identically zero.*

The theorem can also be expressed in terms of $S$-matrix elements:

*the reactions* $n\gamma \to n'\gamma$ *vanish if* $n + n' =$ *an odd number.*

Proof. We use the invariance of the vacuum under $\mathcal{C}$ to obtain:

$$\begin{aligned} G^{\mu_1 \dots \mu_N}(x_1, \dots x_N) &= \langle 0|T\left[A^{\mu_1}(x_1) \dots A^{\mu_N}(x_N)\right]|0\rangle = \\ &= \langle 0|\mathcal{C}T\left[\mathcal{C}A^{\mu_1}(x_1)\mathcal{C}\mathcal{C} \dots \mathcal{C}A^{\mu_N}(x_N)\mathcal{C}\right]\mathcal{C}|0\rangle = \\ &= (-1)^N \langle 0|T\left[A^{\mu_1}(x_1) \dots A^{\mu_N}(x_N)\right]|0\rangle. \end{aligned} \tag{12.80}$$

Clearly, if $N$ is odd the amplitude is equal to its negative and therefore it must vanish.

### 12.5.2 Symmetries of Positronium

*Positronium* is a system formed by an electron and positron bound together by the Coulomb force. Its energy levels are well described, in the non-relativistic approximation, using four quantum numbers:

- the radial quantum number: $n = 1, 2, \dots,$
- the orbital angular momentum: $L = 0, 1, 2, \dots,$
- the total spin: $S = 0, 1,$
- the total angular momentum, $J$, which also takes integer values.

Positronium is similar in every way to the hydrogen atom, except that the *reduced mass*, $\mu$, is about half the reduced mass of hydrogen:

$$\begin{aligned} \mu &= \frac{m_1 m_2}{m_1 + m_2}; \\ \mu_{e^+e^-} &\simeq \frac{1}{2} m_e \simeq \frac{1}{2} \mu_H. \end{aligned} \tag{12.81}$$

Because the reduced mass determines the bound state energy levels, the positronium spectrum is scaled in energy by a factor two compared to that of hydrogen.

Positronium is formed each time a positron comes to rest in matter. At the end of its travel, the positron captures an electron from the surrounding atoms and forms an excited state of positronium. In contrast to hydrogen, positronium is not stable. Once the electron and positron have reached the ground state with $L = 0$, the annihilation probability is significant and the state decays into two or more photons (cf. Landau and Lifshitz [11] for the cross-section calculation).

For the ground state, there are two energy levels with $n = 1$, $L = 0$ and $S = 0$ or 1, known respectively as *parapositronium* and *orthopositronium.* The energy difference between them is very small because it results from magnetic interactions between the spins.

On the basis of the results from the previous section, we can determine the parity and charge conjugation properties of the positronium levels. Because these operations commute with the Hamiltonian, they provide good quantum numbers and determine the selection rules of the decays.

Positronium States. The states of positronium are linear superpositions of the states obtained by applying to the vacuum one creation operator for the electron and another for the positron. The coefficients of the superposition are given by the product of the spherical harmonics corresponding to angular momentum $L$, the Clebsch–Gordan coefficients coupling the electron and positron spins to obtain the total spin $S$ and the radial wave functions corresponding to the quantum number $n$:

$$|n, L, S, J\rangle = \int dp R^{(n,L,S,J)}(p) \int d\Omega_p Y^L_m(\hat{\boldsymbol{p}}) \times \sum_{s,s'} C(S, s_3|1/2, s; 1/2, s') a_s(\boldsymbol{p})^\dagger b_{s'}(-\boldsymbol{p})^\dagger |0\rangle \quad (12.82)$$

where $p = |\boldsymbol{p}|$ and $\hat{\boldsymbol{p}}$ is the unit vector of $\boldsymbol{p}$.

Parity of the Levels. We apply the $\mathcal{P}$ operator to the state (12.82) and use (12.65) and the invariance of the vacuum:

$$\begin{aligned}\mathcal{P}|n, L, S, J\rangle &= \int dp R^{(n,L,S,J)}(p) \int d\Omega_p Y^L_m(\hat{\boldsymbol{p}}) \times \\ &\quad \sum_{s,s'} C(S, s_3|1/2, s; 1/2, s')\ \mathcal{P} a_s(\boldsymbol{p})^\dagger \mathcal{P}\mathcal{P} b_{s'}(-\boldsymbol{p})^\dagger \mathcal{P}\mathcal{P}|0\rangle = \\ &= \int dp R^{(n,L,S,J)}(p) \int d\Omega_p Y^L_m(\hat{\boldsymbol{p}}) \times \\ &\quad \sum_{s,s'} C(S, s_3|1/2, s; 1/2, s')(-1) a_s(-\boldsymbol{p})^\dagger b_{s'}(+\boldsymbol{p})^\dagger |0\rangle \quad (12.83)\end{aligned}$$

where the minus sign originates in the opposite parities of electron and positron. Now we can let $\boldsymbol{p} \to -\boldsymbol{p}$ in the integral and use the parity of the spherical harmonics to obtain:

$$\mathcal{P}|n, L, S, J\rangle = (-1)^{L+1}. \quad (12.84)$$

Charge Conjugation of the Levels. We apply the $\mathcal{C}$ operator and use (12.69) and the invariance of the vacuum:

$$\mathcal{C}|n, L, S, J\rangle = \int dp R^{(n,L,S,J)}(p) \int d\Omega_p Y_m^L(\hat{\mathbf{p}}) \times \sum_{s,s'} \mathcal{C}C(S, s_3|1/2, s; 1/2, s')\, a_s(\mathbf{p})^\dagger \mathcal{C}\mathcal{C} b_{s'}(-\mathbf{p})^\dagger \mathcal{C}\mathcal{C}|0\rangle =$$
$$= \int dp R^{(n,L,S,J)}(p) \int d\Omega_p Y_m^L(\hat{\mathbf{p}}) \times \sum_{s,s'} C(S, s_3|1/2, s; 1/2, s') b_s(\mathbf{p})^\dagger a_{s'}(-\mathbf{p})^\dagger |0\rangle. \quad (12.85)$$

To return to the starting expression we must (i) exchange $a$ with $b$, (ii) let $\mathbf{p} \to -\mathbf{p}$, (iii) exchange $s$ with $s'$.

In the latter operation, we recall that the Clebsch–Gordan coefficients for two spin $\frac{1}{2}$ combinations are *symmetric* for the exchange of $s$ with $s'$ in the case $S = 1$, while they are *antisymmetric* for $S = 0$.

The three operations introduce a factor (i) $-1$; (ii) $(-1)^L$; (iii) $(-1)^{S+1}$ respectively and therefore an overall factor $\eta_C$:

$$\mathcal{C}|n, L, S, J\rangle = \eta_C|n, L, S, J\rangle; \; \eta_C = (-1)(-1)^L(-1)^{S+1} = (-1)^{L+S}. \quad (12.86)$$

Selection Rules. For the positronium ground states we obtain:

$$\text{parapositronium}: \quad J^{PC} = 0^{-+}$$
$$\text{orthopositronium}: \quad J^{PC} = 1^{--} \quad (12.87)$$

As we saw from Furry's theorem, a state with $N$ photons has $C = \pm 1$ according to whether $N$ is even or odd. Therefore the selection rules are:

$$\text{parapositronium} \to 2\gamma; \; \not\to 3\gamma$$
$$\text{orthopositronium} \not\to 2\gamma; \; \to 3\gamma \quad (12.88)$$

The annihilation amplitude into two photons is of order $e^2$, and for three photons of order $e^3$ and the corresponding probabilities of order $\alpha^2$ and $\alpha^3$ respectively, with $\alpha \simeq 1/137$. We therefore expect two positronium components to form in matter, one with a short half-life, parapositronium, and one with a considerably longer half-life, orthopositronium.

The observed values agree well with this rule. They are[1]:

$$\Gamma(para \to 2\gamma)_{expt} = 7990.9(1.7)\ \mu\text{s}^{-1} \quad (12.89)$$
$$\Gamma(ortho \to 3\gamma)_{expt} = 7.0404(10)(8)\ \mu\text{s}^{-1} \quad (12.90)$$

[1] Numbers in parentheses represent the error on the last digits of the quantity; when two numbers are reported, the first is the statistical and the second the systematic error.

a factor of about a thousand between the two decay probabilities, in agreement with the selection rule (12.87) and the QED predictions.

## 12.6 THE CPT THEOREM

We summarise the transformations of the Dirac covariants under $\mathcal{P}$, $\mathcal{C}$ and $\mathcal{T}$, generalising the considerations of sections 12.1, 12.2 and 12.3 to the case of bilinears constructed from two different fermion fields, $\psi_a$ and $\psi_b$. In the case of the scalar density, for example, we find:

$$\begin{aligned}
&\mathcal{P}S_{ab}(\mathbf{x},t)\mathcal{P} = \mathcal{P}\bar{\psi}_a(\mathbf{x},t)\psi_b(\mathbf{x},t)\mathcal{P} = \eta_P(S)S_{ab}(-\mathbf{x},t);\\
&\mathcal{C}S_{ab}(\mathbf{x},t)\mathcal{C} = \eta_C(S)S_{ba}(\mathbf{x},t);\\
&\mathcal{T}S_{ab}(\mathbf{x},t)\mathcal{T} = \eta_T(S)S_{ab}(\mathbf{x},-t)
\end{aligned} \tag{12.91}$$

where $\eta$ are the $\pm$ signs characteristic of the transformation and the specific covariant. In Table 12.1 the values of $\eta$ for the three transformations and for the combined $\theta = CPT$ operation are listed.

The $CPT$ operation is clearly represented by an antilinear operator and acts according to the simple rule:

$$\theta\,[g_S S_{ab}(x)]\,\theta^\dagger = (-1)^N (g_S)^* (S_{ab}(-x))^\dagger \tag{12.92}$$

where $g_S$ is a complex coefficient. $N$ is the number of Lorentz indices of the covariant.

The same rule applies to the vector potential ($N = 1$), and to the Maxwell tensor ($N = 2$) and extends unchanged to operators of scalar or vector fields. In the case of complex fields, the antilinear nature of $\theta$ implies, for example, that

$$\theta\,(\phi_1 + i\phi_2)\,(x)\theta^\dagger = (\phi_1 - i\phi_2)(-x) \tag{12.93}$$

Therefore, for these operators also, $\theta$ implies hermitian conjugation.

So as not to forget .... It is useful to derive the results in Table 12.1 directly from the transformations of the fields under $CPT$. Collecting the formulae (12.59), (12.66) and (12.70) and the corresponding hermitian conjugates, we find:

$$(\mathcal{CPT})\psi(x)(\mathcal{TPC}) = \theta\psi(x)\theta^\dagger = \sigma_2\gamma^0(i\gamma^2)\psi^\dagger(-x) = i\gamma_5\psi^\dagger(-x) \tag{12.94}$$

and, for the hermitian conjugate field:

$$\theta\psi^\dagger(x)\theta^\dagger = -i\gamma_5\psi(-x). \tag{12.95}$$

Table 12.1 Summary of the properties of the Dirac covariants and of the electromagnetic field under $\mathcal{P}$, $\mathcal{C}$, $\mathcal{T}$ transformations, and $\theta = CPT$.

| | S | P | V | A | T | $A^\mu$ | $F^{\mu\nu}$ |
|---|---|---|---|---|---|---|---|
| $\eta_P$ | +1 | - 1 | $g^{\mu\mu}$ | $-g^{\mu\mu}$ | $g^{\mu\mu}\ g^{\nu\nu}$ | $g^{\mu\mu}$ | $g^{\mu\mu}g^{\nu\nu}$ |
| $\eta_C$ | +1 | + 1 | -1 | +1 | -1 | -1 | -1 |
| $\eta_T$ | +1 | - 1 | $g^{\mu\mu}$ | $+g^{\mu\mu}$ | $-g^{\mu\mu}\ g^{\nu\nu}$ | $g^{\mu\mu}$ | $+g^{\mu\mu}g^{\nu\nu}$ |
| $\eta_{CPT}$ | +1 | + 1 | -1 | -1 | +1 | -1 | -1 |

For the general bilinear, we therefore find:

$$\begin{aligned}\theta\bar{\psi}_a\Gamma\psi_b\theta^\dagger &= \theta\psi_a^\dagger(x)\theta^\dagger(\gamma^0\Gamma)^*\theta\psi_b(x)\theta^\dagger = [\gamma_5\psi_a(-x)](\gamma^0\Gamma)^*\gamma_5\psi_b^\dagger(-x) = \\ &= -\psi_b^\dagger(-x)\gamma_5(\gamma^0\Gamma)^\dagger\gamma_5\psi_a(-x) = +\left[\bar{\psi}_b\gamma^0\gamma_5\Gamma^\dagger\gamma_5\gamma^0\psi_a\right](-x) \end{aligned} \quad (12.96)$$

where we have repeatedly made use of the fact that $\gamma_5$ is hermitian and anticommutes with $\gamma^0$.

Combination with $\gamma_5$ produces a minus sign for every factor $\gamma^\mu$ in $\Gamma$, therefore:

$$\begin{aligned}\gamma^0\left(\gamma_5\Gamma^\dagger\gamma_5\right)\gamma^0 &= \gamma^0\left(\gamma_5\Gamma\gamma_5\right)^\dagger\gamma^0 = \\ &= (-1)^N\gamma^0\Gamma^\dagger\gamma^0 = (-1)^N\Gamma \end{aligned} \quad (12.97)$$

where $N$ is the number of vector indices of $\Gamma$. Finally:

$$\theta\left[\bar{\psi}_a\Gamma\psi_b(x)\right]\theta^\dagger = (-1)^N\left[\bar{\psi}_b\Gamma\psi_a(-x)\right] = (-1)^N\left[\bar{\psi}_a\Gamma\psi_b(-x)\right]^\dagger \quad (12.98)$$

as given in Table 12.1.

The CPT Theorem. The $(-1)^N$ factor applies to the *total inversion* of the coordinates:

$$TI: \quad x^\mu \to -x^\mu. \quad (12.99)$$

In a four-dimensional Euclidean space, inversion of the axes has a determinant equal to one and is continuously connected to the identity transformation. In Euclidean space, inversion is a *proper* transformation and therefore a necessary symmetry. This is not true in Minkowski space, in which proper transformations must also have $\Lambda^{00} > 0$, a condition which is not satisfied by TI.

However, as we will see in the discussion of Feynman integrals of [14], quantum theory in Minkowski space is the analytic continuation of the theory defined with a complex time, and there is no obstacle to continue to Minkowski space from purely imaginary time, i.e. from the theory in the four-dimensional Euclidean space. This is the origin of the *CPT Theorem* which, under very general conditions, states that the operation $\theta$, total inversion of the axes supplemented with the operation of hermitian conjugation, is an exact symmetry of any relativistic quantum field theory.

The $CPT$ theorem is due to Luders and Pauli [19, 20]. We will give a demonstration very close to the one of Bjorken and Drell [21]. Subsequently, we will illustrate the most noteworthy consequences of the $CPT$ theorem.

We consider a theory described by a Lagrangian density $\mathcal{L}(x)$. The conditions characterising a relativistic quantum theory, the subject to which the Lagrangian is to be constructed, are as follows.

The Lagrangian must be:

- hermitian,
- a local function of the fields and their derivatives, up to finite order, calculated at the same point,
- a boson operator; fermion fields can appear in even numbers and they can therefore be organised into Dirac bilinears, $\bar{\psi}_a \Gamma \psi_b$, where $\psi_a$ and $\psi_b$ are fields associated with different types of fermion (e.g. electron and neutrino),
- invariant under proper Lorentz transformations,
- a normal product of fields.

Under these conditions, we show that it must necessarily be true that:

$$\theta \mathcal{L}(x) \theta^\dagger = \mathcal{L}(-x) \tag{12.100}$$

from which it follows that the action is invariant under $CPT$:

$$\theta S \theta^\dagger = \int d^4x \; \theta \mathcal{L}(x) \theta^\dagger = \int d^4x \; \mathcal{L}(-x) = \int d^4(-x) \; \mathcal{L}(-x) = S \tag{12.101}$$

and therefore $CPT$ is an exact symmetry.

Proof. The general form of a Lagrangian density which satisfies the conditions above can be written in the following way:

$$\begin{aligned} \mathcal{L}(x) &= \sum_i c_i \; \mathcal{O}_i(x); \\ \mathcal{O}_i(x) &=: \; \ldots A^\mu(x) \ldots (\bar{\psi}_a \Gamma \psi_b)(x) \ldots \partial^\nu \ldots \; : \end{aligned} \tag{12.102}$$

where $c_i$ are complex coefficients. We now apply the $CPT$ operation. On the basis of Table 12.1 and the antilinearity of $\theta$ we obtain:

$$\theta \mathcal{L}(x) \theta^\dagger = \sum_i (c_i)^* (-1)^{N_{tot}} : \; \ldots A^\mu(-x) \ldots \left[ (\bar{\psi}_a \Gamma \psi_b)(-x) \right]^\dagger \ldots \partial^\nu \ldots \; : \tag{12.103}$$

where $N_{tot}$ is the total number of Lorentz indices which appear in the equation (12.102).

Inside the normal product, we can commute the boson operators and put them in the opposite order to which they appear in (12.102), which means we can rewrite (12.103) as:

$$\theta \mathcal{L}(x)\theta^{\dagger} = \sum_i (-1)^{N_{tot}} \left[c_i \ \mathcal{O}_i(-x)\right]^{\dagger}. \tag{12.104}$$

If the Lagrangian should be Lorentz invariant, the Lorentz indices must be summed on invariant tensors. In a four-dimensional space-time, there are only two invariant operations: contraction with $g_{\mu\nu}$ and contraction with $\epsilon_{\mu\nu\rho\sigma}$ (cf. Section 3.3). Both of these operations reduce the free indices to an even number. Therefore, in order that (12.102) should be invariant (i.e. does not contain free indices), the number of vector indices that appear must be even, or $(-1)^{N_{tot}}{=}{+}1$. We conclude therefore that:

$$\theta \mathcal{L}(x)\theta^{\dagger} = \sum_i \left[c_i \ \mathcal{O}_i(-x)\right]^{\dagger} = \left[\mathcal{L}(-x)\right]^{\dagger} = \mathcal{L}(-x) \tag{12.105}$$

where the final step follows from the fact that $\mathcal{L}$ is Hermitian.

Fermi Interaction. It is interesting to apply the considerations just described to the Fermi interaction for neutron $\beta$ decay, which we wrote as (Section 9.2):

$$\begin{aligned}\mathcal{L}_F = &-\frac{G}{\sqrt{2}}\left[\bar{\psi}_P\left(\gamma^\mu + \frac{g_A}{g_V}\gamma^\mu\gamma_5\right)\psi_N \ \bar{\psi}_e\left(\gamma^\mu - \gamma^\mu\gamma_5\right)\psi_{\nu_e}\right] + \\ &-\frac{G}{\sqrt{2}}\left[\bar{\psi}_N\left(\gamma^\mu + (\frac{g_A}{g_V})^*\gamma^\mu\gamma_5\right)\psi_P \ \bar{\psi}_{\nu_e}\left(\gamma^\mu - \gamma^\mu\gamma_5\right)\psi_e\right].\end{aligned} \tag{12.106}$$

We have written explicitly the hermitian conjugate of the first term, necessary to make the Lagrangian real.

The interaction (12.106) clearly is NOT invariant under parity, since it is produced by superpositions of polar and axial vectors, which change relative sign under parity. For example:

$$\mathcal{P}\bar{\psi}_P\left(\gamma^\mu + \frac{g_A}{g_V}\gamma^\mu\gamma_5\right)\psi_N(\boldsymbol{x},t)\mathcal{P} = g^{\mu\mu} \ \bar{\psi}_P\left(\gamma^\mu - \frac{g_A}{g_V}\gamma^\mu\gamma_5\right)\psi_N(-\boldsymbol{x},t). \tag{12.107}$$

Instead, the $CP$ transformation acts in the same way on vector and axial currents:

$$\begin{aligned}&\mathcal{C}\bar{\psi}_P\left(\gamma^\mu + \frac{g_A}{g_V}\gamma^\mu\gamma_5\right)\psi_N\mathcal{C} = (-g^{\mu\mu})\bar{\psi}_N\left(\gamma^\mu + \frac{g_A}{g_V}\gamma^\mu\gamma_5\right)\psi_P \\ &\mathcal{C}\bar{\psi}_e\left(\gamma^\mu - \gamma^\mu\gamma_5\right)\psi_{\nu_e}\mathcal{C} = (-g^{\mu\mu})\bar{\psi}_{\nu_e}\left(\gamma^\mu - \gamma^\mu\gamma_5\right)\psi_e.\end{aligned} \tag{12.108}$$

Invariance under $CP$ can be obtained if the first term of (12.106) transforms into the second, which is its hermitian conjugate. This, in turn, requires that:

$$g_A/g_V = \text{real} \quad \text{(invariance for CP).} \tag{12.109}$$

However, if we apply $\theta$ to the first term of (12.106), $g_A/g_V$ also changes to its complex conjugate and the interaction remains invariant for any value, real or complex, of the constant which appears in (12.106).

### 12.6.1 Equality of Particle and Antiparticle Masses

This is the most direct consequence of the $CPT$ theorem. From the invariance of the Lagrangian, equation (12.100), that of the Hamiltonian easily follows:

$$\theta \boldsymbol{H} \theta^\dagger = \boldsymbol{H}. \tag{12.110}$$

Moreover, from Table 12.1 it follows that $\theta$ changes signs of 3-vectors, like momentum, and changes the sign of each conserved charge that is present in our theory (e.g. electric charge):

$$\theta \boldsymbol{P} \theta^\dagger = -\boldsymbol{P}; \quad \theta Q \theta^\dagger = -Q \tag{12.111}$$

Conversely, angular momentum remains unchanged[2]

$$\theta \boldsymbol{J} \theta^\dagger = \boldsymbol{J} \tag{12.112}$$

where $\boldsymbol{P}$, $\boldsymbol{J}$, $Q$ are the momentum, angular momentum and charge operators.

The Case $M \neq 0$. We consider the case of a particle with non-zero mass. We can choose to be in the rest frame of the particle, in which $\boldsymbol{P} = 0$ and the angular momentum coincides with the spin. In general, other than the mass and the spin, the state is characterised by the value of the conserved charge, which we denote with $q$. Therefore we write the ket which represents the state as:

$$|A\rangle = |M, \boldsymbol{P} = 0, s_z; q\rangle. \tag{12.113}$$

Taking account of (12.111), the state $\theta|A\rangle$ must have the same mass, the same value of $\boldsymbol{s}^2 = s(s+1)$ and of the spin component, but opposite electric charge:

$$\theta|A\rangle = |M, \boldsymbol{P} = 0, s_z; -q\rangle. \tag{12.114}$$

[2]Classically, $\boldsymbol{J} = \boldsymbol{x} \times \boldsymbol{v}$ does not change sign under total inversion since $\boldsymbol{x}$ and $\boldsymbol{v}$ both change sign.

We must distinguish two cases, according to which $q \neq 0$ or $q = 0$.

If $q \neq 0$, as for example it is for the electron, the $CPT$-conjugate state, (12.114), clearly does not coincide with the original state. Another particle of equal mass and spin but the opposite charge must exist: the positron. Since $\theta^2 = 1$, the relationship between particle and antiparticle is perfectly symmetric; the electron is the antiparticle of the positron.

If instead $q = 0$, the conjugate state (12.114) is identical to the original. Naturally, invariance under rotation of the Hamiltonian in the rest system requires the presence of all $2s + 1$ states associated with spin $s$. In this case, we have a spin $s$ particle, absolutely neutral in the sense that it cannot be distinguished from the first particle. This, for example, is the case of the (spin 0) $\pi^0$ meson or of (spin 1) $\rho^0$ and $\omega^0$ mesons.

The Case $M = 0$. In this case, we choose to be in a system in which the particle has its momentum along the $z$-axis and has helicity $\lambda$. We write the state as:

$$|A\rangle = |P_3, \lambda; q\rangle. \tag{12.115}$$

Lorentz invariance, which reduces to invariance under rotation around the $z$-axis, does not require that there should be states other than those in (12.115).

The $CPT$-conjugate state must have the opposite momentum and the same spin components, or opposite helicity:

$$\theta|A\rangle = |-P_3, -\lambda; -q\rangle. \tag{12.116}$$

Here again, we distinguish two cases:

If $q \neq 0$, the conjugate state represents an antiparticle of the original particle, with opposite charge, and opposite helicity. This is the case for particles of the two-component theory (the Weyl neutrino, see Chapter 13). We have a neutrino state with negative helicity and another state of an antineutrino with positive helicity, distinguished by the value of a conserved charge, the lepton number. Other states are not necessary to have a relativistic and $CPT$-invariant theory.

If $q = 0$, we have an absolutely neutral particle which, on the basis of (12.116), must be present with *two helicity states* equal to $\pm\lambda$. This is the case of the photon. On the basis of pure relativistic invariance, it could have only a single state, for example with helicity equal to one. $CPT$-invariance requires both observed states of the photon, with helicity $\pm 1$.

Experimental Verification. Beyond the equality of their masses, it can easily be seen that particle and antiparticle, if unstable, must have opposite magnetic moments and equal half-lives. These relationships have been experimentally tested, with extreme precision in a few fortunate cases. We recall the most important (see [**?**]).

- the mass of the antiproton coincides with that of the proton within a relative precision of $10^{-8}$,
- the masses and half-lives of $K^0$ and anti-$K^0$ mesons agree to within a few parts in $10^{-18}$,
- the electron and positron have equal masses within one part in $10^{-8}$, and their magnetic moments, apart from opposite signs, agree within one part in $10^{-12}$,
- the magnetic moments of muon and antimuon agree within two parts in $10^{-8}$.

The observation of any violation of the $CPT$ theorem would imply the necessity for improvement of the relativistic quantum field theory paradigm and would have enormous conceptual importance.

## 12.7 PROBLEMS FOR CHAPTER 12

### Sect. 12.1

1. The parity transformation $\mathcal{P}$ inverts the signs of the space components leaving the time component unchanged:

$$\mathcal{P}: \quad \boldsymbol{x},\ t \to -\boldsymbol{x},\ t$$

with $\mathcal{P}\mathcal{P} = \mathbb{1}$. Requiring the Dirac action to be invariant under parity, namely

$$\mathcal{L}(x) = \bar{\psi(x)}(i\partial\!\!\!/ - m)\psi(x)$$
$$\mathcal{P}\mathcal{L}(\boldsymbol{x}, t)\mathcal{P} = \mathcal{L}(-\boldsymbol{x}, t)$$

determine the form of the 4×4 matrix $P$ transforming the field $\psi(x)$ according to

$$\mathcal{P}\psi_\alpha(\boldsymbol{x}, t)\mathcal{P} = P_{\alpha\beta}\psi_\beta(-\boldsymbol{x}, t)$$

where $\alpha$ and $\beta$ are spinor indices.
Hint: Consider the term $\bar{\psi}(x)i\partial\!\!\!/\psi(x) = i\bar{\psi(x)}\gamma_\mu\partial^\mu\psi(x)$, treating separately the cases $\mu = 0$ and $\mu = 1,\ 2,\ 3$.

2. Using the result of the previous problem,

   – perform the parity transformation of the Dirac spinors

$$u_s(\boldsymbol{p}) = \sqrt{\frac{E+m}{2m}} \begin{pmatrix} \chi_s \\ \frac{(\boldsymbol{\sigma}\cdot\boldsymbol{p})}{E+m}\chi_s \end{pmatrix}, \quad v_s(\boldsymbol{p}) = \sqrt{\frac{E+m}{2m}} \begin{pmatrix} \frac{(\boldsymbol{\sigma}\cdot\boldsymbol{p})}{E+m}\xi_s \\ \xi_s \end{pmatrix}$$

   $\chi_s$ e $\xi_s$ being Pauli spinors satisfying the relations

$$\sigma_3\chi_s = s\chi_s, \quad (s = \pm 1)$$
$$\sigma_3\xi_s = s\xi_s, \quad (s = \mp 1)\ ;$$

  - verify that $u_s(\mathbf{0})$ and $v_s(\mathbf{0})$ have opposite parity.

**3.** Determine, from the above arguments, the parity of a fermion antifermion pair in $S$-wave, e.g. an electron-positron bound state (positronium) with orbital angular momenum $L = 0$.

**4.** Show that the Weyl equation for the two-component neutrino, eq. (6.171), can be derived from the Lagrangian density

$$\mathcal{L}_{\text{Weyl}}(x) = i\psi^\dagger \left( \frac{\partial}{\partial x^0} - \boldsymbol{\nabla} \cdot \boldsymbol{\sigma} \right) \psi \, .$$

**5.** Show that the Weyl Lagrangian is NOT invariant under parity. On this basis, Pauli rejected Weyl's theory (parity violation had not beeen discovered yet).
Hint: the matrix $P$ should obey $P^2 = 1$ and $P\boldsymbol{\sigma}P = -\boldsymbol{\sigma}$. Does such a matrix exist?

Sect. 12.2

**1.** Define the action of parity and charge conjugation on the two-component Weyl field according to

$$\mathcal{P}\psi(\boldsymbol{x}, t)\mathcal{P} = \sigma_2\psi(-\boldsymbol{x}, t); \;\; \mathcal{C}\psi(\boldsymbol{x}, t)\mathcal{C} = \psi(\boldsymbol{x}, t)^\dagger \, . \qquad (12.117)$$

Prove that the Weyl Lagrangian, $\mathcal{L}_{\text{Weyl}}$ of problem **4** above, is invariant under $\mathcal{PC}$.

Hints:

  - require only invariance of the action, so that derivatives in the Lagrangian density can be freely integrated by parts;
  - use the identity: $\sigma_2\boldsymbol{\sigma}^T\sigma_2 = -\boldsymbol{\sigma}$.

More about Weyl neutrinos in Sect. 13.1.

CHAPTER 13

# WEYL AND MAJORANA NEUTRINOS

We have seen that Dirac spinors transform according to a reducible representation of the Lorentz transformations. On the basis of transformations of $L_{+}^{\uparrow}$ only it should therefore be possible to find *smaller*, truly irreducible representations to describe a spin $\frac{1}{2}$ particle. This reduction leads to two types of theory, of the Weyl neutrino and the Majorana neutrino.

As we will see, the two theories agree for zero mass fermions but differ for particles of non-zero mass giving rise to the Dirac and Majorana theories as physically distinct alternatives.

It should be said immediately that neither of these theories can apply to the electron, or to the proton or neutron. For these particles, the presence of a conserved vector current associated with an electric charge or baryon number makes the Dirac structure essential. The question is open in the case of the neutrino.

The recent discovery of very small neutrino masses has reopened the question of which of the Dirac or Majorana theories is the better candidate to describe the properties of these particles.

## 13.1 THE WEYL NEUTRINO

We consider the Dirac equation in the limit of zero mass:

$$i\partial\!\!\!/\psi = 0. \tag{13.1}$$

Equation (13.1) allows an invariant operator represented by the $\gamma_5$ matrix. If $\psi$ is a solution of (13.1), then $\gamma_5\psi$ also satisfies the equation. Therefore

DOI: 10.1201/9781003436263-13

we can separate the solutions into two invariant sub-spaces by means of the projection operators:

$$a^{(\pm)} = \frac{1 \pm \gamma_5}{2}. \tag{13.2}$$

We define:

$$\psi_L = a^{(-)}\psi; \qquad \psi_R = a^{(+)}\psi. \tag{13.3}$$

If, for example, $\psi_R = 0$ at time $t_0$, it will remain so at later times. The field therefore divides into two irreducible and independent components.

In the Pauli representation of the $\gamma$ matrices:

$$\gamma_5 = \begin{pmatrix} 0 & 1 \\ 1 & 0 \end{pmatrix} \tag{13.4}$$

the eigenvectors of $\gamma_5$ with eigenvalue $h = \pm 1$ have the form:

$$\psi = \begin{pmatrix} \chi \\ h\chi \end{pmatrix}. \tag{13.5}$$

The Dirac equation with $m = 0$ gives rise to two possible equations for the two-dimensional spinors:

$$i\frac{\partial}{\partial t}\chi = \pm(-i\nabla \cdot \sigma)\chi. \tag{13.6}$$

Equation (13.6) is known as the *Weyl equation* and describes a particle of zero mass (because it is compatible with the equation $\Box\chi = 0$) and spin $\frac{1}{2}$[1]. The Weyl equation is used to describe a massless neutrino, which we will return to later.

We examine the solutions to the massless Dirac equation, (13.1), more closely. The spinors with positive energy and momentum $p$ along the $z$-axis take the form:

$$u_+(p) = \frac{1}{\sqrt{2}}\begin{pmatrix} \chi^+ \\ \chi^+ \end{pmatrix}; \qquad u_-(p) = \frac{1}{\sqrt{2}}\begin{pmatrix} \chi^- \\ -\chi^- \end{pmatrix};$$

$$\chi^+ = \begin{pmatrix} 1 \\ 0 \end{pmatrix}; \qquad \chi^- = \begin{pmatrix} 0 \\ 1 \end{pmatrix}. \tag{13.7}$$

We have used the appropriate normalisation for massless spinors in (13.7), i.e.:

$$u_s(p)^\dagger u_r(p) = \delta_{rs}. \tag{13.8}$$

It should be noted that (cf. Section 6.1.4) $\bar{u}(p)u(p) = m/E \to 0$ for $m = 0$.

[1]If we repeat the Dirac construction (Section 6.1) starting directly with $m = 0$, we must introduce only three anticommuting matrices and therefore the solution $\alpha = \sigma$ is acceptable, the minimum dimensionality of the spinor is 2 and instead of the Dirac equation we obtain the Weyl equation.

The spinors with negative energy and momentum $-p$ along the $z$-axis are instead:

$$u_{+}^{(E<0)}(-p) = v_{-}(p) = \frac{1}{\sqrt{2}} \begin{pmatrix} \chi^{+} \\ \chi^{+} \end{pmatrix};$$

$$u_{-}^{(E<0)}(-p) = v_{+}(p) = \frac{1}{\sqrt{2}} \begin{pmatrix} -\chi^{-} \\ \chi^{-} \end{pmatrix}. \tag{13.9}$$

Both in (13.7) and (13.9) the $\pm$ superscripts denote the $\sigma_3$ eigenvalue.

The spinors $u_{+}(p)$ and $v_{-}(p)$ are eigenvectors of $\gamma_5$ with eigenvalues $+1$, while $u_{-}(p)$ and $v_{+}(p)$ belong to the $-1$ eigenvalue.

The exchange of signs between $u$ and $v$ in (13.9) arises from the fact that, in the hole theory, the $u_{\pm}^{(E<0)}(-p)$ spinors are associated with the destruction of a neutrino in a negative energy state with momentum $-p$ and spin $\pm\frac{1}{2}$ along the $z$-axis. This corresponds to the creation of a *hole* (an antineutrino) with momentum $p$ and spin $\mp\frac{1}{2}$, described by the spinor $v(p)$.

The same ideas can be translated into the language of quantised fields as follows. We introduce left-handed, $L$, and right-handed, $R$, fields, defined in (13.3):

$$\psi_L = \sum_{\mathbf{p}} \frac{1}{\sqrt{V}} \left[ a_{-}(\mathbf{p}) u_{-}(\mathbf{p}) e^{-i(px)} + (b_{+}(\mathbf{p}))^{\dagger} v_{+}(\mathbf{p}) e^{+i(px)} \right]$$

$$\psi_R = \sum_{\mathbf{p}} \frac{1}{\sqrt{V}} \left[ a_{+}(\mathbf{p}) u_{+}(\mathbf{p}) e^{-i(px)} + (b_{-}(\mathbf{p}))^{\dagger} v_{-}(\mathbf{p}) e^{+i(px)} \right]. \tag{13.10}$$

From (13.10) it follows that:

- the field $\psi_L$ destroys a neutrino with helicity $-\frac{1}{2}$ and creates an antineutrino with helicity $+\frac{1}{2}$,
- the field $\psi_R$ destroys a neutrino with helicity $+\frac{1}{2}$ and creates an antineutrino with helicity $-\frac{1}{2}$.

The same conclusion is reached by considering the matrix elements of the invariant density $\bar{\psi}(x)\psi(x)$. We write:

$$\psi(x) = \psi(x)_R + \psi(x)_L. \tag{13.11}$$

If we define:

$$\bar{\psi}_L = (a^{-}\psi_L)^{\dagger}\gamma^0 = (\psi_L)^{\dagger} a^{-}\gamma^0 = \bar{\psi}_L a^{+}, \tag{13.12}$$

we see that $\bar{\psi}_L$ should be multiplied by $\psi_R$ and $\bar{\psi}_R$ by $\psi_L$. Thus, we find:

$$\bar{\psi}(x)\psi(x) = \bar{\psi}_L\psi_R + \bar{\psi}_R\psi_L. \tag{13.13}$$

The scalar density has non-zero matrix elements between the vacuum state and the state with a neutrino-antineutrino pair. To calculate these matrix elements, we must examine the spinors with momentum $-p$:

$$
\begin{aligned}
u_\pm(-p) &= \frac{1}{\sqrt{2}}\begin{pmatrix} \chi^\mp \\ \pm\chi^\mp \end{pmatrix}; \\
v_\pm(-p) &= \frac{1}{\sqrt{2}}\begin{pmatrix} \mp\chi^\pm \\ \chi^\pm \end{pmatrix};
\end{aligned}
\tag{13.14}
$$

and helicity $\pm 1$ (the two-dimensional spinors always correspond to the eigenstates of $\sigma_3$).

Keeping only the terms which give matrix elements different from zero, we have:

$$
\begin{aligned}
\langle \nu(p)\bar{\nu}(-p)|\bar{\psi}_R\psi_L|0\rangle &= \bar{u}_+(p)v_+(-p)e^{-i(px)+i(p'x)}; \\
\langle \nu(p)\bar{\nu}(-p)|\bar{\psi}_L\psi_R|0\rangle &= \bar{u}_-(p)v_-(-p)e^{-i(px)+i(p'x)}.
\end{aligned}
\tag{13.15}
$$

The neutrino and antineutrino are created with the same helicity, in agreement with the fact that the state must have zero angular momentum, and must also conform with the rules described above. Fig. 13.1 summarises the states created from the vacuum by the scalar density for each chirality, and those created by the vector density. We leave to the reader the task of showing that the only states which have matrix elements of the vector current different from zero are those illustrated in the figure.

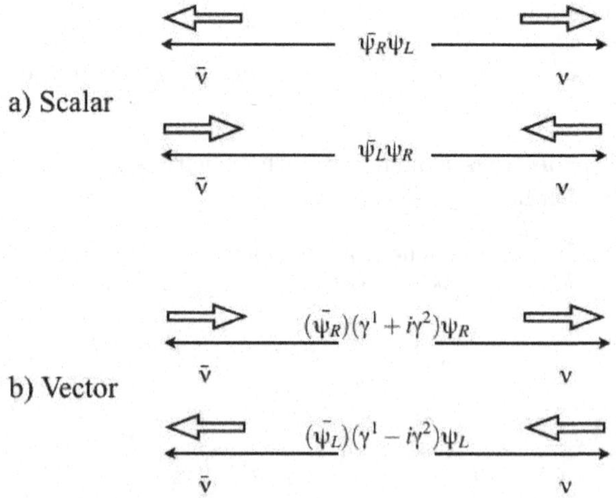

Figure 13.1 States created by the scalar and vector densities acting on the vacuum.

## 13.2 THE MAJORANA NEUTRINO

In the Pauli representation of the $\gamma$ matrices, the Dirac equation (7.23) is a complex equation; if $\psi$ is real at a time $t_0$, in general it will develop an imaginary part at later times. However, a symmetry of the Lagrangian (7.21) exists, *charge conjugation*, which essentially exchanges $\psi$ with $\psi^*$. Using this symmetry we can reduce the Dirac field to two independent components, as pointed out by Majorana [22] in 1937.

The most direct way of studying the question is to start from the fact that representations exist (called Majorana representations, MR in the following), in which all the Dirac matrices are *purely imaginary*.

In general, it is not necessary to know the explicit form of the gamma matrices in the Majorana representation. We give an example here to reassure the reader that they actually do exist. Starting from the Pauli representation, and using (6.25) with $S = (1 + \gamma^2)/2$, we obtain:

$$\widetilde{\gamma}^0 = \alpha_2;\ \widetilde{\gamma}^1 = -i\Sigma^3;\ \widetilde{\gamma}^2 = \gamma^2;\ \widetilde{\gamma}^3 = i\Sigma^1$$
$$\widetilde{\gamma}_5 = \gamma^2\gamma^5 = \begin{pmatrix} \sigma_2 & 0 \\ 0 & -\sigma_2 \end{pmatrix}. \tag{13.16}$$

We note that $\gamma_5$ is also imaginary (and therefore antisymmetric). All other realisations of MR are obtained by applying (6.25) again with $S$ real.

In the Majorana representation, the Dirac equation (7.23) is an equation with real coefficients and therefore allows purely real solutions. For a real field we expect half the number of degrees of freedom compared to a complex field, therefore the field found in this way has only 2 degrees of freedom, exactly as required for a spin $\frac{1}{2}$ particle.

It is interesting to show in detail how the degrees of freedom in a Majorana field are organised. From (6.76) we see that, in the Majorana representation,

$$v(\boldsymbol{p}) = u(\boldsymbol{p})^*$$

therefore the expansion of $\psi$ with the condition that it should be real, takes the form (we also use here the normalisation (13.8) to facilitate the limit $m \to 0$):

$$\psi(x) = \sum_{\boldsymbol{p},r} \frac{1}{\sqrt{V}}[a_r(\boldsymbol{p})u_r(\boldsymbol{p})e^{-i(px)} + a_r(\boldsymbol{p})^\dagger u_r(\boldsymbol{p})^* e^{+i(px)}]. \tag{13.17}$$

The particles created by $\psi^{(-)}$ are identical to those annihilated by $\psi^{(+)}$; a Majorana fermion is the same as its antiparticle and is intrinsically uncharged, like the photon.

Starting from the Lagrangian (7.21) we can quantise the Majorana field without difficulty. The momentum conjugate to $\psi$ is simply $i\psi$. The anticommutation rules are written:

$$\{\psi(\boldsymbol{x},t)_\alpha, i\psi(\boldsymbol{y},t)_\beta\} = i\delta_{\alpha,\beta}\delta^{(3)}(\boldsymbol{x}-\boldsymbol{y}). \tag{13.18}$$

From here the anticommutation relations for the creation and destruction operators $a$ and $a^\dagger$ can be recovered.

The neutrality of the Majorana fermions can also be seen from the form of the conserved current. We find:

$$\begin{aligned} j^\mu = \bar{\psi}\gamma^\mu\psi = \psi^T\gamma^0\gamma^\mu\psi = \\ = (\psi^T\psi, \psi^T\alpha\psi) = \text{complex number} \end{aligned} \tag{13.19}$$

($\psi^T$ denotes a row vector with the same components as $\psi$). That the current should be *a complex number* follows from the fact that, in all the components of $j$, the spinors $\psi_\alpha\psi_\beta$ are multiplied by matrices *symmetric in* $\alpha$ *and* $\beta$, and therefore, in view of the anticommutation relations (13.18), produce a multiple of the identity matrix. If we define the current so as to vanish on the vacuum state, we will have $Q = 0$ in all the states.

A non-trivial result is obtained for the *axial* current:

$$A^\mu = \bar{\psi}\gamma^\mu\gamma_5\psi = \psi^T\gamma^0\gamma^\mu\gamma_5\psi \tag{13.20}$$

because, in this case, the matrices:

$$a^\mu = (\gamma_5, \boldsymbol{\alpha}\gamma_5) \tag{13.21}$$

are antisymmetric. However, the current (13.20) is conserved only in the limit of zero mass. Using the Dirac equation (7.23) we find:

$$\partial_\mu A^\mu = 2m\bar{\psi}i\gamma_5\psi. \tag{13.22}$$

The vanishing of the vector current excludes that the electron or proton, whose electromagnetic current is certainly different from zero can be described by a Majorana field. The neutron too, in view of the conservation of baryon number must be described by a (complex) Dirac field.

In the Majorana representation, the matrices which represent the Lorentz transformations $S(\Lambda)$, equation (6.39), take a special form. In fact, with imaginary gamma matrices, the $\sigma^{\mu\nu}$ matrices are also imaginary and the $S(\Lambda)$ are *real matrices.*

If we examine a complex Dirac field, the spinor:

$$\psi^T\gamma^0 \tag{13.23}$$

transforms like the adjoint spinor, i.e. with $S^{-1}$:

$$\psi'^T\gamma^0(x') = \psi^T S^T(\Lambda)\gamma^0 = \psi^T S^\dagger(\Lambda)\gamma^0 = \psi^T\gamma^0 S^{-1}(\Lambda). \tag{13.24}$$

It then follows that we can construct with $\psi$ a different mass term from what appears in the Dirac Lagrangian:

$$\mathcal{L}_M = \mu\,\psi^T\gamma^0\psi + \mu^*\,\psi^\dagger\gamma^0\psi^\dagger \tag{13.25}$$

where $\mu$ is a *complex* arbitrary parameter. The mass term (13.25) is known as the *Majorana mass*. The Lagrangian:

$$\bar{\psi}(i\partial\!\!\!/)\psi + \mu\psi^T\gamma^0\psi + \mu^*\psi^\dagger\gamma^0\psi^\dagger \tag{13.26}$$

still describes spin $\frac{1}{2}$ fermions with non-zero mass, as we will see, but there is no longer invariant for global phase transformations of the field $\psi$. The corresponding vector current $\bar{\psi}\gamma^\mu\psi$ is not conserved.

## 13.3 RELATIONSHIPS AMONG WEYL, MAJORANA AND DIRAC NEUTRINOS

In this section, we consider the different possibilities to describe the field of a neutrino. In all the formulae which follow we use the Majorana representation.

Starting from a Dirac field, $\nu(x)$, we can separate the left- and right-handed components using the projection operators $a^{(\pm)}$:

$$\nu(x) = \nu(x)_R + \nu(x)_L. \tag{13.27}$$

For massless fields the separation is Lorentz-invariant and the Dirac field divides into irreducible components, each of which describes a Weyl fermion. The matching of degrees of freedom is accounted by:

$$Dirac(4) = 2 \times Weyl(2). \tag{13.28}$$

Again in terms of a massless theory, each Weyl fermion is equivalent to a Majorana fermion. To see this we define:

$$\nu_1(x) = \nu_L(x) + [\nu_L(x)]^\dagger; \qquad \nu_2(x) = \nu_R(x) + [\nu_R(x)]^\dagger \ \text{(MR)} \tag{13.29}$$

Recalling the expansion (13.10) and the conjugation relation (13.17) we find, for example for $\nu_1$:

$$\nu_1(x) = \sum_{\mathbf{p}} \frac{1}{\sqrt{V}} \times$$
$$\left\{[a_-(\mathbf{p})u_-(\mathbf{p}) + b_+(\mathbf{p})u_+(\mathbf{p})]e^{-i(px)} + [a_-(\mathbf{p})^\dagger v_-(\mathbf{p}) + b_+(\mathbf{p})^\dagger v_+(\mathbf{p})]e^{+i(px)}\right\}. \tag{13.30}$$

Equation (13.30) shows that the two spin components of the Majorana fermion are made up of the right-handed antineutrino and the left-handed neutrino, the components of the Weyl field $\nu_L(x)$. Similarly the two spin components of $\nu_2$ are composed of the right-handed neutrino and the left-handed antineutrino of the field $\nu_R$; see Fig. 13.2.

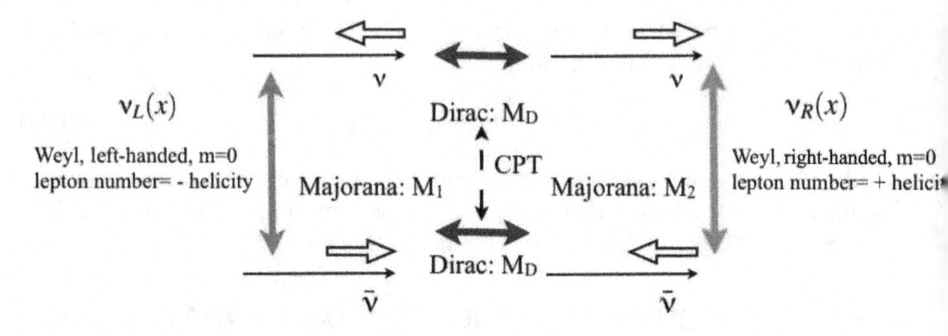


Figure 13.2 Schematic representation of the relationships among Weyl, Majorana and Dirac neutrino states and masses.

The equations in (13.29) can be inverted, recalling that $\gamma_5^T = -\gamma_5$, from which:

$$\left(\frac{1-\gamma_5}{2}\right)(\nu_L)\dagger = [\nu_L\left(\frac{1-\gamma_5}{2}\right)]^\dagger$$
$$= [\left(\frac{1+\gamma_5}{2}\right)\nu_L]^\dagger = 0. \quad (13.31)$$

We therefore obtain:

$$\nu_L = \left(\frac{1-\gamma_5}{2}\right)\nu_1; \qquad (\nu_L)^\dagger = \left(\frac{1+\gamma_5}{2}\right)\nu_1. \quad (13.32)$$

Equations (13.29) and (13.32) show the equivalence of the Weyl and Majorana theories for massless fermions.

The Dirac Lagrangian can be written in terms of Weyl or Majorana fields, according to the equivalence:

$$\mathcal{L}_D = \bar{\nu}i\partial\!\!\!/\nu = \bar{\nu}_R i\partial\!\!\!/\nu_R + \bar{\nu}_L i\partial\!\!\!/\nu_L = \mathcal{L}_W(\nu_R) + \mathcal{L}_W(\nu_L)$$
$$\mathcal{L}_W(\nu_{R(L)}) = \frac{1}{2}\nu_{2(1)}^T\gamma^0\partial\!\!\!/\nu_{2(1)} = \mathcal{L}_M(\nu_{2(1)}) \quad (13.33)$$

The possible terms to be added to (13.33) to give mass to the fermions can be classified as follows:

The Dirac mass term has the form:

$$\mathcal{L}_{mD} = m_D\bar{\nu}\nu = m_D\bar{\nu}_L\nu_R + \text{h.c.} \quad (13.34)$$

where h.c. denotes the hermitian conjugate operator, in this case[2] $m_D\bar{\nu}_R\nu_L$.

[2]We leave it to the reader to show that if we start from a complex $m_D$ we can reduce it to the real case with a redefinition of the relative phase between $\nu_L$ and $\nu_R$ and the consequent redefinition of $\nu_1$ and $\nu_2$.

Again using the antisymmetry of $\gamma_5$, it is easy to see that the Dirac term takes the form, in terms of $\nu_{1,2}$:

$$\mathcal{L}_{mD} = m_D \nu_1^T \gamma^0 \nu_2 = m_D \nu_2^T \gamma^0 \nu_1. \tag{13.35}$$

However, given the Majorana fields $\nu_{1,2}$, we can also consider two new mass terms:

$$\mathcal{L}_{mM} = \frac{1}{2} M_1 \nu_1^T \gamma^0 \nu_1 + \frac{1}{2} M_2 \nu_2^T \gamma^0 \nu_2. \tag{13.36}$$

Individually, the two terms correspond to a Majorana mass for the left-handed and right-handed neutrino, for example:

$$\frac{1}{2} M_1 \nu_1^T \gamma^0 \nu_1 = \frac{1}{2} M_1 (\nu_L^T \gamma^0 \nu_L + \text{h.c.}). \tag{13.37}$$

Overall, the neutrino masses are described by a *symmetric matrix*:

$$\begin{aligned} \mathcal{L}_m &= \frac{1}{2} \zeta^T \gamma^0 \mathcal{M} \zeta \\ \zeta &= \begin{pmatrix} \nu_1 \\ \nu_2 \end{pmatrix}; \quad \mathcal{M} = \begin{pmatrix} M_1 & m_D \\ m_D & M_2 \end{pmatrix}. \end{aligned} \tag{13.38}$$

It is instructive to examine the symmetries of the massless Lagrangian, (13.33), and the Dirac and Majorana mass terms, (13.35) and (13.36).

Equation (13.33) is invariant under two abelian and commuting transformation groups[3]:

$$\begin{aligned} \nu_1 &\to \nu_1' = e^{i\alpha\gamma_5} \nu_1; \\ \nu_2 &\to \nu_2' = e^{i\beta\gamma_5} \nu_2. \end{aligned} \tag{13.39}$$

However, the Majorana mass terms are not invariant, while the Dirac term is invariant for the subgroup of transformations with $\alpha = -\beta$:

$$\begin{aligned} \mathcal{L}_{mM} &\to \frac{1}{2} M_1 \nu_1^T \gamma^0 e^{2i\alpha\gamma_5} \nu_1 + \frac{1}{2} M_2 \nu_2^T \gamma^0 e^{2i\beta\gamma_5} \nu_2; \\ \mathcal{L}_{mD} &\to m_D \nu_1^T \gamma^0 e^{i(\alpha+\beta)\gamma_5} \nu_2. \end{aligned} \tag{13.40}$$

In the specific case $M_1 = M_2 = 0$ a symmetry therefore remains, whatever the value of $m_D$ and a conserved current. In this case, as we see from (13.34), the neutrino is described by the Dirac field $\nu_L + \nu_R$ with conserved current $\bar{\nu}\gamma^\mu \nu$, corresponding to the familiar phase transformations.

To see this more formally, we observe that in the case $M_1 = M_2 = 0$ the mass matrix is proportional to the $\sigma_2$ matrix and is therefore diagonalised

[3]Since $\gamma_5$ is purely imaginary, and the transformations (13.39) are *orthogonal*, as they must be to maintain the Majorana nature of the fields $\nu_{1,2}$.

by the combination $\zeta^{(+)} = \nu_1 + \nu_2$, eigenvalue $+m_D$, and $\zeta^{(-)} = \nu_1 - \nu_2$, eigenvalue $-m_D$. Thus, in the light of (13.29), we can set:

$$\begin{aligned}\nu_{Dirac} &= \frac{1}{2}[\zeta^{(+)} - \gamma_5 \zeta^{(-)}] \\ &= \frac{1}{2}[\nu_L + (\nu_L)^\dagger + \nu_R + (\nu_R)^\dagger] - \frac{1}{2}\gamma_5[\nu_L + (\nu_L)^\dagger - \nu_R - (\nu_R)^\dagger] = \\ &= \nu_L + \nu_R \end{aligned} \tag{13.41}$$

The conserved current in this case is known as the *lepton number* and distinguishes the neutrino from the antineutrino.

In the general case, in which at least one of $M_1$ or $M_2$ is non-zero, the eigenvectors of the mass matrix are two Majorana fields, and there is neither a conserved current nor a difference between neutrino and antineutrino.

The relationships between the Weyl, Dirac and Majorana neutrino states and masses are illustrated in Fig. 13.2. The general particle-antiparticle symmetry requires that the Dirac masses which connect $\nu_L$ with $\nu_R$ or $\bar{\nu}_L$ with $\bar{\nu}_R$ takes the same $m_D$ value.

Comment: the electron case. The requirement that a conserved current should exist, to be identified as the electric current, implies that there should only be the Dirac mass term and that therefore the electron should be described by a pair of Majorana fields as in (13.41), therefore by a four-dimensional Dirac spinor. In turn, this implies the existence of antiparticles (positrons) distinct from particles. The same argument, using the conservation of baryon number, holds for the proton and neutron.

Comment: the see-saw mechanism. In the modern theory of the neutrino, it is supposed both that $M_1 = 0$ and $M_2 >> m_D$. In this case, the mass matrix in (13.38) has as approximate eigenvalues and eigenvectors:

$$\zeta^{(-)} \simeq \nu_1; \quad m \simeq \frac{m_D^2}{M_2}. \tag{13.42}$$

For sufficiently large $M_2$ with fixed $m_D$, the light neutrino has a tiny mass and is a good approximation to a Majorana–Weyl neutrino. As we will see, this situation describes the neutrino of $\beta$-decay very well.

## 13.4 PROBLEM FOR CHAPTER 13

### Sect. 13.1

1. Extend the calculation illustrated in Fig. 13.1 to the axial vector density, $\bar{\psi}\gamma^\mu\gamma_5\psi$.

CHAPTER 14

# APPLICATIONS: QED

The calculation of the elements of the $S$-matrix to a given order of perturbation theory, equation (11.9), is carried out using the general method of *Feynman diagrams*, which will be illustrated in [14].

For processes up to the second order, which we will consider in this and the following chapter, the matrix elements can be calculated by simple inspection and we will not need general formalism, which is essential for the calculation of higher order corrections and renormalisation. The focus here is rather to show how the ideas of second quantisation can be compared with experiments for the simplest processes, which historically had an essential role in the development of the theory of elementary particles.

## 14.1 SCATTERING IN A CLASSICAL COULOMB FIELD

We consider the scattering of an electron in a static, i.e. time-independent, external field. The relevant 4-potential can be written

$$A^\mu(x) = A^\mu(\mathbf{x}) = \int \frac{d^3q}{(2\pi)^3}\, A^\mu(\mathbf{q})\, \mathrm{e}^{i\mathbf{q}\mathbf{x}} \tag{14.1}$$

and the perturbative expansion of the $S$-matrix has the form

$$S = \sum_{n=0}^{\infty} \frac{(ie)^n}{n!} \int d^4x_1 ... d^4x_n\, T\{[: \bar{\psi}(x_1) A\!\!\!/(x_1)\psi(x_1) :] ... [: \bar{\psi}(x_n) A\!\!\!/(x_n)\psi(x_n) :]\}\,, \tag{14.2}$$

where $\psi$ is the field which describes the electron and $T$ is the time-ordering operator.

To first order in the interaction Lagrangian, equation (14.2) reduces to

$$S = 1 + ie \int d^4x : \bar{\psi}(x) A\!\!\!/(x)\psi(x) : \ . \tag{14.3}$$

The physical process described by (14.3) is represented schematically by the Feynman diagram of Fig. 14.1. The interaction with the external field

DOI: 10.1201/9781003436263-14

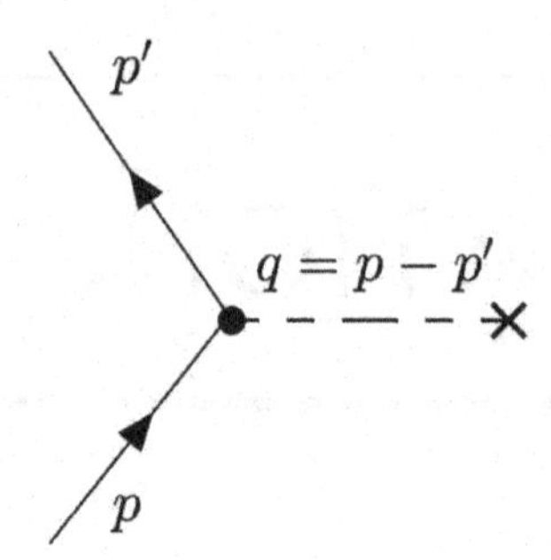


Figure 14.1 Diagrammatic representation of electron scattering in a static field.

causes the transition of the electron from the initial state $|i\rangle = |pr\rangle$ to the final state $|f\rangle = |p'r'\rangle$, where $p \equiv (E, \mathbf{p})$ and $p' \equiv (E', \mathbf{p}')$ are the 4-momenta and $r$ and $r'$ the spin projections.

The corresponding $S$-matrix has the form

$$\begin{aligned} S_{if} = \langle f|S|i\rangle &= ie\left(\frac{m}{VE}\right)^{1/2}\left(\frac{m}{VE'}\right)^{1/2}\int d^4x\, \mathrm{e}^{i(p'-p)x}\, \bar{u}_{r'}(\mathbf{p}')A\!\!\!/(x)u_r(\mathbf{p}) \\ &= ie\left(\frac{m}{VE}\right)^{1/2}\left(\frac{m}{VE'}\right)^{1/2}\int d^3x\, \mathrm{e}^{i(\boldsymbol{q}+\boldsymbol{p}-\boldsymbol{p}')\boldsymbol{x}} \\ &\quad\times \int dt\, e^{i(E'-E)t}\int \frac{d^3q}{(2\pi)^3}\bar{u}_{r'}(\mathbf{p}')A\!\!\!/(\boldsymbol{q})u_r(\mathbf{p}) \\ &= (2\pi)\,\delta(E-E')\left(\frac{m}{VE}\right)^{1/2}\left(\frac{m}{VE'}\right)^{1/2}\mathcal{M}_{if}\,, \end{aligned} \tag{14.4}$$

where $V$ is the normalisation volume and

$$\mathcal{M}_{if} = ie\, \bar{u}_{r'}(\mathbf{p})A\!\!\!/(\mathbf{p}'-\mathbf{p})u_r(\mathbf{p})\,. \tag{14.5}$$

We note that in (14.4) only the $\delta$-function associated with the conservation of energy appears. Momentum is not conserved, because the static source, which generates the field, breaks translational invariance. The relation which expresses the conservation of energy, which implies

$$|\boldsymbol{p}| = |\boldsymbol{p}'|\,, \tag{14.6}$$

clearly shows that the momentum absorbed by the source is neglected.

The differential cross section for the process is obtained from the formulae given in Chapter 11:

$$d\sigma = \frac{V}{v}\, W_{if}\, \frac{V d^3p'}{(2\pi)^3}\,, \tag{14.7}$$

where $W_{if}$ is the transition probability per unit time

$$W_{if} = \frac{|S_{if}|^2}{T} = (2\pi)\,\delta(E-E')\left(\frac{m}{VE}\right)^2|\mathcal{M}_{if}|^2\,, \tag{14.8}$$

and $T$ is the duration of the interaction. Using equations (14.7) and (14.8), and the relation:

$$|\boldsymbol{p}|^2 d|\boldsymbol{p}| = |\boldsymbol{p}|E'dE' \ , \tag{14.9}$$

from (14.7) we obtain

$$d\sigma = \frac{VE}{|\mathbf{p}|}\,(2\pi)\;\delta(E-E')\;\left(\frac{m}{VE}\right)^2 |\mathcal{M}_{if}|^2 \frac{V}{(2\pi)^3}\;|\mathbf{p}|E'dE'd\Omega_{p'} \ . \tag{14.10}$$

To arrive at the expression for the differential cross-section, the probability that the electron is scattered into a solid angle element $d\Omega_{p'}$, we must integrate over the energy using the $\delta$-function, with the result

$$\frac{d\sigma}{d\Omega_{p'}} = \left(\frac{m}{2\pi}\right)^2 |\mathcal{M}_{if}|^2 = \left(\frac{me}{2\pi}\right)^2 |\bar{u}_{r'}(\boldsymbol{p}')\not{A}(\boldsymbol{q})u_r(\boldsymbol{p})|^2 \ , \tag{14.11}$$

where $\boldsymbol{q} = \boldsymbol{p}' - \boldsymbol{p}$.

We now consider the case in which the external field is the Coulomb field generated by an atomic nucleus, which we suppose to be pointlike. Therefore we have:

$$A^\mu(\boldsymbol{x}) \equiv \left(\frac{Z}{4\pi}\frac{e}{|\boldsymbol{x}|}, 0, 0, 0\right) \ , \tag{14.12}$$

with the Fourier transform (FT) given by (cf. Problem 1):

$$A^\mu(\boldsymbol{q}) \equiv \left(Z\frac{e}{|\boldsymbol{q}|^2}, 0, 0, 0\right) \ . \tag{14.13}$$

Substituting (14.13) into (14.11), summing over the projections of the electron spin in the final state and averaging over those of the initial state of the electron we obtain

$$\begin{aligned}\frac{d\sigma}{d\Omega_{p'}} &= \left(\frac{me}{2\pi}\right)^2 Z^2 \frac{e^2}{|\boldsymbol{q}|^4}\frac{1}{2}\sum_{rr'}|\bar{u}_{r'}(\boldsymbol{p}')\gamma^0 u_r(\boldsymbol{p})|^2 \\ &= \frac{(2m\alpha Z)^2}{|\boldsymbol{q}|^4}\frac{1}{(2m)^2}\frac{1}{2}Tr\left[(\not{p}'+m)\gamma^0(\not{p}+m)\gamma^0\right] \ ,\end{aligned} \tag{14.14}$$

where $\alpha = e^2/4\pi$ and the sums over $r$ and $r'$ are carried out using the completeness relations satisfied by the Dirac spinors. The calculation of the trace in (14.14) is easily done using the result

$$\begin{aligned}Tr\left[(\not{p}'+m)\gamma^0(\not{p}+m)\gamma^0\right] &= Tr(\not{p}'\gamma^0\not{p}\gamma^0) + m^2 Tr(\gamma^0\gamma^0) \\ &= 4p'_\mu p_\nu(2g^{\mu 0}g^{\nu 0} - g^{\mu\nu}g^{00}) + 4m^2 \\ &= 4[2EE' - (pp') + m^2] = 4[EE' + (\boldsymbol{p}'\cdot\boldsymbol{p}) + m^2] \ .\end{aligned} \tag{14.15}$$

In this way we obtain

$$\begin{aligned}\frac{d\sigma}{d\Omega_{p'}} &= 2\,\frac{(\alpha Z)^2}{|\boldsymbol{q}|^4}\,[E^2+(\boldsymbol{p}\cdot\boldsymbol{p}')+m^2] \\ &= 2\,\frac{(\alpha Z)^2}{|\boldsymbol{q}|^4}\,2E^2\left(1-\frac{|\boldsymbol{p}|^2}{E^2}\sin^2\frac{\theta}{2}\right) \\ &= 2\,\frac{(\alpha Z)^2}{|\boldsymbol{q}|^4}\,2E^2\left(1-v^2\sin^2\frac{\theta}{2}\right).\end{aligned} \tag{14.16}$$

where $v$ and $\theta$ are, respectively, the velocity and scattering angle of the electron, i.e. the angle between the vectors $\boldsymbol{p}$ and $\boldsymbol{p}'$. Using the relation

$$|\boldsymbol{q}|^2 = |\boldsymbol{p}'-\boldsymbol{p}|^2 = 2|\boldsymbol{p}|^2(1-\cos\theta) = 4E^2v^2\sin^2\frac{\theta}{2}\,, \tag{14.17}$$

we can rewrite equation (14.17) in the form originally obtained by Mott

$$\frac{d\sigma}{d\Omega_{p'}} = \frac{(\alpha Z)^2}{4E^2v^4\sin^4(\theta/2)}\left(1-v^2\sin^2\frac{\theta}{2}\right). \tag{14.18}$$

The Mott cross-section describes the elastic scattering of electrons by nuclei to the lowest order in the fine structure constant $\alpha$. Obviously, this approximation is not applicable in the case of very heavy nuclei, for which the coupling constant $Z\alpha$ becomes too large.

In the non-relativistic limit, i.e. for $v \ll 1$ and $E \sim m$, from (14.18) we recover the famous Rutherford cross-section

$$\frac{d\sigma}{d\Omega_{p'}} = \frac{(\alpha Z)^2}{4m^2\sin^4(\theta/2)}\,, \tag{14.19}$$

whose experimental measurement led to the development of the planetary model of the atom.

In the ultrarelativistic limit, $v \simeq 1$, the Mott cross section becomes:

$$\left.\frac{d\sigma}{d\Omega_{p'}}\right|_{v=1} = \frac{(\alpha Z)^2\cos^2(\theta/2)}{4E^2v^4\sin^4(\theta/2)}. \tag{14.20}$$

The ultrarelativistic Mott cross section vanishes for back-scattering of the electron, $\theta = \pi$.

Note. We have succeeded in deriving the Rutherford cross-section from a calculation to first order in $\alpha$. This is surprising because the same expression is obtained from classical mechanics as an exact result. The explanation of this apparent contradiction is due to Dalitz, who showed that, in the non-relativistic limit, the inclusion of higher-order corrections only results in the appearance of a phase factor in the amplitude, which leaves the cross-section unchanged.

Coulomb Divergence. The integral of the Mott cross section, (14.20), over solid angle diverges for $\theta = 0$, preventing definition of the total cross-section. To be exact, for small values of $\theta$

$$\frac{d\sigma}{d\Omega_{p'}}\, d\cos\theta \simeq 4\frac{\alpha Z^2}{4E^2 v^4 \theta^4}\, \theta d\theta\ . \tag{14.21}$$

The origin of this divergence can be traced to the trend towards singularity, for $|\mathbf{q}| \to \mathbf{0}$, of the FT of the Coulomb field, equation (14.13), or to the too slow fall off of the Coulomb field itself, as $|\mathbf{x}| \to \infty$, equation (14.12). Due to this slow decrease, even particles which pass very far from the charge at the origin feel its effect; the asymptotic states, in the interaction representation, do not truly tend to a constant and this prevents us from rigorously defining the $S$-matrix.

Fortunately, isolated electric charges do not exist in Nature. The positive charge of each nucleus, in ordinary matter, is screened by the negative charges of the atomic electrons. Also in an ionised plasma, due to the long range of the electrostatic forces, the positive charges attract negative charges around them so that, on average, the plasma is electrically neutral. The result is that the integrand in (14.21) is suppressed for small values of $\theta$.

We recall that the FT at $|\boldsymbol{q}|$ is sensitive to the values of the function for $|\boldsymbol{x}| \simeq |\boldsymbol{q}|^{-1}$. Therefore, if we consider scattering on an atom, the presence of the external electrons is felt at values of $|\boldsymbol{q}|$ such that:

$$|\boldsymbol{q}|R \leq 1\ , \tag{14.22}$$

where $R$ is the atomic radius, or for:

$$\theta \leq \frac{1}{REv}\ . \tag{14.23}$$

In this $\theta$ region the differential cross section is suppressed by the fact that the electron sees a gradually decreasing total charge and the total cross-section is finite.

Non-relativistic Form Factor. We can improve the approximation of the pointlike charge of the nucleus by introducing a *form factor*.

We suppose that the charge density inside the nucleus is described by a function $Ze\rho(\boldsymbol{x})$. The charge density must decrease rapidly with increasing $|\mathbf{x}|$, vanish outside the nucleus, i.e. for $|\boldsymbol{x}| > R_N$, and satisfy the normalisation condition:

$$\int \rho(\boldsymbol{x}) d^3x = 1 \tag{14.24}$$

so that the total charge is still given by $Ze$. The electrostatic potential of the nucleus is now given by:

$$A^0(\boldsymbol{x}) = \frac{Ze}{4\pi}\int d^3y \frac{\rho(y)}{|x-y|}\ , \tag{14.25}$$

and its FT is given by (cf. Problem 1):

$$A^0(\boldsymbol{q}) = \frac{Ze}{|q|}\, F(q);$$
$$F(\boldsymbol{q}) = \int d^3x\; e^{i\boldsymbol{q}\cdot\boldsymbol{x}}\rho(\boldsymbol{x}). \tag{14.26}$$

Correspondingly the Mott cross section becomes:

$$\frac{d\sigma}{d\Omega_{p'}} = \frac{\alpha Z^2}{4E^2v^4\sin^4(\theta/2)}\,|F(\boldsymbol{q})|^2\left(1 - v^2\sin^2\frac{\theta}{2}\right). \tag{14.27}$$

## 14.2 ELECTROMAGNETIC FORM FACTORS

We consider the matrix element $\langle p'|J^\mu_p|p\rangle$ of the proton current, equation (9.16), between states with momenta $p$ and $p'$.

Substituting the expansion in plane waves of the fields, we find ($M$ denotes the proton mass ):

$$\langle p'|J^\mu_p|p\rangle = \frac{M}{V\sqrt{E'E}}\bar{u}(p')\left[\gamma^\mu + \frac{\kappa}{2M}i\sigma_{\mu\nu}q^\nu\right]u(p) = \frac{M}{V\sqrt{E'E}}J^\mu(p',p)\,, \tag{14.28}$$

where we have introduced the *momentum transfer*:

$$q^\mu = p'^\mu - p^\mu\,. \tag{14.29}$$

We can ask what would be the most general form of $J^\mu(p',p)$ compatible with a conserved polar 4-vector:

$$q_\mu J^\mu(p',p) = 0\,. \tag{14.30}$$

Clearly the current must have the form:

$$J^\mu(p',p) = \bar{u}(p')\Gamma^\mu(p',p)u(p)\,, \tag{14.31}$$

with $\Gamma^\mu$ a linear combination of the available 4-vectors, which we can choose as:

$$\gamma_\mu,\ \sigma_{\mu\nu}q^\nu,\ q_\mu,\ p'_\mu + p_\mu,\ \sigma_{\mu\nu}(p'^\nu + p^\nu)\,. \tag{14.32}$$

However, taken between the two spinors $\bar{u}(p')$ and $u(p)$, these vectors are not independent of each other, as a result of the equation of motion which $u(p)$ satisfies, (6.76), and the corresponding equation for $\bar{u}(p')$. Multiplying the first equation of (6.76) by $\bar{u}(p')\gamma_\mu$, we obtain:

$$\bar{u}(p')\left[p_\mu - i\sigma_{\mu\nu}p^\nu\right]u(p) = M\,\bar{u}(p')\gamma_\mu u(p)\,.$$

Proceeding symmetrically with the equation for $\bar{u}(p')$, we also obtain:

$$\bar{u}(p')\left[p'_\mu + i\sigma_{\mu\nu}p'^\nu\right]u(p) = M\,\bar{u}(p')\gamma_\mu u(p)\ .$$

Summing and subtracting the two equations, we find the relations[1]:

$$\frac{p'^\mu + p^\mu}{2M}\bar{u}(p')u(p) + \frac{i}{2M}\bar{u}(p')\sigma_{\mu\nu}q^\nu u(p) = \bar{u}(p')\gamma_\mu u(p) \quad (14.33)$$

$$q_\mu \bar{u}(p')u(p) + \frac{i}{2M}\bar{u}(p')\sigma_{\mu\nu}(p'^\nu + p^\nu)u(p) = 0\ , \quad (14.34)$$

which shows that we can limit ourselves to the first three vectors:

$$\Gamma_\mu(p', p) = A\gamma_\mu + B\frac{i}{2M}\sigma_{\mu\nu}q^\nu + Cq_\mu\ . \quad (14.35)$$

Moreover, from the conservation equation (14.30) we obtain:

$$q_\mu J^\mu(p', p) = Cq^2\bar{u}(p')u(p) = 0\ , \quad (14.36)$$

or $C = 0$. We can therefore write:

$$J^\mu(p', p) = \bar{u}(p')\left[A\gamma_\mu + B\frac{i}{2M}\sigma_{\mu\nu}q^\nu\right]u(p)\ . \quad (14.37)$$

The requirement that (14.37) transforms like a 4-vector implies, in general, that the coefficients $A$ and $B$ are invariant functions of 4-momenta. Because $p^2 = p'^2 = m^2$, the only non-trivial invariant combination is $q^2$ and we obtain, finally, the form of the current:

$$J^\mu(p', p) = \bar{u}(p')\left[F_1(q^2)\gamma_\mu + F_2(q^2)\frac{i}{2M}\sigma_{\mu\nu}q^\nu\right]u(p). \quad (14.38)$$

It is easily seen that condition (9.17) implies

$$F_1(0) = 1\ , \quad (14.39)$$

while from (9.10) we find:

$$F_2(0) = \kappa\ . \quad (14.40)$$

Similar formulae naturally hold for the neutron, with $F_{1,n}(0) = 0$.

The functions $F_1(q^2)$ and $F_2(q^2)$ are known as Dirac and Pauli form factors, respectively, and can be measured, as we will see, by elastic scattering of electrons on hydrogen (for the proton) and on deuterium (obtaining the neutron form factor by subtraction).

The form (14.38) is obviously not unique because we can choose the basic 4-vectors in different ways. For example, we can write the current as:

$$J^\mu(p', p) = \bar{u}(p')\left[F_a(q^2)\frac{p'^\mu + p^\mu}{2M} + F_b(q^2)\frac{i}{2M}\sigma_{\mu\nu}q^\nu\right]u(p)\ . \quad (14.41)$$

[1] The first of these relations is well known as *Gordon decomposition.*

Using equation (14.34), this choice corresponds to:

$$F_1(q^2) = F_a(q^2); \quad F_2(q^2) = F_b(q^2) - F_a(q^2) \tag{14.42}$$

or, for $q^2 = 0$:

$$F_a(0) = 1; \quad F_b(0) = 1 + F_2(0). \tag{14.43}$$

$F_b(0)$ immediately gives the total magnetic moment (in Bohr magneton units).

The choice (14.41) is best suited for the calculation of radiative corrections to the anomalous magnetic moment of the electron; see [14].

A particularly convenient choice to describe the elastic electron–nucleon cross section is given by the *Sachs form factors*

$$\begin{aligned} G_E &= F_1 - \tau F_2, \quad G_M = F_1 + F_2 \\ \tau &= -\frac{q^2}{4M^2}, \end{aligned} \tag{14.44}$$

which are known, respectively, as the electric and magnetic form factors. The normalisation conditions follow from equation (14.44):

$$G_E(0) = 1, \quad G_M(0) = 1 + \kappa\,. \tag{14.45}$$

## 14.3 THE ROSENBLUTH FORMULA

In the scattering of electrons on protons, when the electron energy becomes of the order of GeV, we must include a fully quantum description of the proton. To do this, we extend the interaction Lagrangian of the electromagnetic field according to the approach illustrated in Chapter 9, by introducing the electromagnetic current of the proton:

$$\mathcal{L}_{e.m.} = eA_\mu(x)J^\mu_{tot}(x); \quad J^\mu_{tot}(x) = J^\mu_e(x) + J^\mu_p(x)\,. \tag{14.46}$$

For the electron current, we take, as before:

$$J^\mu_e(x) = -:\bar{\psi}_e(x)\gamma^\mu\psi_e(x):\,. \tag{14.47}$$

The matrix element for the proton current between proton states is parameterised in terms of the form factors introduced in the previous section, equation (14.38):

$$\begin{aligned} &< p'|J^\mu_p(0)|p> = \sqrt{\frac{M^2}{V^2E_{p'}E_p}}\bar{u}(p')\Gamma^\mu(p',p)u(p)\,, \\ &\Gamma^\mu(p',p) = F_1(q^2)\gamma_\mu + F_2(q^2)\frac{i}{2M}\sigma_{\mu\nu}q^\nu\,. \end{aligned} \tag{14.48}$$

To lowest order of perturbation theory, electron-proton scattering is described by terms of second order:

$$S^{(2)} = \frac{(ie)^2}{2} \int d^4x\, d^4y\, T\left[A_\mu(x)J^\mu_{tot}(x)A_\nu(y)J^\nu_{tot}(y)\right] =$$
$$= (ie)^2 \int d^4x\, d^4y\, T\left[A_\mu(x)J^\mu_e(x)A_\nu(y)J^\nu_p(y)\right] + \ldots \qquad (14.49)$$

We have written only the terms of the $T$-product which have the right annihilation and creation operators to destroy the initial particles and create the final particles, summarising the others, which are not essential, in the ellipsis dots.

Matrix Elements. As explained earlier, the operators which appear in equation (14.49) satisfy the free particle equations of motion. Therefore fields which correspond to different particles commute or anticommute among themselves, according to which type of statistics they obey. Furthermore, the external states are tensor products of the states of the different particles, for example:

$$|e, p, 0_\gamma >= a^\dagger(e)a^\dagger(p)|0 >= |e > |p > |0_\gamma >= a^\dagger(e)|0_e > a^\dagger(p)|0_p > |0_\gamma > ,$$

where we have denoted the different momenta using the names of the relevant particles, with, for example, the state with zero electrons being $|0_e >$, and, naturally:

$$|0 >= |0_e, 0_p, 0_\gamma >= |0_e > |0_p > |0_\gamma > .$$

The current of each particle operates only on the corresponding states.

Finally, we can take the currents outside the $T$-product and write:

$$< f|S|i > = (ie)^2 \int d^4x\, d^4y < e'|J^\mu_e(x)|e > .$$
$$. < 0_\gamma|T\left[A_\mu(x)A_\nu(y)\right]|0_\gamma >< p'|J^\nu_p(y)|p > . \qquad (14.50)$$

Fig. 14.2 (a) gives a space-time picture of the scattering process. The initial electron propagates up to a point $x$ where it is destroyed by the current, which creates the final electron at the same point. Also, at $x$ a photon which propagates from point $x$ to point $y$, or from $y$ to $x$, is emitted or absorbed, while at point $y$ the proton current changes the initial proton into the final one. The amplitude of the propagation of the photon between $x$ and $y$ is given by the $T$-product on the vacuum of the electromagnetic fields and the product of the three factors in the integral of (14.50) represents the overall amplitude of the *history* which corresponds to values of $x$ and $y$.

The observation of the final particles does not determine the points $x$ and $y$ where the interaction took place, therefore, according to the principles of quantum mechanics, we must sum over the amplitudes of each *history* to have the total amplitude for the process, integrating over the values of $x$ and $y$.

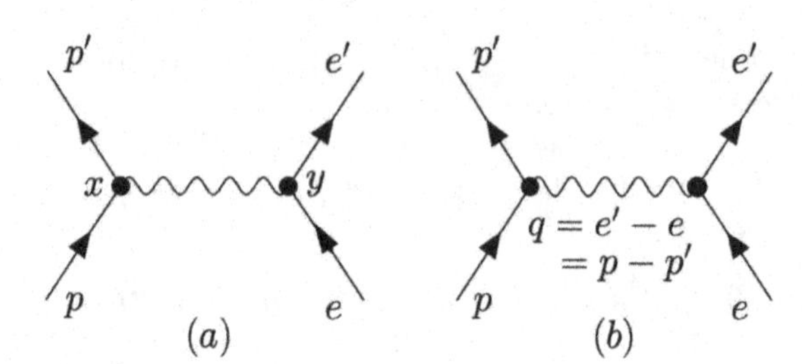


Figure 14.2 Graphical representations of the electron-proton scattering amplitudes in (a) space-time and (b) momentum space.

Proceeding as in the previous section, we find (the minus sign corresponds to the negative electron charge; it is here as a reminder only, as it disappears when computing probabilities):

$$< e'|J_e^\mu(x)|e >= e^{i(e'-e)x}\sqrt{\frac{m_e^2}{V^2 E_e E_{e'}}}(-1)\bar{u}(e')\gamma^\mu u(e) \ , \tag{14.51}$$

and similarly:

$$< p'|J_p^\nu(y)|p >= e^{i(p'-p)y}\sqrt{\frac{M^2}{V^2 E_p E_{p'}}}\bar{u}(p')\Gamma^\nu(p',p)u(p) \ . \tag{14.52}$$

Using the translational invariance of the vacuum in the propagation function of the electromagnetic field, we finally find:

$$\begin{aligned} < f|S|i >= (2\pi)^4\delta^{(4)}(p'+e'-p-e)\sqrt{\frac{m_e^2}{V^2 E_e E_{e'}}}\sqrt{\frac{M^2}{V^2 E_p E_{p'}}} \\ \times \ e^2 \ \bar{u}(e')\gamma^\mu u(e) \ i(D_F)_{\mu\nu}(p-p') \ \bar{u}(p')\Gamma^\nu(p',p)u(p) \ , \end{aligned} \tag{14.53}$$

where we have set, according to equation (8.27):

$$i(D_F)_{\mu\nu}(q) = \int d^4x \ e^{-iqx} < 0|T\left[A_\mu(x)A_\nu(0)\right]|0 >= \frac{-ig^{\mu\nu}}{q^2+i\epsilon} \ . \tag{14.54}$$

The result (14.53) can be graphically represented in Fig. 14.2(b). The initial and final particles emerge from the vertices, each one represented by a Dirac spinor ($u$ for an incoming fermion, $\bar{u}$ for an outgoing fermion) and by the corresponding normalisation factor. The interaction is represented by a factor $-ie\gamma$ or $ie\Gamma$ for electron and proton. The momentum $q = p - p' = e' - e$ is associated with the photon propagator so that 4-momentum is conserved at each vertex.

Differential Cross Section. Following the formulae of the previous chapter, the differential cross section is written:

$$d\sigma = \frac{1}{4}\,(2\pi)^4\delta^{(4)}(p'+e'-p-e)\times$$
$$\times\ \frac{m_e^2M^2}{E_eE_{e'}E_pE_{p'}}\,\frac{d^3p'd^3e'}{(2\pi)^6}\,\frac{e^4}{(q^2)^2}\,H_{\mu\nu}L^{\mu\nu}\ . \qquad (14.55)$$

The factor $\frac{1}{4}$ comes from the averaging over the initial spins and we have set:

$$H_{\mu\nu} = \sum_{r,s}\bar{u}_r(p')\Gamma^\mu u_s(p)\bar{u}_s(p)\Gamma^\nu u_r(p') =$$
$$= Tr\left[\frac{(\not{p}+M)}{2M}\Gamma^\nu\frac{(\not{p}'+M)}{2M}\Gamma^\mu\right] = \frac{h_{\mu\nu}}{4M^2}\ , \qquad (14.56)$$

$$L_{\mu\nu} = \sum_{p,q}\bar{u}_p(e')\gamma^\mu u_q(e)\bar{u}_q(e)\gamma^\nu u_p(e') =$$
$$= Tr\left[\frac{\not{e}}{2m_e}\gamma^\nu\frac{\not{e}'}{2m_e}\gamma^\mu\right] = \frac{l_{\mu\nu}}{4m_e^2}\ . \qquad (14.57)$$

Here and in what follows we neglect the electron mass in the numerators.

To obtain the differential cross section in the variables describing the final electron, which are those which are normally observed, we must integrate over the momentum of the final proton using the three-dimensional $\delta$-function for momentum conservation, which fixes:

$$\mathbf{p}' = \mathbf{e} - \mathbf{e}' \qquad (14.58)$$

In addition, conservation of energy fixes the energy of the final electron. In the proton rest frame, the argument of the $\delta$-function for the energy in (14.55) is:

$$f(E_{e'}) = E_{e'} + E_{p'} - E_e - M =$$
$$= E_{e'} + \sqrt{M^2 + E_{p'}^2 - 2E_{e'}E_e\cos\theta + E_e^2} - E_e - M\ , \qquad (14.59)$$

with the derivative with respect to $E_{e'}$ equal to:

$$\frac{\partial f}{\partial E_{e'}} = \frac{ME_e}{E_{e'}E_{p'}}\ . \qquad (14.60)$$

Using the relation:

$$\delta(f(x)) = \frac{1}{|(\partial f/\partial x)_{x_0}|}\,\delta(x-x_0)\ , \quad f(x_0)=0\ , \qquad (14.61)$$

and introducing the fine structure constant:

$$\alpha = \frac{e^2}{4\pi} , \tag{14.62}$$

we find:

$$d\sigma = \frac{\alpha^2}{16q^4} \left( \frac{E_{e'}}{E_e} \right)^2 \frac{X}{M^2} \, d\Omega , \tag{14.63}$$

where

$$X = l_{\mu\nu} h^{\mu\nu} , \tag{14.64}$$

and d$\Omega$ is the solid angle of the final electron.

Kinematic Variables. In the proton rest frame, the process is described by the energy of the initial electron and the scattering angle, $\theta$, of the final electron. As an alternative we can use the squared momentum transfer,

$$q^2 = (p - p')^2 = (e' - e)^2 , \tag{14.65}$$

or the energy, $E_{e'}$, of the final electron. From its definition, we find:

$$q^2 = 2E_e E_{e'}(1 - \cos\theta) = 4E_e E_{e'} \sin^2 \frac{\theta}{2} . \tag{14.66}$$

In general $W^2$ is used to denote the squared mass of the system recoiling against the final electron. In the elastic scattering process which we are considering we must have:

$$W^2 = M^2 = (p + e - e')^2 = M^2 + q^2 + 2M(E_e - E_{e'}) ,$$

or:

$$E_{e'} = E_e + \frac{q^2}{2M} ,$$

and therefore, using equation (14.66):

$$E_{e'} = \frac{E_e}{1 + \frac{2E_e}{M} \cos\theta} . \tag{14.67}$$

Traces. We must calculate:

$$h_{\mu\nu} = Tr \left[ (\not{p} + M) \Gamma^\nu (\not{p}' + M) \Gamma^\mu \right] . \tag{14.68}$$

If we use equation (14.48), in the trace there are up to six $\gamma$ matrices. As an alternative we can again use the Gordon decomposition, (14.34), and express the proton current as a combination of $\gamma^\mu$ and of:

$$Q^\mu = (p + p')^\mu . \tag{14.69}$$

We find:

$$\Gamma(p',p) = A(q^2)\,\gamma^\mu + B(q^2)\,\frac{Q^\mu}{2M}\,,$$
$$A = F_1 + F_2\,,\quad B = -F_2\,. \tag{14.70}$$

With this arrangement, the trace (14.68) is easily calculated, with the result

$$h_{\mu\nu} = 2\left\{Q^\mu Q^\nu\left[(A+B)^2 - \frac{q^2}{4M^2}B^2\right] - \Pi^{\mu\nu}(q)A^2\right\}\,,$$
$$\Pi^{\mu\nu}(q) = q^\mu q^\nu - q^2 g^{\mu\nu}\,. \tag{14.71}$$

The trace of the electron is obtained from (14.71) with the obvious substitutions:

$$Q^\mu \;\to\; E^\mu = (e+e')^\mu,\quad M \;\to m_e \simeq 0\,,$$
$$A \;\to\; 1,\quad B \;\to\; 0\,, \tag{14.72}$$

yielding

$$l_{\mu\nu} = 2\left[Q^\mu Q^\nu - \Pi^{\mu\nu}(q)\right]\,.$$

Finally, we find:

$$X = 4\left\{\left[(Q_\mu E^\mu)^2 + q^2 Q^2\right]\,\left[(A+B)^2 - \frac{q^2}{4M^2}B^2\right] + 2(q^2)^2\,A^2\right\}\,.$$

From the preceding equations, we can explicitly calculate[2]:

$$\left[(Q_\mu E^\mu)^2 + q^2 Q^2\right] = 4M^2(E_e + E_{e'})^2 + q^2(4M^2 - q^2)$$
$$= 16M^2 E_e(E_e + \frac{q^2}{2M}) + 4M^2 q^2 = 16M^2 E_e E_{e'}\cos^2\frac{\theta}{2}\,.$$

Finally setting:

$$2(q^2)^2 = 8(-q^2)E_e E_{e'}\sin^2\frac{\theta}{2}\,,$$

we find:

$$\frac{X}{M^2} = 64 E_e E_{e'}\cos^2\frac{\theta}{2}\left\{\left(F_1^2 - \frac{q^2}{4M^2}\,F_2^2\right) + \frac{-q^2}{2M^2}\,(F_1+F_2)^2\,\tan^2\frac{\theta}{2}\right\}\,. \tag{14.73}$$

Substituting into (14.63), we obtain the *Rosenbluth formula*:

$$\frac{d\sigma}{d\Omega} = \frac{\alpha^2\cos^2\frac{\theta}{2}}{4E_e^2\sin^4\frac{\theta}{2}}\left(\frac{E_{e'}}{E_e}\right)$$
$$\times\left\{\left(F_1^2 - \frac{q^2}{4M^2}\,F_2^2\right) + \frac{-q^2}{2M^2}\,(F_1+F_2)^2\,\tan^2\frac{\theta}{2}\right\}. \tag{14.74}$$

[2] Using the relation $(p'e) = (pe')$.

Expressing $F_{1,2}$ in terms of the Sachs form factors, equation (14.44), we finally obtain:

$$\frac{d\sigma}{d\Omega} = \left[\frac{d\sigma}{d\Omega}\right]_0 \left(\frac{E_{e'}}{E_e}\right) \times \left[\frac{G_E^2 + \tau G_M^2}{1+\tau} + 2\tau G_M^2 \tan^2\frac{\theta}{2}\right] , \quad (14.75)$$

where, as before, $\tau = -\frac{q^2}{4M^2}$ and:

$$\left[\frac{d\sigma}{d\Omega}\right]_0 = \frac{\alpha^2 \cos^2\frac{\theta}{2}}{4E_e^2 \sin^4\frac{\theta}{2}} , \quad (14.76)$$

is the Mott cross section.

The Sachs form factors are convenient since equation (14.75) does not contain interference terms. The combinations $G_E^2 + \tau G_M^2$ and $G_M^2$ are obtained separately from the angular dependence of the cross section at fixed $q^2$. The relative sign of $G_{E,M}$ is obtained from the cross section on polarised protons.

The neutron form factor is obtained by subtraction from scattering on deuterium.

The measurement of the proton form factor by Hofstadter and collaborators at the end of the 1950s provided an important test of the non-pointlike, and presumably non-elementary, nature of the proton and, more generally, of other hadronic particles.

## 14.4 COMPTON SCATTERING

We consider the process

$$e + \gamma \to e' + \gamma' \quad (14.77)$$

which to the lowest order of perturbation theory is described by Feynman diagrams (1) and (2) in Fig. 14.3.

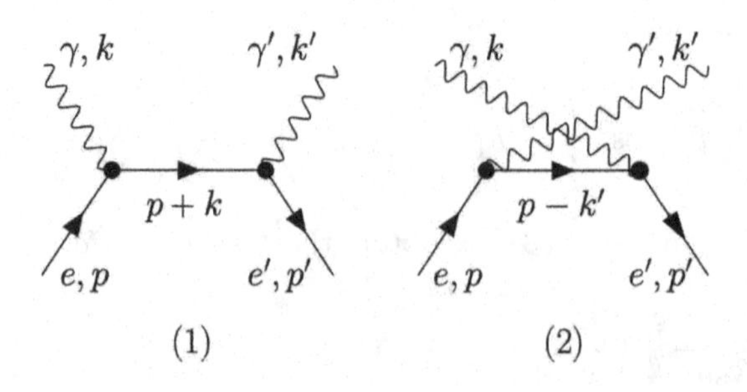


Figure 14.3 Feynman diagrams describing the amplitude of $e + \gamma \to e' + \gamma'$ to order $\alpha$.

To write down the corresponding amplitudes it is useful to introduce the Mandelstam variables[3]

$$s = (p+k)^2 = (p'+k')^2 = E_{cm}^2 \ , \tag{14.78}$$

$$t = (p'-p)^2 = (k-k')^2 \ , \tag{14.79}$$

$$u = (p-k')^2 = (k-p')^2 \ , \tag{14.80}$$

where $E_{cm}$ is the total energy of the system in the centre of mass frame, defined by the relation $\mathbf{p}+\mathbf{k}=0$, and $\mathbf{p}$ and $\mathbf{k}$ are the respective momenta of the electron and photon present in the initial state.

The cross section for the process is proportional to the squared modulus of the sum of the amplitudes associated with diagrams (1) and (2), averaged over the spin states of the particles in the initial state and summed over the spin states of the particles in the final state.

$$\sigma \propto \overline{|M_1+M_2|^2} = \overline{|M_1|^2} + \overline{|M_2|^2} + \overline{M_1 M_2^*} + \overline{M_1^* M_2} \ . \tag{14.81}$$

The amplitudes of diagrams (1) and (2) are easily computed to be:

$$M_1 = \bar{u}'\epsilon_\mu'^{\,*}(ie\gamma^\mu)i\frac{\not{p}+\not{k}+m}{(p+k)^2-m^2}(ie\gamma^\nu)\epsilon_\nu u \ , \tag{14.82}$$

and

$$M_2 = \bar{u}'\epsilon_\mu^{\,*}(ie\gamma^\mu)i\frac{\not{p}-\not{k}'+m}{(p-k')^2-m^2}(ie\gamma^\nu)\epsilon_\nu' u \ , \tag{14.83}$$

where $u = u_s(\boldsymbol{p})$ and $u' = u_{s'}(\boldsymbol{p}')$ are the four-spinors associated with the electrons $e$ and $e'$, and $\epsilon^\mu = \epsilon_r^\mu(\boldsymbol{k}')$ and $\epsilon' = \epsilon_{r'}^\mu(\mathbf{k}')$ are the polarisation vectors of the photons $\gamma$ and $\gamma'$.

Before going into the detailed calculations, we comment on some properties of amplitudes with external photons implied by invariance under gauge transformations, which are useful for the calculation of the sum over polarization states of the photons.

Implications of Gauge Invariance. The amplitude of any process which involves a photon with polarisation $r$ in the initial or final state can be put in the form

$$M = \epsilon_{r,\mu}\mathcal{M}^\mu \ . \tag{14.84}$$

The polarisation vector $\epsilon_r$ depends on the gauge, as is easily seen by considering the transformation

$$A_\mu(x) \to A_\mu{}'(x) = A_\mu(x) + \partial_\mu \Lambda(x) \ , \tag{14.85}$$

[3]For completeness, we also give the definition of the variable $t$, which will not be used in the calculations discussed in this section.

where

$$A_\mu(x) = \epsilon_{r,\mu} e^{ikx} \ . \tag{14.86}$$

Choosing $\Lambda(x) = C e^{ikx}$ we find

$$A_\mu{}'(x) = (\epsilon_{r,\mu} + iCk_\mu) e^{ikx} = \epsilon_{r,\mu}{}' e^{ikx} \tag{14.87}$$

i.e. the gauge transformation changes the polarisation vector $\epsilon_r$ into $\epsilon_r{}'$.

The condition that the amplitude (14.84) remains invariant under the transformation of the polarisation vector is

$$\epsilon_{r,\mu} \mathcal{M}^\mu = \epsilon_{r,\mu}{}' \mathcal{M}^\mu = (\epsilon_{r,\mu} + iCk_\mu) \mathcal{M}^\mu \ , \tag{14.88}$$

that is:

$$k_\mu \mathcal{M}^\mu = 0 \ . \tag{14.89}$$

This result allows us to make the summation

$$\sum_r |\mathcal{M}|^2 = (\mathcal{M}^\mu)^* \mathcal{M}^\nu \sum_{r=1}^{2} \epsilon^*_{r,\mu} \epsilon_{r,\nu} \ , \tag{14.90}$$

where (see the discussion of Chapter 5)

$$\sum_{r=1}^{2} \epsilon^*_{r,\mu} \epsilon_{r,\nu} = -g_{\mu\nu} - \frac{k_\mu k_\nu - (kn)(k_\mu n_\nu + k_\nu n_\mu)}{(kn)^2} \ , \tag{14.91}$$

with $n^\mu \equiv (1,0,0,0)$. From (14.89) it follows that

$$\sum_{r=1}^{2} |\mathcal{M}|^2 = (\mathcal{M}^\mu)^* \mathcal{M}^\nu \sum_{r=1}^{2} \epsilon^*_{r,\mu} \epsilon_{r,\nu} = -(\mathcal{M}^\mu)^* \mathcal{M}^\nu \ g_{\mu\nu} = -|\mathcal{M}|^2 \ . \tag{14.92}$$

The result is that:

- when multiplied by gauge invariant amplitudes, the sum over polarizations of the photon can be replaced by $-g_{\mu\nu}$.

**Calculation of $|M_1|^2$ and $|M_2|^2$.** To sum over the spins of the electrons, we use the relation equation (6.84), Section 6.1.4:

$$\sum_s u_s(\boldsymbol{p}) \overline{u}_s(\boldsymbol{p}) = \frac{\not{p} + m}{2m} \ , \tag{14.93}$$

where $m$ is the mass of the electron. In this way we obtain:

$$\begin{aligned} \overline{|M_1|^2} &= \frac{e^4}{(s-m^2)^2} \frac{1}{4} \sum_{s,s'} \bar{u}' \gamma^\mu (\not{p} + \not{k} + m) \gamma^\nu u \bar{u} \gamma_\nu (\not{p} + \not{k} + m) \gamma_\mu u' \qquad (14.94) \\ &= \frac{e^4}{(s-m^2)^2} \frac{1}{16m^2} \mathrm{Tr}\ (\not{p}' + m) \gamma^\mu (\not{p} + \not{k} + m) \gamma^\nu (\not{p} + m) \gamma_\nu (\not{p} + \not{k} + m) \gamma_\mu . \end{aligned}$$

To calculate the trace of the right-hand side of equation (14.94), we apply the rules summarised in Section 6.3.

First, using the invariance property of the trace under cyclic permutations of the arguments, we can move the $\gamma_\mu$ matrix to the left of the factor $(\not{p}'+m)$. Then exploiting the relation, valid for any four-vector $a$,

$$\gamma_\mu \not{a} \gamma^\mu = a_\rho \gamma_\mu \gamma^\rho \gamma^\mu = a_\rho \gamma_\mu (2g^{\rho\mu} - \gamma^\mu \gamma^\rho) = -2\not{a} \ , \tag{14.95}$$

we can put the trace to be calculated in the form

$$\text{Tr } (4m - 2\not{p}')(\not{p} + \not{k} + m)(4m - 2\not{p})(\not{p} + \not{k} + m) = A + B + C + D \ , \tag{14.96}$$

with

$$A = 16m^2 \text{Tr } (\not{p} + \not{k} + m)(\not{p} + \not{k} + m) \tag{14.97}$$
$$B = 4\text{Tr } \not{p}'(\not{p} + \not{k} + m)\not{p}(\not{p} + \not{k} + m) \tag{14.98}$$
$$C = -8m\text{Tr } \not{p}'(\not{p} + \not{k} + m)(\not{p} + \not{k} + m) \tag{14.99}$$
$$D = -8m\text{Tr } (\not{p} + \not{k} + m)\not{p}(\not{p} + \not{k} + m) \ . \tag{14.100}$$

Defining $\tilde{s} = p+k = p'+k'$, from which it follows that $\tilde{s}^2 = s$, and recalling that the trace of the product of an odd number of $\gamma$-matrices vanishes, we can immediately calculate the term $A$, with the result

$$A = 16m^2 \text{Tr } (\not{\tilde{s}}\not{\tilde{s}} + m^2) = 64m^2(s + m^2) \ , \tag{14.101}$$

which is easily obtained from the identity, valid for any two 4-vectors $a$ and $b$:

$$\text{Tr } \not{a}\not{b} = a_\rho b_\sigma \text{Tr } (\gamma^\rho \gamma^\sigma) = 4a_\rho b_\sigma g^{\rho\sigma} = 4(ab) \ . \tag{14.102}$$

The contribution of the terms $C$ and $D$ is obtained in a similar manner. For example, for $C$ we find the expression

$$C = -8m\text{Tr } \not{p}'(\not{\tilde{s}} + m)(\not{\tilde{s}} + m) = -16m^2 \text{Tr } \not{p}'\not{\tilde{s}} = -64m^2(p'\tilde{s}) \ , \tag{14.103}$$

which substituting

$$(p'\tilde{s}) = p'(p' + k') = m^2 + (p'k') \ , \tag{14.104}$$

with $(p'k') = (s - m^2)/2$ becomes

$$C = -32m^2(s + m^2) \ . \tag{14.105}$$

The calculation of term $D$ is similar and the result is the same

$$D = -64m^2(p\tilde{s}) = -32m^2(s + m^2) \ . \tag{14.106}$$

Summing (14.101), (14.105) and (14.106) we find

$$A + C + D = 0 \ , \tag{14.107}$$

from which it follows that the only term which contributes to the trace (14.96) is $B$, which we can rewrite as

$$B = \text{Tr}\,(\not{p}'(\not{\tilde{s}}\not{p}\not{\tilde{s}} + m^2\not{p}'\not{p}) \,. \tag{14.108}$$

The calculation is carried out by using the relation

$$\begin{aligned} \text{Tr}\,(\not{a}\not{b}\not{c}\not{d}) &= a_\lambda b_\mu c_\nu d_\rho \text{Tr}\,(\gamma^\lambda\gamma^\mu\gamma^\nu\gamma^\rho) \\ &= 4a_\lambda b_\mu c_\nu d_\rho (g^{\lambda\mu}g^{\nu\rho} + g^{\lambda\rho}g^{\mu\nu} - g^{\lambda\nu}g^{\mu\rho}) \\ &= 4[(ab)(cd) + (ad)(bc) - (ac)(bd)] \,. \end{aligned} \tag{14.109}$$

The result is

$$\begin{aligned} B &= 16[2(p'\tilde{s})(p\tilde{s}) - (pp')s + m^2(pp')] \\ &= 32\{(pk)(pk') + m^2[(pp') + (p'k)]\} \,, \end{aligned} \tag{14.110}$$

which can be written in terms of the variables $s$ and $u$ by noting that

$$(pk') = \frac{m^2 - u}{2} \,, \quad [(pp') + (p'k)] = \frac{s + m^2}{2} \,. \tag{14.111}$$

Finally we find

$$B = 8\left[4m^4 + (s - m^2)(m^2 - u) + 2m^2(s - m^2)\right] \,. \tag{14.112}$$

From equations (14.94) and (14.96) we find, in conclusion:

$$\overline{|M_1|^2} = \frac{1}{4m^2}\frac{2e^4}{(s - m^2)^2}\left[4m^4 - (s - m^2)(u - m^2) + 2m^2(s - m^2)\right] \,. \tag{14.113}$$

The expression for $|M_2|^2$, see equation (14.83), is obtained from (14.113) with the substitution $s \rightleftharpoons u$:

$$\overline{|M_2|^2} = \frac{1}{4m^2}\frac{2e^4}{(u - m^2)^2}\left[4m^4 - (u - m^2)(s - m^2) + 2m^2(u - m^2)\right] \,. \tag{14.114}$$

Interference. The calculation of the interference terms is carried out in a similar way. We obtain

$$\overline{M_1 M_2^*} + \overline{M_1^* M_2} = \frac{1}{4m^2}\frac{4e^4}{(s - m^2)(u - m^2)}\left[4m^4 + m^2(s - m^2) + m^2(u - m^2)\right] \,. \tag{14.115}$$

Summing Up. Putting together the results (14.113), (14.114) and (14.115) we find, finally:

$$\begin{aligned} \overline{|M_1 + M_2|^2} &= \frac{2}{m^2}e^4\left[\left(\frac{m^2}{s - m^2} + \frac{m^2}{u - m^2}\right)^2\right. \\ &+ \left.\left(\frac{m^2}{s - m^2} + \frac{m^2}{u - m^2}\right) - \frac{1}{4}\left(\frac{u - m^2}{s - m^2} + \frac{s - m^2}{u - m^2}\right)\right] \,. \end{aligned} \tag{14.116}$$

Invariance under Gauge Transformations. As we saw in the previous section, the gauge transformation (14.85) does not leave the polarisation vectors of the electromagnetic field unchanged. We can write the total scattering amplitude, $M_1 + M_2$, in the form[4]

$$M_1 + M_2 = \epsilon_\mu \epsilon_\nu (\mathcal{M}_1^{\mu\nu} + \mathcal{M}_2^{\mu\nu}) \,, \tag{14.117}$$

and we see from equation (14.88) that it must satisfy the condition

$$k_\mu k'_\nu (\mathcal{M}_1^{\mu\nu} + \mathcal{M}_2^{\mu\nu}) = 0 \,. \tag{14.118}$$

We consider the first term on the left-hand side of (14.118), which we can rewrite using the definitions (14.82) and (14.83), with $k^2 = 0$ and $p^2 = m^2$,

$$k_\mu k'_\nu \mathcal{M}_1^{\mu\nu} = \frac{\bar{u}' \not{k}' (\not{p} + \not{k} + m) \not{k} u}{2(pk)} \,. \tag{14.119}$$

The numerator of this expression can be put in a very simple form by noting that

$$(\not{p} + \not{k} + m)\not{k} u = (\not{p} + m)\not{k} u = [2(pk) - \not{k}(\not{p} - m)]u = 2(pk)u \,, \tag{14.120}$$

from which it follows that

$$k_\mu k'_\nu \mathcal{M}_1^{\mu\nu} = \bar{u}' \not{k}' u \,. \tag{14.121}$$

The second term, thanks to the relation $p - k' = p' - k$, can be put in the form

$$k_\mu k'_\nu \mathcal{M}_2^{\mu\nu} = -\frac{\bar{u}' \not{k} (\not{p}' - \not{k} + m) \not{k}' u}{2(p'k)} \,. \tag{14.122}$$

Since

$$-\bar{u}' \not{k} (\not{p}' - \not{k} + m) = -\bar{u}' \not{k} (\not{p}' + m) = -\bar{u}' [2(p'k) - (\not{p}' - m)\not{k}] = -2(p'k)\bar{u}' \,, \tag{14.123}$$

we obtain

$$k_\mu k'_\nu \mathcal{M}_2^{\mu\nu} = -\bar{u}' \not{k}' u = -k_\mu k'_\nu \mathcal{M}_1^{\mu\nu} \,, \tag{14.124}$$

confirming that the scattering amplitude for $e + \gamma \to e' + \gamma'$ is invariant under gauge transformations. We note that only the sum $M_1 + M_2$ is invariant, while the amplitudes corresponding to the two processes illustrated in the Feynman diagrams of the figure, considered separately, are not.

---

[4]Without loss of generality, we may choose real polarisation vectors.

Klein–Nishina Cross Section. Now we will use the result (14.116) to obtain the cross section in the laboratory frame, defined by the relations

$$p \equiv (m, \mathbf{0}), \quad p' \equiv (E', \mathbf{p}') ,$$

$$k \equiv (\omega, \boldsymbol{k}), \quad k' \equiv (\omega', \boldsymbol{k}') ,$$

with $|\mathbf{k}| = \omega$ and $|\mathbf{k}|' = \omega'$, from which it follows that

$$s = (p+k)^2 = m^2 + 2m\omega, \quad u = (p-k')^2 = m^2 - 2m\omega' . \tag{14.125}$$

Furthermore, $(p-p')^2 = (k'-k)^2$ implies that

$$1 - \cos\theta = m\left(\frac{1}{\omega'} - \frac{1}{\omega}\right) , \tag{14.126}$$

where $\theta$ is the angle between the vectors $\boldsymbol{k}$ and $\boldsymbol{k}'$.

Substituting equations (14.125) and (14.126) into (14.116), we obtain

$$\overline{|M_1 + M_2|^2} = \frac{1}{4m^2} 2e^4 \left(\frac{\omega'}{\omega} + \frac{\omega}{\omega'} - \sin^2\theta\right) . \tag{14.127}$$

The cross section is defined as

$$d\sigma = \frac{1}{F}\frac{W}{T}\frac{V}{(2\pi^3)} d^3p' \frac{V}{(2\pi^3)} d^3k' , \tag{14.128}$$

where the flux of incident photons, $F$, is

$$F = \frac{1}{V}\frac{|\mathbf{k}|}{\omega} = \frac{1}{V} , \tag{14.129}$$

while the transition probability per unit time is

$$\begin{aligned} \frac{W}{T} &= \frac{1}{T} VT(2\pi)^4 \delta^{(4)}(p + k - p' - k') \\ &\quad \times \frac{1}{2V\omega}\frac{1}{2V\omega'}\frac{m}{Vm}\frac{m}{VE'} \overline{|M_1 + M_2|^2} . \end{aligned} \tag{14.130}$$

After substituting equations (14.127), (14.129) and (14.130) into (14.128), we may carry out the integration over $\mathbf{p}'$ exploiting the $\delta$-function associated with conservation of momentum, to obtain

$$d\sigma = \frac{1}{64\pi^2}\frac{1}{\omega\omega' m E'} 2e^4 \left(\frac{\omega'}{\omega} + \frac{\omega}{\omega'} - \sin^2\theta\right) \delta(m + \omega - E' - \omega') d^3k' , \tag{14.131}$$

or, recalling that $d^3k' = d\Omega'\omega'^2 d\omega'$, where $\Omega'$ is the solid angle which identifies the direction of the vector $\boldsymbol{k}'$,

$$\frac{d\sigma}{d\Omega'} = \frac{e^4}{32\pi^2}\frac{\omega'}{\omega}\frac{1}{mE'} \left(\frac{\omega'}{\omega} + \frac{\omega}{\omega'} - \sin^2\theta\right) \delta(m + \omega - E' - \omega') d\omega' . \tag{14.132}$$

The integration over $\omega'$ with the $\delta$-function is carried out using the rule

$$\delta[F(\omega')] = \left|\frac{dF}{d\omega'}\right|^{-1} \delta(\omega' - \omega_0') \ , \tag{14.133}$$

with $F(\omega_0') = 0$. Recalling that $E' = \sqrt{m^2 + |\mathbf{k} - \boldsymbol{k}'|^2}$, we find

$$\begin{aligned} \left|\frac{d}{d\omega'}(m + \omega - E' - \omega')\right| &= \frac{1}{E'}(\omega' - \omega\cos\theta) + 1 \\ &= \frac{(p'k')}{E'\omega'} = \frac{(pk)}{E'\omega'} = \frac{m\omega}{E'\omega'} \ , \end{aligned} \tag{14.134}$$

and therefore

$$\frac{d\sigma}{d\Omega'} = \frac{e^4}{32\pi^2}\frac{\omega'}{\omega}\frac{1}{mE'}\frac{m\omega}{E'\omega'}\left(\frac{\omega'}{\omega} + \frac{\omega}{\omega'} - \sin^2\theta\right) \ , \tag{14.135}$$

or

$$\frac{d\sigma}{d\Omega'} = \frac{\alpha^2}{2m^2}\left(\frac{\omega'}{\omega}\right)^2\left(\frac{\omega'}{\omega} + \frac{\omega}{\omega'} - \sin^2\theta\right) \ . \tag{14.136}$$

The cross section (14.136) was obtained by Klein and Nishina in 1929 and it provides an accurate description of the Compton effect, observed experimentally for the first time in 1923.

We now consider equation (14.136) in the non-relativistic limit: $\omega/m \ll 1$. Solving equation (14.126) for $\omega'$ we obtain the relation

$$\omega' = \frac{\omega}{1 + \frac{\omega}{m}(1 - \cos\theta)} \ , \tag{14.137}$$

which shows that, in the limit which interests us here, $\omega/\omega' \to 1$ and equation (14.136) becomes

$$\frac{d\sigma}{d\Omega'} = \frac{\alpha^2}{2m^2}\left(1 + \cos^2\theta\right) \ . \tag{14.138}$$

To obtain the total cross section we carry out the angular integral

$$\sigma = 2\pi\,\frac{\alpha^2}{2m^2}\int d\cos\theta\,\left(1 + \cos^2\theta\right) = \frac{8\pi}{3}\frac{\alpha^2}{m^2} \ . \tag{14.139}$$

Equation (14.139) is the Thomson cross section, which describes the interaction of the classical electromagnetic field with an electron.

## 14.5 INVERSE COMPTON SCATTERING

In the previous section we considered the photon-electron scattering cross section in the laboratory frame, in which the electron is at rest. The expression we obtained for the squared modulus of the transition amplitude is relativistically invariant, however, and can also be used to describe collisions involving electrons in flight, in which case the kinematical variables are defined as

$$p \equiv (E, \boldsymbol{p}), \quad p' \equiv (E', \boldsymbol{p}') \; , \tag{14.140}$$

$$k \equiv (\omega, \boldsymbol{k}), \quad k' \equiv (\omega', \boldsymbol{k}') \; . \tag{14.141}$$

We will now consider relativistic electrons, with $E \gg m$, and see how in these conditions it is possible to obtain a final state photon with energy $\omega' \gg \omega$. From conservation of total four-momentum, which implies

$$(pk) = (p'k') = (p + k - k')k' = (pk') + (kk') \; , \tag{14.142}$$

we obtain the relation

$$E\omega - \mathbf{p} \cdot \mathbf{k} = E\omega' - \mathbf{p} \cdot \mathbf{k}' + \omega\omega'(1 - \cos\theta) \; , \tag{14.143}$$

that we can rewrite as

$$\omega(1 - \beta\cos\phi) = \omega'(1 - \beta\cos\phi') + \frac{\omega\omega'}{E}(1 - \cos\theta) \; , \tag{14.144}$$

where $\phi$ and $\phi'$ are the angles contained between the direction of $\mathbf{p}$ and, respectively, those of $\boldsymbol{k}$ and $\boldsymbol{k}'$, and $\beta = |\mathbf{p}|/E$ is the electron velocity (in units with $c = 1$).

Solving equation (14.144) for $\omega'$, we obtain

$$\omega' = \omega \, \frac{1 - \beta\cos\phi}{1 - \beta\cos\phi' + \frac{\omega}{E}(1 - \cos\theta)} \; , \tag{14.145}$$

which obviously reduces to equation (14.137) in the limit $\beta \to 0$. For relativistic electrons $\beta \approx 1$ and we can use the expansion, valid for $1/\gamma^2 \ll 1$,

$$\beta = \frac{|\mathbf{p}|}{E} = \sqrt{\frac{E^2 - m^2}{E^2}} = \sqrt{1 - \frac{1}{\gamma^2}} \approx 1 - \frac{1}{2\gamma^2} \; , \tag{14.146}$$

where $\gamma = E/m = (1 - \beta^2)^{-1/2}$ is the Lorentz factor.

In this regime, the energy of the final state photon has a maximum for $\phi = \pi$ and $\phi' = 0$, implying $\theta = \phi - \phi' = \pi$:

$$\omega'_{max} = \omega \frac{1 + \beta}{1 - \beta + \frac{2\omega}{E}} \approx \omega \frac{2}{\frac{m^2}{2E^2} + \frac{2\omega}{E}} = E \frac{z}{1 + z}, \tag{14.147}$$

with

$$z = \frac{4\omega E}{m^2} \; . \tag{14.148}$$

Now we consider the case in which $\boldsymbol{p}$ and $\boldsymbol{k}$ are still oppositely directed, i.e. $\phi = \pi$, but the momenta of the particles in the final state are such that

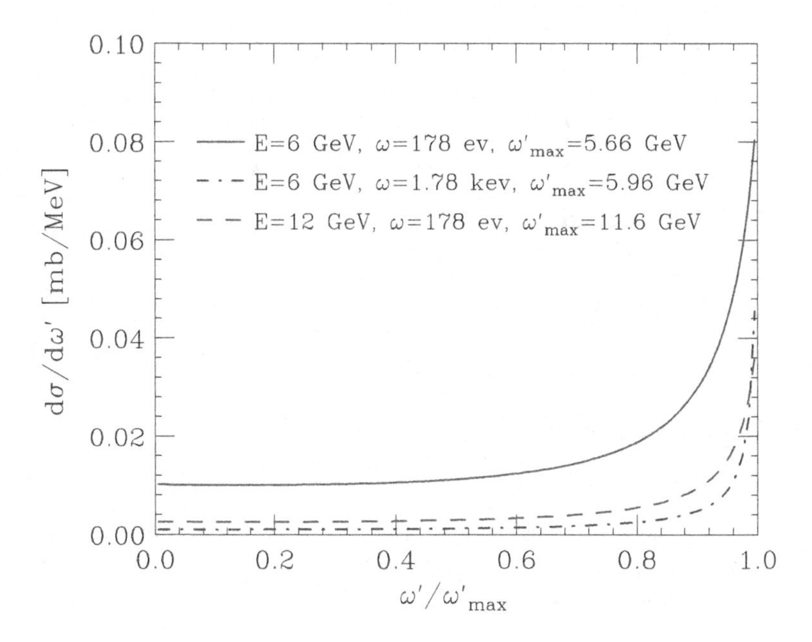


Figure 14.4 Energy distribution of photons produced by Compton scattering of relativistic electrons.

$\phi' \approx 0$ and $\theta \approx \pi$. In these conditions equation (14.145) can be rewritten in the form

$$\omega' \approx E \frac{z}{1+x+z} \,, \tag{14.149}$$

with

$$x = \frac{E^2}{m^2} \phi'^2 \,. \tag{14.150}$$

In Compton scattering of electrons produced by a particle accelerator on photons with energies of order 1 eV, obtained with a laser, it is possible to produce photons with maximum energies of order GeV. The corresponding differential cross section shown in Fig. 14.4 exhibits a peak at $\omega \approx \omega'_{max}$ whose width decreases with the increase of $E$ and $\omega$.

This technique is used for the production of beams of photons for nuclear physics experiments. The first beam of this type, of energy ∼80 MeV, was obtained at the end of the 1970s at the Frascati Laboratory, using electrons from the *Adone storage ring* with energy $E = 1.5$ GeV and photons of energy $\omega = 2.45$ eV. A photon beam obtained in this way possesses the important property of having a high degree of polarisation. This is a consequence of the fact that for relativistic electrons helicity is a good quantum number.

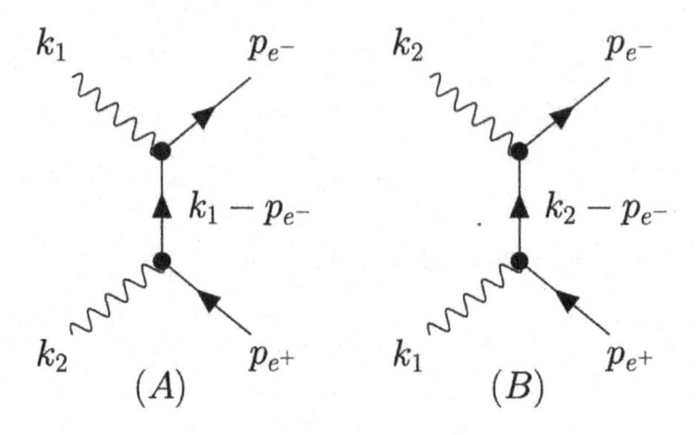


Figure 14.5 Feynman diagrams describing the amplitude of $\gamma + \gamma \to e^+ + e^-$ to order $\alpha$.

Therefore, in the scattering process there is very little transfer of angular momentum, and the polarisation of the scattered photon is very close to that of the incoming laser light.

## 14.6 THE PROCESSES $\gamma\gamma \to e^+e^-$ AND $e^+e^- \to \gamma\gamma$

The discussion of photon-electron scattering in Section 14.4 can be easily generalised to describe the process

$$\gamma_1 + \gamma_2 \to e^- + e^+ \ , \tag{14.151}$$

in which the collision between the two photons results in the creation of an electron-positron pair. Comparing the Feynman diagrams which describe this process to order $\alpha^2$, represented in Fig. 14.5, to those for Compton scattering, Fig. 14.3, one sees that the only difference consists in the substitution

$$k \to k_1 \ , \ k' \to -k_2 \ , \ p \to -p_{e^-} \ , \ p' \to p_{e^+} \ . \tag{14.152}$$

From this it follows that the expression for $\overline{|M_1 + M_2|^2}$ written in terms of Mandelstam variables $s$, $t$ and $u$, equation (14.116), can also be used to obtain the transition probability for the process (14.151) by setting

$$s = (k_1 - p_{e^-})^2 = (p_{e^+} - k_2)^2 \ , \tag{14.153}$$

$$t = (p_{e^+} + p_{e^-})^2 = (k_1 + k_2)^2 \ , \tag{14.154}$$

$$u = (p_{e^-} - k_2)^2 = (k_1 - p_{e^+}^2)^2 \ . \tag{14.155}$$

where $k_1 \equiv (\omega_1, \mathbf{k}_1)$, $k_2 \equiv (\omega_2, \mathbf{k}_2)$, $p_{e^+} \equiv (E_{e^+}, \mathbf{k}_{e^+})$ and $p_{e^-} \equiv (E_{e^-}, \mathbf{k}_{e^-})$ represent, respectively, the 4-momenta of the particles in the initial and final states.

We note that, compared to the case of Compton scattering, the roles of the variables $s$ and $t$ are exchanged. One says therefore that the two processes (14.77) and (14.151) correspond to *crossed channels* of the same reaction.

The calculation of the differential cross section, which can easily be carried out starting from equation (14.116), gives the result

$$\begin{aligned} \frac{d\sigma}{ds} &= 8\pi\frac{\alpha^2}{t^2}\left[\left(\frac{m^2}{s-m^2}+\frac{m^2}{u-m^2}\right)^2\right. \\ &+ \left.\left(\frac{m^2}{s-m^2}+\frac{m^2}{u-m^2}\right)-\frac{1}{4}\left(\frac{u-m^2}{s-m^2}+\frac{s-m^2}{u-m^2}\right)\right] . \end{aligned} \quad (14.156)$$

In the centre of mass frame, in which $\boldsymbol{p}_{e^+} = -\boldsymbol{p}_{e^+}$, $\boldsymbol{k}_1 = -\boldsymbol{k}_2$, $\omega_1 = \omega_2 = \omega$ and $E_{e^+} = E_{e^-} = \omega$, the total cross section, which is obtained by integration of (14.156), has the form

$$\sigma = \frac{\pi}{2}\frac{\alpha^2}{m^2}(1-\beta^2)\left[(3-\beta^4)\ln\frac{1+\beta}{1-\beta} - 2\beta(2-\beta^2)\right] , \quad (14.157)$$

where

$$\beta = \left(1 - \frac{m^2}{\omega^2}\right)^{1/2} , \quad (14.158)$$

and, obviously, the energy of the photons must satisfy the condition $\omega > m$.

Equation (14.157) can easily be generalised to the case of any reference frame in which the photons move along opposing paths in the same overall direction, with energies $\omega_1$ and $\omega_2$, by making the substitution

$$\beta \to \left(1 - \frac{m^2}{\omega_1\omega_2}\right)^{1/2} , \quad (14.159)$$

which implies the threshold condition

$$\omega_1 \geq \frac{m^2}{\omega_2} . \quad (14.160)$$

The cross section for the process (14.151) applies to the collisions of high energy photons in cosmic rays with the thermal background radiation (*cosmic microwave background*, CMB) which permeates the universe. The mean free path of a photon with energy $\omega_1$ can be estimated from the relation

$$\lambda(\omega_1) = \frac{1}{\sigma\rho_\gamma} , \quad (14.161)$$

where $\rho_\gamma$ is the density of thermal photons, of energy $\omega_2$, and $\sigma$ is the total cross section defined by equations (14.157) and (14.159).

Comment. Setting $\omega_2 \approx 6 \times 10^{-4}$ eV, the value which is obtained from the spectral distribution of a black body at a temperature of $T = 2.7$ K, and using the result of recent measurements which give $\rho_\gamma = 410$ cm$^{-3}$, a threshold energy of $\omega_1 \approx 4 \times 10^{14}$ eV $= 400$ TeV is obtained. At energies just above the threshold the cross section has a value $\sigma \approx 10^{-26}$ cm$^2$, and the mean free path of $\lambda \approx 31.5$ kpc turns out to be comparable with the dimensions of our galaxy.

We conclude this section by noting that, as well as photon-electron scattering (14.77) and $e^+e^-$ pair production (14.151), there is another process described by the transition probability (14.116), which is annihilation:

$$e^+ + e^- \to \gamma_1 + \gamma_2 \,. \tag{14.162}$$

The cross section calculation is carried out using the Mandelstam variables

$$s = (p_{e^-} - k_1)^2 = (k_2 - p_{e^+})^2 \,, \tag{14.163}$$

$$t = (p_{e^+} + p_{e^-})^2 = (k_1 + k_2)^2 \,, \tag{14.164}$$

$$u = (k_2 - p_{e^-})^2 = (p_{e^+}^2 - k_1) \,, \tag{14.165}$$

which are obtained immediately from (14.155) by changing the signs of all the four-momenta.

In the centre of mass frame the resulting cross section has the form

$$\sigma = \frac{\pi\alpha^2}{m^2(\lambda+1)}\left[\frac{\lambda^2+4\lambda+1}{\lambda^2-1}\ \ln(\lambda+\sqrt{\lambda^2-1}) \ - \ \frac{\lambda+3}{\sqrt{\lambda^2-1}}\right] \,, \tag{14.166}$$

with $\lambda = E_{cm}/m$. In the non-relativistic limit equation (14.166) becomes

$$\sigma \approx \frac{1}{v_{rel}}\frac{\pi\alpha^2}{m^2} \,, \tag{14.167}$$

where $v_{rel} = 2\lambda\sqrt{\lambda^2-1}$ is the relative velocity of the electron–positron system.

From the cross section of the process (14.162) the mean lifetime of parapositronium (i.e. the bound state of an electron and positron with spin zero) can be obtained, for which the decay into two photons dominates[5] (cf. Chapter 12).

The mean lifetime is obtained from the product of the probability of the annihilation process and the flux, equal to $v_{rel}|\psi(0)|^2$, where $\psi(r)$ is the normalised wave function of the positronium ground state. Taking into account

[5]As shown in Chapter 12, the spin one state, orthopositronium, has negative charge conjugation and cannot annihilate into two photons, resulting in a much longer lifetime.

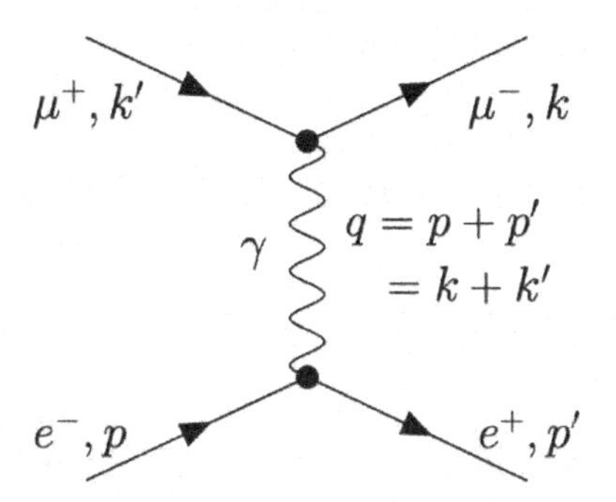


Figure 14.6 Graphical representation of the amplitude for electron-positron annihilation into a $\mu^+ \ \mu^-$ pair. The arrows indicate the flux of negative electrical charge, therefore they are in the opposite direction to the momenta of $e^+$ and $\mu^+$; $q^\mu$ is the momentum associated with the line which represents the propagation of the photon; momentum is conserved at the vertices.

that only one of the four spin states of the $e^+e^-$ system is available for annihilation into two photons, we obtain

$$\Gamma = 4|\psi(0)|^2 \ v_{rel}\sigma \approx |\psi(0)|^2 \frac{4\pi\alpha^2}{m^2} = \frac{1}{2} m\alpha^5 \tag{14.168}$$

which corresponds to a lifetime

$$\tau = \Gamma^{-1} \approx 1.2 \times 10^{-10} \text{ s} \tag{14.169}$$

in agreement with the experimental result quoted in Chapter 12.

## 14.7 $e^+e^- \to \mu^+\mu^-$ ANNIHILATION

We consider the process of annihilation of an electron–positron pair and creation of a muon–antimuon pair

$$e^+ \ e^- \to \mu^+ \ \mu^- \ , \tag{14.170}$$

represented schematically by the Feynman diagram of Fig. 14.6.

To obtain the cross section to second order in the interaction Lagrangian we must calculate the matrix element

$$\langle f|S^{(2)}|i\rangle \ , \tag{14.171}$$

where the initial and final states are[6] $|i\rangle = |pr, p'r'\rangle = a^\dagger_{pr} b^\dagger_{p'r'}|0\rangle$ and $|f\rangle = |ks, k's'\rangle = a^\dagger_{ks} b^\dagger_{k's'}|0\rangle$ and the $S$-matrix has the form

$$S^{(2)} = -\frac{e^2}{2} \, 2 \int d^4x \int d^4y \ T\{: \bar{\psi}_e(x) \not{A}(x) \psi_e(x) : : \bar{\psi}_\mu(y) \not{A}(y) \psi_\mu(y) :\} \ , \tag{14.172}$$

[6]For simplicity of notation, where no confusion can arise, we use the same symbols for the creation and destruction operators of the electrons and muons.

with the factor 2 in front of the integrals to allow for the fact that two identical contributions are obtained with the exchange $x \rightleftharpoons y$. Obviously the field operators $\psi_e$ and $\psi_\mu$, which describe the electron and muon (and relevant antiparticles) commute, both with each other and the electromagnetic field. We can therefore rewrite the time-ordered product in the form

$$T\{j_e^\lambda(x)A_\lambda(x)j_\mu^\nu(y)A_\nu(y)\} = j_e^\lambda(x)j_\mu^\nu(y)T\{A_\lambda(x)A_\nu(y)\} \ , \tag{14.173}$$

where $j_e^\lambda =: \bar{\psi}_e\gamma^\lambda\psi_e :$ and $j_\mu^\nu =: \bar{\psi}_\mu\gamma^\nu\psi_\mu :$ are the electromagnetic currents of the electron and the muon. Because neither the initial nor the final state contain photons, the contribution of the electromagnetic field to the transition matrix element is given by

$$\langle 0|T\{A^\lambda(x)A^\nu(y)\}|0\rangle = iD_F^{\lambda\nu}(x-y) \ , \tag{14.174}$$

where $D_F^{\lambda\nu}$ is the photon propagator which, as we showed in Section 8.4, is given by

$$iD_F^{\lambda\nu}(x-y) = \int \frac{d^4q}{(2\pi)^4}\, e^{iq(x-y)}\, \frac{-ig^{\lambda\nu}}{q^2+i\epsilon} \ . \tag{14.175}$$

The other matrix elements to be calculated are

$$\langle 0|j_e^\lambda(x)|pr, p'r'\rangle \ , \quad \langle ks, k's'|j_\mu^\nu(y)|0\rangle \ . \tag{14.176}$$

We consider the electron current (spin indices are omitted to simplify the notation)

$$j_e^\lambda(x) = \sum_{q,q'} N_q N_{q'} : \left[a_q^\dagger \bar{u}_e(q)e^{iqx} + b_q\bar{v}_e(q)e^{-iqx}\right]\gamma^\lambda \times \left[a_{q'}u_e(q')e^{-iq'x} + b_{q'}^\dagger v_e(q')e^{iq'x}\right] : \ , \tag{14.177}$$

with $N_q = \sqrt{m/VE_q}$. The only term which gives a non-zero contribution to the matrix element (14.176) is the one containing the destruction operators $a_p$ and $b_{p'}$, i.e.

$$N_pN_{p'}\ b_{p'}a_p\ \bar{v}_e(p')\gamma^\lambda u_e(p)\ e^{-i(p+p')} \ . \tag{14.178}$$

Similarly, the only term which contributes to the muon current matrix element is

$$N_kN_{k'}\ a_k^\dagger b_{k'}^\dagger\ \bar{u}_\mu(k)\gamma^\nu v_\mu(k')\ e^{+i(k+k')} \ . \tag{14.179}$$

We therefore obtain ($u_e = u_e(p)$, $\bar{u}'_e = \bar{u}_e(p')$, ...)

$$\langle f|S^{(2)}|i\rangle = -e^2\ N_pN_{p'}N_kN_{k'}(\bar{v}'_e\gamma^\lambda u_e)(\bar{u}_\mu\gamma^\nu v'_\mu) \times \int d^4x \int d^4y\ e^{-i[(p+p')x-(k+k')y]}(-ig_{\lambda\nu})\int \frac{d^4q}{(2\pi)^4}\frac{e^{iq(x-y))}}{q^2} \ . \tag{14.180}$$

The $+i\epsilon$ term in the denominator of the photon propagator is irrelevant, because $q^2 > 0$. Carrying out the integrations and substituting the expressions for the spinor normalisation constants, we obtain the result

$$\langle f|S^{(2)}|i\rangle = (2\pi)^4\delta^{(4)}(p+p'-k-k')\frac{m}{V\sqrt{E_pE_{p'}}}\,\frac{M}{V\sqrt{E_kE_{k'}}}\mathcal{M}_{if}\ , \quad (14.181)$$

where $m$ and $M$ are the masses of the electron and the muon, respectively. The invariant amplitude $\mathcal{M}_{if}$ is defined by

$$\mathcal{M}_{if} = ie^2\ (\bar{v}'_e\gamma^\lambda u_e)\frac{1}{(p+p')^2}(\bar{u}_\mu\gamma_\lambda v'_\mu)\ , \quad (14.182)$$

from which it follows that

$$|\mathcal{M}_{if}|^2 = \frac{e^4}{(p+p')^2}(\bar{v}'_e\gamma^\lambda u_e\bar{u}_e\gamma^\nu v'_e)(\bar{u}_\mu\gamma_\lambda v'_\mu\bar{v}'_\mu\gamma_\nu u_\mu)\ . \quad (14.183)$$

The average over spins of the particles in the initial state, and the sum over spins of particles in the final state, is carried out by using the completeness relations for the Dirac spinors:

$$\sum_s u_e\bar{u}_e = \frac{\not{p}+m}{2m}\ , \quad -\sum_{s'} v'_e\bar{v}'_e = \frac{\not{p}'+m}{2m}\ , \quad (14.184)$$

and

$$\sum_k u_\mu\bar{u}_\mu = \frac{\not{k}+M}{2M}\ , \quad -\sum_{k'} v'_\mu\bar{v}'_\mu = \frac{\not{k}'+M}{2M}\ . \quad (14.185)$$

From the term corresponding to the electron current, we obtain

$$\sum_{s,s'}\bar{v}'_e\gamma^\lambda u_e\bar{u}_e\gamma^\nu v'_e = -\frac{1}{4m^2}\mathrm{Tr}\left[(\not{p}'-m)\gamma^\lambda(\not{p}+m)\gamma^\nu\right]\ , \quad (14.186)$$

which, substituted together with the analogous expression for the muon current in (14.183), gives the result

$$\overline{|\mathcal{M}_{if}|^2} = \frac{1}{4}\sum_{s,s'}\sum_{k,k'}|\mathcal{M}_{if}|^2 = \frac{e^4}{4}\,\frac{1}{(p+p')^4}\times$$
$$\frac{1}{4m^2}\mathrm{Tr}\left[(\not{p}'-m)\gamma^\lambda(\not{p}+m)\gamma^\nu\right]\frac{1}{4M^2}\mathrm{Tr}\left[(\not{k}+m)\gamma_\lambda(\not{k}'-m)\gamma_\nu\right]\ . \quad (14.187)$$

Using the rules for the calculation of products of $\gamma$-matrix traces, we find

$$\begin{aligned}\mathrm{Tr}\left[(\not{p}'-m)\gamma^\lambda(\not{p}+m)\gamma^\nu\right] &= \mathrm{Tr}(\not{p}'\gamma^\lambda\not{p}\gamma^\nu) - m^2\mathrm{Tr}(\gamma^\lambda\gamma^\nu)\\ &= p'^\rho p^\sigma\mathrm{Tr}(\gamma^\rho\gamma^\lambda\gamma^\sigma\gamma^\nu) - 4m^2g^{\lambda\nu}\\ &= 4\left\{p'^\lambda p^\nu + p'^\nu p^\lambda - g^{\lambda\nu}\left[(pp')+m^2\right]\right\}\ ,\end{aligned} \quad (14.188)$$

and, similarly

$$\mathrm{Tr}\left[(\not{k}+M)\gamma_\lambda(\not{k}'-M)\gamma_\nu\right] = 4\left\{k_\lambda k'_\nu + k_\nu k'_\lambda - g_{\lambda\nu}\left[(kk')+M^2\right]\right\} . \tag{14.189}$$

From now on we set to zero the mass of the electron; $m \sim 0.5$ MeV, which is negligible compared to the mass of the muon, $M \sim 105$ MeV and to the 4-momenta of the particles which participate in the process. With this approximation, we obtain

$$\begin{aligned}\mathrm{Tr}&\left[(\not{p}'-m)\gamma^\lambda(\not{p}+m)\gamma^\nu\right]\mathrm{Tr}\left[(\not{k}+M)\gamma_\lambda(\not{k}'-M)\gamma_\nu\right]\\ &= 32\left[(pk')(p'k)+(pk)(p'k')+(pp')M^2\right] ,\end{aligned} \tag{14.190}$$

from which it follows that

$$|\overline{\mathcal{M}}_{if}|^2 = \frac{e^4}{(p+p')^4}\frac{1}{4m^2}\frac{1}{4M^2}\, 8\left[(pk')(p'k)+(pk)(p'k')+(pp')M^2\right] . \tag{14.191}$$

The explicit calculation of $|\overline{\mathcal{M}}_{if}|^2$ is easily carried out in the centre of mass frame, defined by the condition $\mathbf{p}+\mathbf{p}'=0$. Choosing the $z$-axis in the direction of $\mathbf{p}$ and again neglecting the electron mass, we can write

$$p \equiv (E,0,0,E), \quad p' \equiv (E,0,0,-E), \tag{14.192}$$

$$k \equiv (E,0,|\boldsymbol{k}|\sin\theta,|\boldsymbol{k}|\cos\theta), \quad k' = (E,0,-|\boldsymbol{k}|\sin\theta,-|\boldsymbol{k}|\cos\theta), \tag{14.193}$$

where $\theta$ is the angle between the momenta of the electron and muon and $|\mathbf{k}| = \sqrt{E^2-M^2}$. From the definitions it follows that

$$(pk') = (p'k) = E^2\left(1+\frac{|\boldsymbol{k}|}{E}\cos\theta\right) , \tag{14.194}$$

$$(pk) = (p'k') = E^2\left(1-\frac{|\boldsymbol{k}|}{E}\cos\theta\right) , \tag{14.195}$$

$$(p+p')^2 = 2(pp') = 4E^2 . \tag{14.196}$$

Substituting into (14.191) we obtain the expression

$$|\overline{\mathcal{M}}_{if}|^2 = \frac{1}{4m^2}\frac{1}{4M^2}e^4\left[\left(1+\frac{M^2}{E^2}\right)+\left(1-\frac{M^2}{E^2}\right)\cos^2\theta\right] , \tag{14.197}$$

which implies (compare with (14.181))

$$\begin{aligned}|\langle f|S^{(2)}|i\rangle|^2 &= \frac{TV}{V^4}(2\pi)^4\delta^{(4)}(p+p'-k-k')\times\\ &\frac{e^4}{16E^4}\left[\left(1+\frac{M^2}{E^2}\right)+\left(1-\frac{M^2}{E^2}\right)\cos^2\theta\right] .\end{aligned} \tag{14.198}$$

$$\frac{|\boldsymbol{v}-\boldsymbol{v}'|}{V} = 2\,\frac{|\boldsymbol{v}|}{V} = \frac{2}{V}\frac{|p|}{E} = \frac{2}{V}\,. \tag{14.199}$$

The result is

$$d\sigma = \frac{V}{2}\frac{V}{V^4}(2\pi)^4\delta^{(4)}(p+p'-k-k')\frac{e^4}{16E^4}\left[\ldots\ldots\right]\frac{V}{(2\pi)^3}d^3k\frac{V}{(2\pi)^3}d^3k'\,. \tag{14.200}$$

The integration over $\mathbf{k}$ can be carried out thanks to the function $\delta^{(3)}(\mathbf{p}+\mathbf{p}'-\mathbf{k}-\mathbf{k}')$, while to integrate over $\mathbf{k}'$ we use the relation

$$d^3k' = d\Omega_{k'}|\boldsymbol{k}'|^2 d|\boldsymbol{k}'| = d\Omega_{k'}|\boldsymbol{k}'|E'dE'\,, \tag{14.201}$$

which, substituted into (14.200), gives the result

$$d\sigma = \frac{1}{2}\frac{1}{(2\pi)^2}\,\delta(2E-2E')\,\frac{e^4}{16E^4}\left[\ldots\ldots\right]\,d\Omega_{k'}|\boldsymbol{k}'|E'dE'\,, \tag{14.202}$$

or (with $\alpha = e^2/4\pi$)

$$\frac{d\sigma}{d\Omega_{k'}} = \frac{\alpha^2}{16E^2}\frac{|\boldsymbol{k}|}{E}\left[\ldots\ldots\right], \tag{14.203}$$

or

$$\frac{d\sigma}{d\Omega_{k'}} = \frac{\alpha^2}{4E^2_{\rm cm}}\left(1-\frac{M^2}{E^2}\right)^{1/2}\left[\left(1+\frac{M^2}{E^2}\right)+\left(1-\frac{M^2}{E^2}\right)\cos^2\theta\right]\,, \tag{14.204}$$

where the total energy in the centre of mass frame is $E_{\rm cm} = 2E$. Finally, carrying out the angular integration we obtain the total cross section

$$\sigma = 2\pi\int_{-1}^{1}\frac{d\sigma}{d\Omega_{k'}}\,d\cos\theta = 2\pi\frac{\alpha^2}{4E^2_{\rm cm}}\left(1-\frac{M^2}{E^2}\right)^{1/2}\frac{8}{3}\left(1+\frac{1}{2}\frac{M^2}{E^2}\right)\,. \tag{14.205}$$

Obviously, equation (14.205) makes sense only if

$$1-\frac{M^2}{E^2} > 0\,, \tag{14.206}$$

or if the threshold condition for $\mu^+\mu^-$ pair production, $E^2 > M^2$ or $E^2_{\rm cm} > 4M^2$, is satisfied. In the limit $E \to \infty$

$$\sigma \to \frac{4}{3}\pi\frac{\alpha^2}{E^2_{\rm cm}}\,. \tag{14.207}$$

The trend of equation (14.207) with energy was easily predictable, because the cross section has dimensions of an area, i.e. in natural units with $\hbar = c = 1$ the inverse of energy squared. In the limit $E \to \infty$ the only energy scale available is $E$, all the masses being negligible. To second order in the interaction Lagrangian, we therefore find

$$\sigma \propto \frac{\alpha^2}{E^2_{\rm cm}}\,, \tag{14.208}$$

in agreement with equation (14.207).

## 14.8 PROBLEMS FOR CHAPTER 14

### Sect. 14.1

**1.** Write the expression for the Fourier transform of the electrostatic potential due to the charge distribution: $Ze\rho(\mathbf{x})$.

### Sect. 14.4

**1.** Starting from the Dyson formula and using translation invariance, show that the Compton scattering amplitude to order $\alpha$ can be concisely written as:

$$\mathcal{A} = (2\pi)^4 \delta^{(4)}(p' + q' - p - q)\ \epsilon^\mu(q)\epsilon^\nu(q')\mathcal{M}_{\mu\nu}$$
$$\mathcal{M}_{\mu\nu} = \langle p', r'| \int d^4x\ e^{-iqx}\ T\left\{J_\mu(x)J_\nu(0)\right\}|p, r\rangle\ .$$

**2.** Recalling the result of problem **2.** to Sect. 8.3 show that:

$$q^\mu \int d^4x\ e^{iqx}\ T\left\{J_\mu(x)J_\nu(0)\right\} = 0\ .$$

This identity is used in the text as a check that gauge invariance is respected in the calculation of $\mathcal{M}_{\mu\nu}$ from the Feynman diagrams. It belongs to the large class of *Ward identities* embodying the consequences of electric charge conservation.

**3.** Using equations (14.116), (14.140) and (14.141), show that the differential cross section for Compton scattering on a moving electron can be cast in the form

$$\frac{d\sigma}{d\Omega'} = \frac{\alpha^2}{(s-m^2)^2}\, 2\omega'^2 \left[4\left(\frac{m^2}{s-m^2} + \frac{m^2}{u-m^2}\right)^2 + 4\left(\frac{m^2}{s-m^2} + \frac{m^2}{u-m^2}\right) - \left(\frac{u-m^2}{s-m^2} + \frac{s-m^2}{u-m^2}\right)\right], \tag{14.209}$$

where $\Omega'$ is the differential solid angle specifying the direction of the scattered photon.

**4.** Using equations (14.209) and (14.145) with $\phi = \pi$, show that, in the limit $\omega \ll E$, the differential cross sections of Fig. 14.4 can be written in the form

$$\frac{d\sigma}{d\omega'} = \frac{\pi\alpha^2}{2\omega E^2}\left[\frac{m^4}{4\omega^2 E^2}\left(\frac{\omega'}{E-\omega'}\right)^2 - \frac{m^2}{\omega E}\left(\frac{\omega'}{E-\omega'}\right) + \frac{E-\omega'}{E} + \frac{E}{E-\omega'}\right].$$

CHAPTER 15

# APPLICATIONS: WEAK INTERACTIONS

## 15.1 NEUTRON DECAY

$$n \to p + e^- + \bar{\nu} \; . \tag{15.1}$$

The interaction Lagrangian which describes the process (15.1) and (9.24) is recalled here for convenience:

$$\mathcal{L}_F = -\frac{G_F}{\sqrt{2}} \left[ \bar{\psi}_p \gamma^\mu (1 + \frac{g_A}{g_V} \gamma_5) \psi_n \right] \left[ \bar{\psi}_e \gamma_\mu (1 - \gamma_5) \psi_\nu \right] = -\frac{G_F}{\sqrt{2}} H^\mu L_\mu \; . \tag{15.2}$$

All the particles are represented by Dirac fields and $G_F$ is the Fermi constant.

The Lagrangian is the product of two operators: the nuclear current $H^\mu$, which induces the transition between heavy particles, $n \to p$, and the lepton current $L^\mu$, which creates the lepton pair from the vacuum. Assuming the validity of (15.2) we calculate the mean lifetime of the neutron and the electron asymmetry with respect to the spin of the neutron. Comparing with experimental values we can determine the two constants which appear in (15.2).

To first order of perturbation theory:

$$\begin{aligned} \langle p, e, \bar{\nu} | S | n \rangle &= \frac{-iG_F}{\sqrt{2}} \langle p, e, \bar{\nu} | \int d^4x H^\mu(x) L_\mu(x) | n \rangle \\ &= (2\pi)^4 \delta^{(4)}(P_f - P_i) \frac{-iG_F}{\sqrt{2}} \langle p | H^\mu(0) | n \rangle \langle e, \bar{\nu} | L_\mu(0) | 0 \rangle \; . \end{aligned} \tag{15.3}$$

The nuclear matrix element can be calculated in the limit in which the proton is non-relativistic, given the small $n$-$p$ mass difference compared to the mass of the proton. In this limit, only the currents which correspond to diagonal Dirac matrices survive, i.e.

$$\gamma^0 \; , \quad \gamma^i \gamma_5 \; .$$

DOI: 10.1201/9781003436263-15

The corresponding matrix elements are known respectively as *Fermi* and *Gamow–Teller transitions.* We find

$$\langle p|\bar{\psi}_p\gamma^0\psi_n|n\rangle = \langle p|\bar{\psi}_p^{(-)}\gamma^0\psi_n^{(+)}|n\rangle = \chi_p^\dagger\chi_n = h^0 \text{ (Fermi)} ,$$
$$\langle p|\bar{\psi}_p\gamma^i\gamma_5\psi_n|n\rangle = \langle p|\bar{\psi}_p^{(-)}\gamma^i\gamma_5\psi_n^{(+)}|n\rangle = \chi_p^\dagger\sigma^i\chi_n = h^i \text{ (Gamow–Teller)} ,$$

where $\chi_p$ and $\chi_n$ are two-dimensional spinors.

For the leptons, no approximation can be used:

$$\langle e,\bar{\nu}|\bar{\psi}_e\gamma^\mu(1-\gamma_5)\psi_\nu|0\rangle = \langle e,\bar{\nu}|\bar{\psi}_e^{(-)}\gamma^\mu(1-\gamma_5)\psi_\nu^{(-)}|0\rangle$$
$$= \sqrt{\frac{m_e m_\nu}{E_e E_\nu}}\bar{u}_e(p_e)\gamma^\mu(1-\gamma_5)v_\nu(p_\nu) = \sqrt{\frac{m_e m_\nu}{E_e E_\nu}}\, l^\mu .$$

The squared modulus of the Feynman matrix element is:

$$|M|^2 = \frac{G_F^2}{2}[h^\mu(h^\nu)^*][l_\mu l_\nu^*] . \tag{15.4}$$

**The Nuclear Part.** In the sum over the proton spin in the nuclear part of $|M|^2$ the projector of the two spin states is used:

$$\sum_{spin\ p}(\chi_p)_a(\chi_p^\dagger)_b = \delta_{ab}(a,b=1,2) . \tag{15.5}$$

Therefore, assuming the neutron is in a given spin state:

$$\sum_{spin\ p} h^\mu(h^\nu)^* = \mathrm{Tr}\left[a^\mu(\chi_n\chi_n^\dagger)a^\nu\right] ,$$

with

$$a^\mu = (1, \frac{g_A}{g_V}\cdot\boldsymbol{\sigma}) . \tag{15.6}$$

**Neutron Polarisation.** The state of a neutron with polarisation $P$ is described using a *density matrix* in place of the projector of spin states in (15.5). If $A$ and $B$ are, respectively, the probability to find the neutron with spin *up* or *down* along the 3-axis, the density matrix is given by:

$$\rho_{ab} = (\chi_n)_a A(\chi_n^\dagger)_b + (\chi_n)_a B(\chi_n^\dagger)_b$$
$$= \left(\frac{1+\sigma_3}{2}\right)A + \left(\frac{1-\sigma_3}{2}\right)B = \frac{1+P\sigma_3}{2} , \tag{15.7}$$

where we have used $A+B=1$ and the polarisation along the 3-axis is:

$$P = \langle\sigma_3\rangle = Tr(\rho\sigma_3) = A - B .$$

For a non-polarised neutron, $P = 0$ and the insertion of $\rho$ into equation (15.6) simply gives the average over the initial spin states. In any case, we find:

$$\sum_{spin\ p} h^\mu (h^\nu)^* = Tr\left[a^\mu \rho a^\nu\right] = H^{\mu\nu} \ ,$$
$$(H^{\mu\nu})^* = H^{\nu\mu} \ .$$

Explicitly:

$$H^{00} = 1 \ , \ \ H^{i0} = H^{0i} = \frac{g_A}{g_V} P \delta^{i3} \ ,$$
$$H^{ij} = \left(\frac{g_A}{g_V}\right)^2 \left(\delta^{ij} - iP\epsilon^{ij3}\right) \ .$$

Lepton Part. In the lepton part, we sum over all the spins, which are usually not observed, using the formulae (6.84) and (6.85) for the projectors of the solutions for positive (electron) and negative (antineutrino) energy. We obtain

$$\sum_{spin\ e,\nu} (l_\mu l_\nu^*) = \sum_{spin\ e,\nu} \left[\bar{u}_e(p_e)\gamma_\mu(1-\gamma_5)v_\nu(p_\nu)\right]\left[v_\nu(p_\nu)^\dagger(1-\gamma_5){\gamma_\nu}^\dagger\gamma^0 u_e(p_e)\right]$$
$$= \sum_{spin\ e,\nu} \left[\bar{u}_e(p_e)\gamma_\mu(1-\gamma_5)v_\nu(p_\nu)\right]\left[\bar{v}_\nu(p_\nu)\gamma_\nu(1-\gamma_5)u_e(p_e)\right] \ ,$$

because from (6.17) it follows that:

$$\gamma^0(1-\gamma_5){\gamma_\nu}^\dagger\gamma^0 = \gamma_\nu(1-\gamma_5) \ ,$$

and therefore:

$$\sum_{spin\ e,\nu} (l_\mu l_\nu^*) = \frac{1}{4m_e m_\nu} Tr\left[(\not{p}_e + m_e)\gamma_\mu(1-\gamma_5)(\not{p}_\nu - m_\nu)\gamma_\nu(1-\gamma_5)\right]$$
$$\simeq 2\frac{1}{4m_e m_\nu} Tr\left[\not{p}_e\gamma_\mu\not{p}_\nu\gamma_\nu(1-\gamma_5)\right]$$
$$= 8\frac{1}{4m_e m_\nu}\left[(p_e)_\mu(p_\nu)_\nu + (p_e)_\nu(p_\nu)_\mu - g_{\mu\nu}(p_e\cdot p_\nu) + i\epsilon_{\alpha\beta\mu\nu}(p_e)^\alpha(p_\nu)^\beta\right]$$
$$= 8\frac{1}{4m_e m_\nu} L_{\mu\nu} \ . \tag{15.8}$$

We have used the results of Section 6.3 and the relations:

$$(1-\gamma_5)(1+\gamma_5) = 0; \ \ (1-\gamma_5)(1-\gamma_5) = 2(1-\gamma_5) \ ,$$

which imply, among other things, that the terms proportional to the masses in the numerator give zero contribution. In components:

$$L_{00} = E_e E_\nu + (\boldsymbol{p}_e\cdot\boldsymbol{p}_\nu); \ L_{0i} = -\left[E_e(p_\nu)^i + (p_e)^i E_\nu\right] + i\epsilon^{ijk}(p_e)^j(p_\nu)^k \ ,$$
$$L_{ij} = \left[(p_e)^i(p_\nu)^j + (p_e)^j(p_\nu)^i + \delta^{ij}(p_e p_\nu)\right] + i\epsilon^{ijk}\left[E_e(p_\nu)^k - (p_e)^k E_\nu\right] \ .$$

Phase Space. Comparing with (11.41), we obtain:

$$d\Gamma = \frac{G_F^2}{(2\pi)^5}\delta^{(4)}(P_f - P_i)\frac{d^3p_e d^3p_\nu}{4E_e E_\nu}d^3p_p \cdot [4H^{\mu\nu}L_{\mu\nu}] \ .$$

We can integrate the proton momentum using the three-dimensional $\delta$-function. In addition, conservation of energy, in the non-relativistic limit for the proton, can be written:

$$m_n - m_p - \left[\frac{(\boldsymbol{p}_e + \boldsymbol{p}_\nu)^2}{2m_p} + E_e + E_\nu\right] = 0 \ .$$

The kinetic energy of the proton is negligible, from which we find:

$$E_e + E_\nu = \Delta m = m_n - m_p \ . \tag{15.9}$$

The result (15.9) explains the continuous spectrum of $\beta$ rays; the energy released in the transition is fixed, as in atomic or nuclear transitions with the emission of a photon, but this energy is divided in a random way between the electron and the neutrino, which is not observed.

Equation (15.9) has the practical consequence that we can integrate freely over the direction of the neutrino momentum, with its energy fixed by $E_\nu = \Delta m - E_e$. Overall, we can substitute

$$\delta^{(4)}(P_f - P_i)\frac{d^3p_e d^3p_\nu}{4E_e E_\nu}d^3p_p(\dots)$$
$$\to \frac{\pi}{2}p_\nu p_e dE_e d\cos\theta_{en}\int d\Omega_\nu(\dots) \ .$$

From (15.4) and (15.8) it can be seen that $|M|^2$ depends linearly on $(p_\nu)^\mu$. We can therefore omit from $L_{\mu\nu}$ the terms proportional to $\boldsymbol{p}_\nu$ which integrate to zero over the solid angle.

Lifetime and Spin Asymmetries. Putting everything together, we find:

$$d\Gamma = \frac{G_F^2}{4\pi^3}(1+3\lambda^2)\left[1 + A_{en}(Pv_e\cos\theta_{en})\right]p_\nu E_\nu p_e E_e dE_e d\cos\theta_{en} \ , \tag{15.10}$$

$$A_{en} = -2\frac{\lambda(1+\lambda)}{(1+3\lambda^2)} \ , \tag{15.11}$$

where $\theta_{en}$ is the angle between the momentum of the electron and the neutron spin direction. We have set

$$\lambda = \frac{g_A}{g_V} \ ,$$

and in addition:

$$p_e = \sqrt{E_e^2 - m_e^2} \ , \quad v_e = \frac{p_e}{E_e} \ ,$$
$$E_\nu = \Delta m - E_e \ , \quad p_\nu = \sqrt{E_\nu^2 - m_\nu^2} \ .$$

Table 15.1 Observables in the decay of the neutron.

| $\Delta m$ (MeV) | Lifetime (s) | $A_{en}$ | $A_{\nu n}$ | $g_A/g_V$ |
|---|---|---|---|---|
| 1.293 | 885.7 $\pm$0.8 | $-0.1173$ $\pm$0.0013 | 0.983 $\pm$0.004 | $-1.2695$ $\pm$0.0029 |

We can also express $d\Gamma$ as a function of the electron energy and the angle between the neutrino and the neutron spin direction. The neutrino momentum is reconstructed by measuring the proton momentum as well as that of the electron, using $\boldsymbol{p}_\nu = -\boldsymbol{p}_e - \boldsymbol{p}_p$. It is easily seen that the terms in $P\lambda$ are symmetric for exchange of $e \to \nu$ while terms in $P\lambda^2$ change sign. Therefore, with $v_\nu = 1$, we obtain:

$$d\Gamma = \frac{G_F^2}{4\pi^3}(1+3\lambda^2)\left[1 + A_{\nu n}(P cos\theta_{\nu n})\right] p_\nu E_\nu p_e E_e dE_e dcos\theta_{\nu n} \ , \tag{15.12}$$

$$A_{\nu n} = -2\frac{\lambda(1-\lambda)}{(1+3\lambda^2)} \ . \tag{15.13}$$

From measurements of $A_{en}$ and $A_{\nu n}$ we can determine $\lambda$ and therefore obtain $G_F$ by comparison of the experimental value of the lifetime with the expression derived from (15.10) or (15.12):

$$\Gamma = \frac{1}{\tau} = \frac{G_F^2 \Delta m^5}{60\pi^3}\left[1 + 3\left(\frac{g_A}{g_V}\right)^2\right] \cdot I(\frac{m_e}{\Delta m}) \ ,$$

$$I(x) = 30 \int_x^1 dt \ t(1-t)^2\sqrt{t^2 - x^2} \simeq 0.473 \ . \tag{15.14}$$

The integral $I$ is normalised to give 1 in the limit $m_e = 0$ where we have again neglected the mass of the neutrino. The numerical value is obtained using the mass values tabulated in Tables 15.1 and 15.2.

Table 15.2 Properties of charged leptons are determined by the weak interactions, from the Particle Data Group [12]. The numbers in brackets denote the error on the last digit of each quantity. In the last two columns $B(l)$, denotes the fraction of lepton decays, with the emission of a $l^- \bar{\nu}_l$ pair, with $l = e, \mu$.

| | $m$ (MeV) | lifetime | $B(e)(\%)$ | $B(\mu)(\%)$ |
|---|---|---|---|---|
| $e$ | 0.510998902(21) | $> 4.6 \cdot 10^{26}$ y | 0 | 0 |
| $\mu$ | 105.658357(5) | $2.19703(4) \cdot 10^{-6}$ s | 100 | 0 |
| $\tau$ | 1776.99(28) | $2.906(11) \cdot 10^{-13}$ s | 17.84(6) | 17.37(6) |

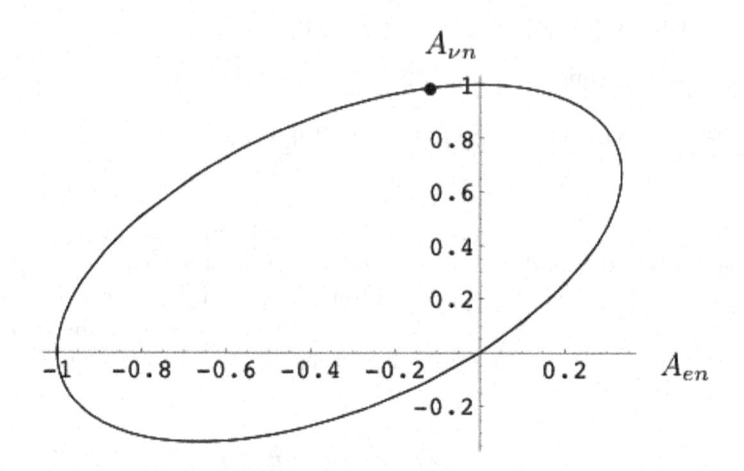


Figure 15.1 Eliminating $\lambda$ from equations (15.11) and (15.13) one finds the second degree consistency relation in the asymmetries, equation (15.17), which is represented by the ellipse shown in the figure, in the plane $(A_{en}, A_{\nu n})$. The experimental point is indicated by the dot and, indeed, it lies on the ellipse, showing that the consistency condition is quite well obeyed.

Numerical Analysis Considering $\lambda$ and $\lambda^2$ as independent quantities, from (15.11) and (15.13) and the values in Table 15.1, we find the value of $g_A/g_V$:

$$\begin{aligned}
\Delta &= A_{en} - A_{\nu n} \simeq -1.10, \quad \Sigma = A_{en} + A_{\nu n} \simeq 0.866 \, , \\
\lambda^2 &= \frac{-\Delta}{4+3\Delta} \simeq 1.57 \, , \\
\lambda &= \frac{-\Sigma}{4+3\Delta} \simeq -1.24 = \frac{g_A}{g_V} \, ,
\end{aligned} \tag{15.15}$$

with a consistency check that requires:

$$\left[\frac{-\Sigma}{4+3\Delta}\right]^2 = \frac{-\Delta}{4+3\Delta} \, ,$$

or:

$$1.53 \simeq 1.57 \, . \tag{15.16}$$

Another way of expressing the consistency is to say that the experimental point in the $(A_{en}, A_{\nu n})$ plane should be on the ellipse given by the equation:

$$\Sigma^2 + 3(\Delta + \frac{2}{3})^2 = \frac{4}{9} \, . \tag{15.17}$$

Fig. 15.1 shows that this is true to an excellent approximation.

Finally, from this simple analysis we find a value for $g_A/g_V$ in (15.15) very close to that adopted by the Particle Data Group [?] listed in Table 15.1. Substituting the latter value into (15.14), we find the Fermi constant to be:

$$G_F = 1.18 \cdot 10^{-5}\ \mathrm{GeV}^{-2}\ . \tag{15.18}$$

Comment. The existence of a correlation between the neutron spin and the electron direction of flight shows that the weak interaction violates spatial reflection symmetry. Under this operation, the electron momentum (polar vector) changes sign while the neutron spin (axial vector) does not change. If the final state is an eigenstate of parity, as happens if $P$ commutes with the Hamiltonian, we must have $\langle \mathbf{v}_e \cdot \boldsymbol{\sigma}_n \rangle = 0$, and therefore an average value of $cos\theta_{en}$ equal to zero.

The existence of parity violation in weak interactions, and therefore in $\beta$ decays, was hypothesised by Lee and Yang [23] in 1956 to resolve the so-called $\theta - \tau$ puzzle, the decay of $K$ mesons into both two and three $\pi$ mesons. The first experimental observation of the spin asymmetry was in $\beta$ decays of nuclei by Wu and collaborators in 1957 [24].

Limits on the Mass of the Neutrino. The electron energy distribution is potentially sensitive to the mass of the neutrino in the highest range of energies, the *endpoint* of the spectrum. For this reason, in equation (15.10) we have kept the neutrino mass different from zero. After integration over angle, we can write the spectrum as:

$$\frac{d\Gamma_e}{dE_e} = f(E_e) = C \cdot p_\nu E_\nu E_e p_e \simeq C' p_\nu E_\nu\ ,$$

with $C$ and $C'$ constants. In the region in which $E_e \simeq \Delta m$, we can keep only the neutrino terms, which vary rapidly, and approximate the others with their value at the end point $E_e \simeq p_e \simeq \Delta m$.

It is useful to consider the *square root* of the spectrum:

$$g(E_e) = \sqrt{f(E_e)} = \sqrt{\left[(E_e - \Delta m)\sqrt{(E_e - \Delta m)^2 - m_\nu^2}\right]}\ . \tag{15.19}$$

For a neutrino mass exactly equal to zero, $g(E_e)$ vanishes *linearly* at the endpoint, while if $m_\nu \neq 0$ the curve vanishes with an *infinite derivative*. The effect allows an estimate of the neutrino mass, or at least an upper limit, with greater sensitivity the smaller the endpoint value, which in the case of the neutron is $\Delta m \simeq 1.29$ MeV. A particularly favourable nuclide is tritium, which decays according to the scheme:

$$^3\mathrm{H} \to {}^3\mathrm{He} + e^- + \bar{\nu}$$

with $\Delta m$ equal to 18.6 keV.

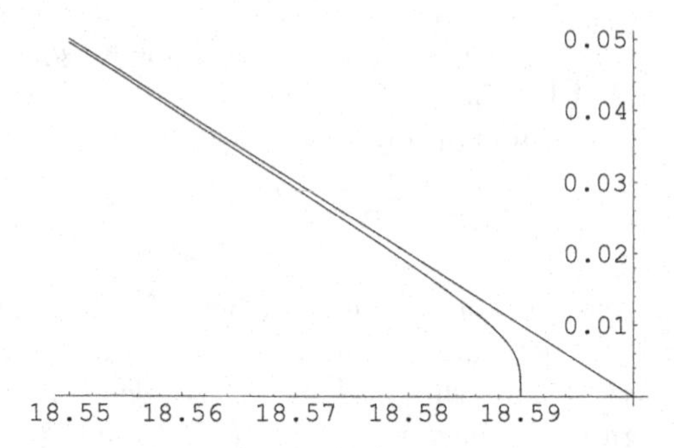


Figure 15.2 Calculated shape of $g(E_e)$ in the decay of tritium, near to the endpoint. The upper curve corresponds to a zero neutrino mass, the lower curve to $m = 10$ eV.

Curves which represent the electron energy spectrum corresponding to $m_\nu = 0$ and $m_\nu = 10$ eV are shown in Fig. 15.1. At present no effect has been observed at the endpoint of tritium which positively indicates the existence of a neutrino mass in $\beta$ decay, but only an upper limit [?]:

$$m_{\nu_e} \leq 3 \text{ eV}.$$

## 15.2 MUON DECAY

The muon decays via a process analogous to $\beta$ decay of the neutron:

$$\mu^-(p) \to \nu_\mu(p') + e^-(q) + \bar{\nu}_e(q') \ . \tag{15.20}$$

We have introduced two different types of neutrino, associated with two charged leptons, in accord with experimental and theoretical evidence accumulated since the 1960s.

The decay of the muon can be described with a Lagrangian of the Fermi type. In view of the strong similarity between electron and muon displayed by the electromagnetic interaction, we assume a Lagrangian in which the $V$–$A$ structure of the $\nu_e$–$e$ pair is extended to the $\nu_\mu$–$\mu$ pair:

$$\mathcal{L}_{\mu-dec} = -\frac{G^{(\mu)}}{\sqrt{2}} \left[\bar{\psi}_{\nu_\mu}\gamma^\lambda(1-\gamma_5)\psi_\mu\right] \left[\bar{\psi}_e\gamma_\lambda(1-\gamma_5)\psi_{\nu_e}\right] + \text{hermitian conjugate.} \tag{15.21}$$

$G^{(\mu)}$ is a new constant analogous to the Fermi constant introduced for the neutron. In the muon rest frame, the $S$-matrix element for the decay is:

$$\langle \nu_\mu, e, \bar{\nu}_e|S|\mu\rangle = (2\pi)^4\delta^{(4)}(p - p' - q - q')\sqrt{\frac{m_e m_{\nu_\mu} m_{\nu_e}}{E(p')E(q)E(q')V^4}} M(i \to f) \ ,$$

where we have introduced the invariant Feynman amplitude, $M(i \to f)$. The decay probability is calculated starting from:

$$< \sum_{spin\ fin} |M(i \to f)|^2 >= \frac{(G^{(\mu)})^2 [8 M^{\mu\nu} L_{\mu\nu}]}{8 m_e m_{\nu_\mu} m_{\nu_e}} ,$$

where $< .. >$ denotes the average over the muon spin. The tensor $L_{\mu\nu}$ is the same as introduced in (15.8), while, for a state of definite spin:

$$M^{\mu\nu} = Tr\left[(\chi\chi^\dagger)\gamma^\nu \not{p}' \gamma^\mu (1-\gamma_5)\right] .$$

In general, we must replace the spinor product with a density matrix, similarly to what was done for the neutron:

$$< (\chi)_\alpha (\chi^\dagger)_\beta > \to \begin{pmatrix} \rho(P) & 0 \\ 0 & 0 \end{pmatrix} , \tag{15.22}$$

$$\rho(P) = \frac{1 + P\sigma_3}{2} ,$$

where $P$ is the polarisation of the muon along the 3-axis. To calculate $M^{\mu\nu}$ we must use the identity (6.170):

$$\begin{aligned} M^{\mu\nu} &= p'^\mu S^\nu + p'^\nu S^\mu - g^{\mu\nu} p'_\alpha S^\alpha - i\epsilon^{\nu\alpha\mu\rho} p'_\alpha S^\rho , \\ S^\mu &= Tr\left[\rho\gamma^\mu (1-\gamma_5)\right] . \end{aligned}$$

Multiplying by $L_{\mu\nu}$ we find:

$$M^{\mu\nu} L_{\mu\nu} = 4(qp') S^\alpha q'_\alpha , \tag{15.23}$$

where the momenta are attributed as in (15.20). From this:

$$d\Gamma = \frac{(G^{(\mu)})^2}{(2\pi)^5} \cdot 32 \cdot q_\rho S_\sigma \frac{d^3 q}{2E_e} \left[ \delta^{(4)}(p - q - p' - q') \frac{d^3 p' d^3 q}{4E_{\nu_e} E_{\nu_\mu}} (p'^\rho q'^\sigma) \right] .$$

If we do not observe the neutrinos, we must integrate the quantity in brackets. The size of the integration volume being invariant, the result is a tensor in the indices $\rho, \sigma$ constructed with $g^{\rho\sigma}$ and the vector $Q = p - q$, the only variable which remains after the integration. We therefore set:

$$\begin{aligned} \langle p'^\rho q'^\sigma \rangle &= \int \delta^{(4)}(p - q - p' - q') \frac{d^3 p' d^3 q}{4E_{\nu_e} E_{\nu_\mu}} p'^\rho q'^\sigma \\ &= A(Q^2) g^{\rho\sigma} + B(Q^2) Q^\rho Q^\sigma . \end{aligned}$$

To determine $A$ and $B$, we note the relations $(p'^2 = q'^2 = 0)$:

$$\begin{aligned} g_{\rho\sigma} \cdot \langle p'^\rho q'^\sigma \rangle &= \langle \frac{1}{2} (p' + q')^2 \rangle = \frac{Q^2}{2} \cdot I , \\ Q_\rho Q_\sigma \cdot \langle p'^\rho q'^\sigma \rangle &= \frac{1}{4} (Q^2)^2 \cdot I , \\ I = \langle 1 \rangle &= \int \delta^{(4)}(p - q - p' - q') \frac{d^3 p' d^3 q}{4E_{\nu_e} E_{\nu_\mu}} . \end{aligned}$$

In this way we find two equations for $A$ and $B$, which give:

$$\langle p'^{\rho} q'^{\sigma} \rangle = \frac{1}{6} I \left( \frac{1}{2} g^{\rho\sigma} Q^2 + Q^{\rho} Q^{\sigma} \right) .$$

An easy calculation, in addition, provides[1]:

$$I = \frac{\pi}{2} ,$$

and therefore, finally, neglecting as usual terms of order $m_e^2$, we find

$$\frac{d\Gamma}{dE_e dcos\theta_e} = \frac{G^{(\mu)2}}{24\pi^3} m_\mu^2 E_e^2 \left\{ (3 - \frac{4E_e}{m_\mu}) + (1 - \frac{4E_e}{m_\mu}) \frac{Tr\left[\rho \not{a}(1-\gamma_5)\right]}{E_e} \right\} .$$

The trace is easily calculated, since:

$$\not{a}(1-\gamma_5) = \begin{pmatrix} E_e + \boldsymbol{p}_e \cdot \sigma & \dots \\ \dots & \dots \end{pmatrix} ,$$

from which:

$$Tr\left[\rho \not{a}(1-\gamma_5)\right] = E_e (1 + v_e P cos\theta_e) .$$

We normalise the electron energy to the maximum value it can have in the decay, $E_{max} = m_\mu/2$, putting:

$$x = \frac{E_e}{E_{max}} = \frac{2x}{m_\mu} ,$$

and we find, finally:

$$\frac{d\Gamma}{dx dcos\theta_e} = \frac{G^{(\mu)2} m_\mu^5}{192\pi^3} \left[3 - 2x + (1-2x) \boldsymbol{v}_e \cdot \boldsymbol{P}\right] . \tag{15.24}$$

Integrating over the remaining variables, we find from (15.24) the total decay rate:

$$\Gamma = \frac{1}{\tau} = \frac{G^{(\mu)2} m_\mu^5}{192\pi^3} . \tag{15.25}$$

Both the electron spectrum and the spin asymmetry are in perfect agreement with the experimental data. Comparing equation (15.25) with the observed value of the lifetime, Table 15.2, we find in addition:

$$G^{(\mu)} = 1.16 \cdot 10^{-5} \; GeV^2 . \tag{15.26}$$

---

[1] In the rest system of $Q^\mu$ the three-dimensional $\delta$-function eliminates the integration over $\mathbf{p}'$, therefore $I = \int d^3q'/(4E'^2)\delta(2E' - Q^0) = (4\pi)/(2 \cdot 4) = \pi/2$.

Kinematic Limits. Taking account of conservation of momentum, the conservation of energy in the decay is written:

$$m_\mu = E_e + E_{\nu_\mu} + |\mathbf{p}_e + \mathbf{p}_{\nu_\mu}| = E_e + E_{\nu_\mu} + \sqrt{p_e^2 + p_{\nu_\mu}^2 + 2p_e p_{\nu_\mu} cos\theta_{e\nu_\mu}} \ .$$

The kinematic limits of the decay in the $E_e$–$E_{\nu_\mu}$ plane are given by the condition $cos\theta_{e\nu_\mu} = \pm 1$. Simplifying to the case of particles of zero mass, we find:

$$m_\mu = E_e + E_{\nu_\mu} + |E_e \pm E_{\nu_\mu}| \ .$$

With the positive sign, we have:

$$\begin{aligned} &E_e + E_{\nu_\mu} = \frac{m_\mu}{2} \qquad \text{therefore :} \\ &E_e = x\,\frac{m_\mu}{2} \quad (0 \le x \le 1)\ ,\ E_{\nu_\mu} = (1-x)\,\frac{m_\mu}{2}\ ,\ E_{\bar{\nu}_e} = \frac{m_\mu}{2}\ . \end{aligned} \tag{15.27}$$

With the negative sign, we have two solutions:

$$\begin{aligned} &E_e = \frac{m_\mu}{2} \qquad \text{therefore :} \\ &E_{\nu_\mu} = y\,\frac{m_\mu}{2} \quad (0 \le y \le 1)\ ,\ E_{\bar{\nu}_e} = (1-y)\,\frac{m_\mu}{2}\ , \end{aligned} \tag{15.28}$$

or:

$$\begin{aligned} &E_{\nu_\mu} = \frac{m_\mu}{2} \qquad \text{therefore :} \\ &E_e = x\,\frac{m_\mu}{2} \quad (0 \le x \le 1)\ ,\ E_{\bar{\nu}_e} = (1-x)\frac{m_\mu}{2}\ . \end{aligned} \tag{15.29}$$

Finally, the extreme configurations are three collinear configurations, in which one particle takes the maximum energy, $m_\mu/2$, and the other two divide the remaining half between them (see Fig. 15.3).

Forbidden Configurations in the Decay. The results in equations (15.23) and (15.24) show that, if we set $m_e = 0$, the decay probability vanishes for certain configurations of the final state particles. This arises from the $V$–$A$ structure of the Lagrangian, for which $e^-$ and $\nu_\mu$ are produced with negative helicity and the antineutrino, $\bar{\nu}_e$, has positive helicity.

According to (15.23), the decay rate is proportional to the $e\nu_\mu$ invariant mass:

$$(p_e p_{\nu_\mu}) \simeq \frac{1}{2}(p_e + p_{\nu_\mu})^2 = \frac{1}{2}M^2_{e\nu_\mu}\ ,$$

which vanishes in the collinear configuration, Fig. 15.3. In this configuration, the spins of $e^-$, $\nu_\mu$ and $\bar{\nu}_e$ are all parallel and similarly oriented, therefore this state has projection $-\frac{3}{2}$ along the electron direction of flight. Because the initial state only has projection $\pm\frac{1}{2}$, the probability of the configuration must vanish because of conservation of angular momentum.

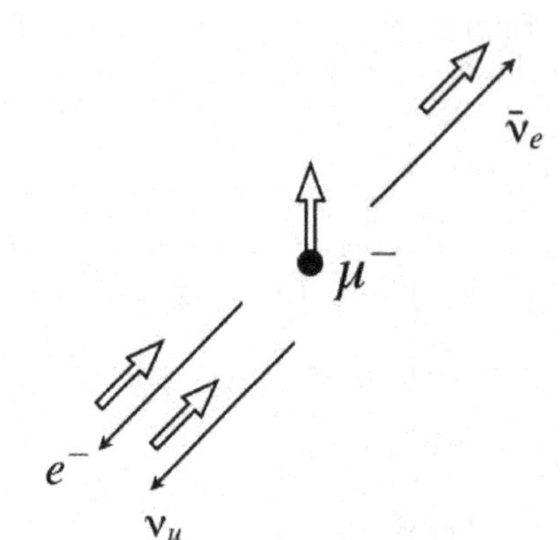


Figure 15.3 The $V$–$A$ interaction implies that, in the limit of zero mass, particles have negative helicity and antiparticles positive helicity. Therefore in the configuration in which $M_{e\nu_\mu} = 0$ the amplitude has to vanish because of conservation of angular momentum.

According to equation (15.24), the probability vanishes for $x = 1$, $P = 1$, $cos\theta_e = 1$, i.e. when the electron travels in the direction of the muon spin with maximum energy, Fig. 15.1. In this situation, $\nu_\mu$ and $\bar{\nu}_e$ are also collinear, in opposite directions, with opposite helicities and therefore with zero component of the total spin. The final state has $S_3 = (s_e)_3 = -\frac{1}{2}$ while the initial state has $S_3 = (s_\mu)_3 = +\frac{1}{2}$.

## 15.3 UNIVERSALITY, CURRENT × CURRENT THEORY

The Fermi constants (15.18) and (15.26) are surprisingly similar. This fact suggests that in the $\beta$ decay Lagrangian an overall weak current appears, analogous to what happens in electromagnetic interactions:

$$\begin{aligned}
\mathcal{L}_W &= -\frac{G}{\sqrt{2}} \left(J_N^\lambda + J_\mu^\lambda\right) (J_e^\dagger)_\lambda \,, \\
J_N^\lambda + J_\mu^\lambda &= g_V \bar{\psi}_p \gamma_\lambda (1 - \frac{g_A}{g_V}\gamma_5)\psi_n + \bar{\psi}_{\nu_\mu}\gamma_\lambda(1-\gamma_5)\psi_\mu \,, \\
(J_e^\dagger)^\lambda &= \bar{\psi}_e \gamma_\lambda (1-\gamma_5)\psi_{\nu_e} \,,
\end{aligned}$$

where we have set $G^{(\mu)} = G$ and $g_V \simeq 1$ represents an eventual scaling factor between the Fermi constants for neutron $\beta$ decay and that of the muon.

If we wish to maintain the universality between electron and muon we must make another step and include the contributions of the two particles in the same current. In this way, we arrive at the *current* × *current* expression:

$$\mathcal{L}_W = -\frac{G}{\sqrt{2}} J_W^\lambda (J_W^\dagger)_\lambda \ ,$$
$$J_W^\lambda = J_N^\lambda + J_e^\lambda + J_\mu^\lambda \ . \tag{15.30}$$

The analogy between electromagnetic interactions is now much deeper (see for example the expression for the total electromagnetic current which appears in equation (9.17)). In addition, we expect that, by adding appropriate terms to the nuclear current, we can describe the weak interactions of *all* hadronic particles.

The current × current Lagrangian describes new processes, compared to those considered up to now, in particular:

- a weak interaction between nucleons, which introduces a *parity-violating* component into the nuclear forces. In view of the value of $G$ we expect a small effect. The eigenstates of the complete nuclear Hamiltonian can be a superposition of states with opposite parity. The consequence is a polarisation asymmetry in the $\gamma$ decay of these states. This type of effect *is actually observed with the correct order of magnitude* (asymmetry $\simeq 10^{-5}$),
- interactions of muon neutrinos with nuclear matter identical to those of electron neutrinos, except for a kinematic effect connected to the difference between the masses of electron and muon, which, however, is a negligible effect at high energy.

The reaction products of each type of neutrino include the corresponding lepton:

$$\nu_e + \text{nucleus} \to e + \dots \ , \ \not\to \mu + \dots \tag{15.31}$$
$$\nu_\mu + \text{nucleus} \to \mu + \dots \ , \ \not\to e + \dots \tag{15.32}$$

The selection rules follow from the invariance of the Lagrangian (15.30) for global phase transformations carried out *separately* on the $\nu_e, e$ fields and on the $\nu_\mu, \mu$ fields. The symmetry implies the conservation of two types of *lepton charge*: electron number and muon number

$$N_e = N(e^-) + N(\nu_e) - N(e^+) - N(\bar{\nu}_e) \ , \tag{15.33}$$
$$N_\mu = N(\mu^-) + N(\nu_\mu) - N(\mu^+) - N(\bar{\nu}_\mu) \ . \tag{15.34}$$

The observation of the selection rule (15.32) at the beginning of the 1960s allowed the existence of the muon neutrino, $\nu_\mu \neq \nu_e$, to be established.

The current × current theory finds spectacular confirmation in the decay of the $\tau$ lepton. If we add to the weak current (15.30) the term which corresponds to the $\tau \to \nu_\tau$ transition, with the same $V$–$A$ structure as the others, we conclude that this particle must have three types of decay[2]

$$\tau^- \to \begin{cases} \nu_\tau + e + \nu_e \\ \nu_\tau + \mu + \nu_\mu \\ \nu_\tau + hadrons \end{cases} .$$

The theoretical prediction of the probability of the semi-hadronic decay mode requires elements of the Standard Model which will be developed in [13]. Concerning instead the first two decay modes, we can use the formula (15.25) with the substitution $m_\mu \to m_\tau$, in the limit in which we neglect the mass of the muon or electron. We then obtain the prediction:

$$\begin{aligned}\Gamma(\tau \to \nu_\tau + l^- + \nu_l) &= \frac{G^2 \cdot m_\tau^5}{192\pi^3} \\ &= (\frac{m_\tau}{m_\mu})^5 \Gamma_\mu \qquad (l = e, \mu) ,\end{aligned} \tag{15.35}$$

or:

$$B(e) = B(\mu);$$

$$\tau(\tau) = \frac{1}{\Gamma_\tau} = \frac{B(l)}{\Gamma(\tau \to \nu_\tau + l^- + \nu_l)} = B(l)(\frac{m_\mu}{m_\tau})^5 \tau(\mu) = 2.86 \cdot 10^{-13}\text{s} ,$$

in excellent agreement with the values for the leptonic decay branching ratios, $B(e), B(\mu)$ and the lifetime given in Table 15.2.

Alternatively, from the experimental values of the $\tau$ lifetime and the leptonic branching ratios we can derive a new value of the Fermi constant. We find:

$$G^{(\tau)} = 1.15 \cdot 10^{-5}\ \text{GeV}^2 , \tag{15.36}$$

in excellent agreement with the value determined from the muon lifetime, (15.26).

Universality: Latest Developments. The situation that we have shown, concerning the decays of the neutron and muon, is that anticipated in the classic work of Feynman and Gell–Mann [25] in which they proposed the $V$–$A$ form of the Fermi interaction and universality of the vector current. For a precision measurement of the Fermi constant, however, it is necessary to take into account electromagnetic corrections to the lifetime. A precise calculation of these corrections shows that in reality, $G_F$ is slightly less than $G^\mu$ by about 3%. The most precise measurement of $G_F$ is obtained from $\beta$ transitions between nuclei of isotopic spin 1 and spin-parity $J^P = 0^+$, the so-called *superallowed* Fermi transitions. The most recent results give [26]:

$$\frac{G_F}{G^{(\mu)}} = 0.9739 \pm 0.0005 . \tag{15.37}$$

[2]For the decays of the $\tau^+$ we must exchange particles with antiparticles and vice versa.

Still more critical is the situation in $\beta$ decays of strange particles in which the corresponding Fermi constant results in a value equal to $\frac{1}{5}$ of the Fermi constant $G_F$.

The reconciliation of these facts with the universality of the weak interactions is due to Cabibbo [27].

The study of the structure of the weak hadronic current is of crucial importance in understanding the nature and properties of the constituents of hadrons. The extension of the Feynman–Gell–Mann–Cabibbo theory to particles with *charm* is due to Glashow, Iliopoulos and Maiani [28], and the subsequent extension to particles formed of $b$ and $t$ quarks to Kobayashi and Maskawa [29].

## 15.4 TOWARDS A FUNDAMENTAL THEORY

Since Fermi's original paper it has been suspected that the four-fermion interaction represented by (9.23) and the two subsequent modifications are only a low energy approximation to a more fundamental interaction, in which the force is transmitted by an intermediate particle, as happens for electromagnetic interactions by the photon. This particle must be a boson and has been given the name *intermediate boson.* If this is the case, an even stronger link between the weak and electromagnetic interactions can be hypothesised, with a symmetry which connects the electromagnetic current to the weak current and, at the same time, the mediator boson to the photon.

The success of the $V$–$A$ theory implies that the eventual intermediate boson should be described by a *vector field*, in which case the name *intermediate vector boson* is used. The first unified theories of the weak and electromagnetic interactions were due to Schwinger [30] and, subsequent to the $V$–$A$ theory, Glashow [31].

To see how this idea works, we suppose to add to the Lagrangian of ordinary matter (leptons, nucleons, etc.) the intermediate vector boson Lagrangian:

$$\begin{aligned}
\mathcal{L}_{IVB} &= \mathcal{L}^{(0)}_{IVB} + \mathcal{L}_{int} \ , \\
\mathcal{L}^{(0)}_{IVB} &= -\frac{1}{2} W^{\mu\nu} W^{\dagger}_{\mu\nu} - M^2 W_\mu (W^\mu)^\dagger \ , \ W^{\mu\nu} = \partial^\nu W^\mu - \partial^\mu W^\nu \\
\mathcal{L}_{int} &= -g \left( W^{\dagger}_\mu J^\mu_W + W_\mu J^{\dagger\,\mu}_W \right) \ , \qquad (15.38)
\end{aligned}$$

where we have described the intermediate boson with a complex vector field; $\mathcal{L}^{(0)}_{IVB}$ is the free Lagrangian and $g$ is a new coupling constant that we assume to be small.

Treating $\mathcal{L}_{int}$ as a perturbation, we have:

$$S = 1 + i \int d^4x \mathcal{L}_{int} + \frac{(i)^2}{2} \int d^4x d^4y T \left[ \mathcal{L}_{int}(x) \mathcal{L}_{int}(y) \right] + \dots \ . \qquad (15.39)$$

The first order term contributes to processes in which an intermediate boson is emitted or absorbed. If the mass $W$ is large enough, these processes are forbidden by conservation of energy, for example in the neutron decay. The second-order term can be expanded in the following way:

$$\begin{aligned}
&\frac{(i)^2}{2}\int d^4x d^4y\, T\left(\mathcal{L}_{int}(x)\mathcal{L}_{int}(y)\right) \qquad (15.40)\\
&= \frac{(ig)^2}{2}\int d^4x d^4y T\left(W_\mu^\dagger(x)J_W^\mu(x)W_\nu(y)J_W^{\dagger\,\nu}(y) + (x\to y, y\to x)\right) + \ldots\\
&= (ig)^2\int d^4x d^4y\, J_W^{\dagger\,\nu}(y)\left[T\left(W_\nu(y)W_\mu^\dagger(x)\right)\right]J_W^\mu(x) + \ldots\ ,
\end{aligned}$$

where the ellipsis dots represent terms with the emission or absorption of two intermediate bosons (still more forbidden) and we have used the fact that in the Dyson formula we use free fields and therefore the currents and fields commute.

In equation (15.41) the intermediate boson propagator appears which, because of its large mass, can be approximated by a $\delta$-function, cf. equation (8.14) and Problem 1 below. In this limit we obtain the product of the currents:

$$\begin{aligned}
S &= 1 + i\int d^4x\left[\frac{-G}{\sqrt{2}}\left(J_W^{\dagger\,\mu}(x)J_{W\mu}(x)\right)\right] + \ldots \qquad (15.41)\\
&= 1 + i\int d^4x\, \mathcal{L}_W(x) + \cdots\ ,
\end{aligned}$$

where again the ellipsis dots represent terms irrelevant at low energy and we have put:

$$\frac{G}{\sqrt{2}} = \frac{g^2}{M_W^2} > 0\ . \qquad (15.42)$$

In the limit of large mass of the intermediate boson and with a definite sign[3] for $G$, the terms of order $g^2$ in the $S$-matrix agree with the first order term in the Fermi constant of the current $\times$ current Lagrangian (15.30)!

As well as giving the physical dimension of $G$ (recall that $g$ is dimensionless, like the electric charge, in natural units), equation (15.42) gives a valuable clue for constructing a unified theory; the *true coupling constant* $g$ can be of the same order of magnitude as the electric charge if $M_W$ is sufficiently large. We require

$$g^2 \simeq e^2 = 4\pi\alpha \simeq \frac{4\pi}{137} \simeq 0.091 \qquad (15.43)$$

and we use the value of the Fermi constant (15.26). We find:

$$M_W^2 = \frac{\sqrt{2}g^2}{G} \simeq (100\ \text{GeV})^2\ . \qquad (15.44)$$

[3]This is irrelevant for the applications just illustrated but essential to give the correct sign for the interaction energy of neutrinos with matter, cf. Chapter 16.

This is the order of magnitude of the mass of the intermediate boson in a theory in which the fundamental weak interactions are described by equation (15.38) are unified with electromagnetic interactions.

As will be shown in the next volume [13], to have a unified theory in agreement with $\beta$ decay it is necessary to introduce further interactions mediated by an electrically neutral vector boson, $Z^0$, with a new coupling constant and a mass of the same order of magnitude as $g$ and $M_W$. In this theory, the overall interaction Lagrangian is written:

$$\mathcal{L}_{int} = -eA_\mu J^\mu_{e.m.} - g(W^\dagger_\mu J^\mu_W + W^\mu J^\dagger_{W\mu}) - g_1 J^\mu_Z Z_\mu \,, \tag{15.45}$$

with $g, g_1$ of order $e$.

## 15.5 PROBLEMS FOR CHAPTER 15

### Sect. 15.4

**1.** Show that the Lagrangian $\mathcal{L}^{(0)}_{IVB}$

$$\mathcal{L}^{(0)}_{IVB} = -\frac{1}{2} W^{\mu\nu} W^\dagger_{\mu\nu} - M^2 W_\mu (W^\mu)^\dagger$$
$$W^{\mu\nu} = \partial^\nu W^\mu - \partial^\mu W^\nu$$

leads to the equations of motion:

$$-\left(\partial_\rho \partial^\rho + M^2\right) W^\mu + \partial^\mu \left(\partial_\rho W^\rho\right) = 0 \,.$$

**2.** Show that the Feynman propagator of the intermediate boson $W$ is

$$D^{\mu\nu}_F(x) =< 0|T\left[W^\mu(x) W^{\dagger\nu}(0)\right]|0>=$$
$$= \frac{1}{(2\pi)^4} \int d^4k \frac{i}{k^2 - M_W^2 + i\epsilon} \left(-g^{\mu\nu} + \frac{k^\mu k^\nu}{M_W^2}\right) e^{-ikx} \,.$$

**3.** Show that in the limit $M_W \to \infty$

$$D^{\mu\nu}_F(x) \to +\frac{i g^{\mu\nu}}{M_W^2} \delta^{(4)}(x) \,.$$

CHAPTER 16

# NEUTRINO OSCILLATIONS

Since the 1970s the systematic observation of neutrinos of both natural and artificial origin has been made possible by the development of experimental apparatus of large dimensions (tons or thousands of tons) which signal, with different methods, the occurrence of a neutrino interaction.

Fig. 16.1 shows the general principle of these measurements. A source produces neutrinos or antineutrinos from $\beta$ decay and the detector, at a distance $L$ from the source, signals the interaction of the neutrino by means of the observation of a charged lepton produced in the inverse $\beta$ process; see Table 16.1 for the sources and decays utilised.

In the historic experiment of Cowan and Reines [32] in 1956, the source was the Savannah River nuclear reactor, which produced a calculable flux of antineutrinos from neutron decays. The detector was a tank containing water. The antineutrinos produced positrons via the reaction:

$$\bar{\nu} + p \to e^+ + n \, , \tag{16.1}$$

on protons of the water, followed by the annihilation of the positron with an electron of the medium, which gives rise to two $\gamma$ rays, each of energy 0.5 MeV. To reduce the uncertainty level, the water contained a certain amount of cadmium chloride, cadmium being capable of absorbing the neutron with the emission, after a delay of around 5 $\mu$s, of another $\gamma$ ray with a characteristic energy, following the reaction:

$$n + {}^{108}\text{Cd} \to {}^{109m}\text{Cd} \to {}^{109}\text{Cd} + \gamma \, , \tag{16.2}$$

where ${}^{109m}$Cd denotes a metastable excited state of ${}^{109}$Cd. A scintillator material dissolved in the water transformed the $\gamma$ rays into light flashes detected with a system of photomultipliers.


DOI: 10.1201/9781003436263-16

Table 16.1 Sources and detection methods for naturally occurring and artificially produced neutrinos (for more detailed information see Ref. [16]). In atmospheric neutrinos, the complete cascade of decays: $\pi \to \mu\nu_\mu$, $\mu \to \nu_\mu e \nu_e$ gives a 2 : 1 ratio between the fluxes of $\nu_\mu$ and $\nu_e$.

| Source | Production | $E_\nu$(MeV) | $L$(km) | Reaction in detector | Method |
|---|---|---|---|---|---|
| Sun (Be-B) | $\nu_e$ | $1-10$ | $1.4 \cdot 10^8$ | $\nu_e\, ^{37}\mathrm{Cl} \to e\, ^{37}\mathrm{Ar}$ | radioch. |
| Sun (p-p) | $\nu_e$ | $0.2-0.7$ | $1.4 \cdot 10^8$ | $\nu_e\, ^{71}\mathrm{Ga} \to e\, ^{71}\mathrm{Ge}$ | radioch. |
| Sun (B) | $\nu_e$ | $5.5-10$ | $1.4 \cdot 10^8$ | $\nu_e\, p \to e\, n$ | Cherenkov |
| Sun (B) | $\nu_e$ | $6-10$ | $1.4 \cdot 10^8$ | $\nu\, d \to \nu\, p\, n$ | Cherenkov |
| Supernova 1987 | $e\, p \to n\, \nu_e$ | 1 | $1.7 \cdot 10^{18}$ | $\nu_e\, Nucleus \to e + \cdots$ | Cherenkov |
| Atmosphere (zenith) | $\begin{pmatrix} \pi \to \mu\nu_\mu \\ \mu \to \nu_\mu e \nu_e \end{pmatrix}$ | $10^3$ | $\sim 20$ | $\nu_{\mu/e}\, Nucleus \to \mu/e + \cdots$ | Cherenkov |
| Atmosphere (nadir) | $\begin{pmatrix} \pi \to \mu\nu_\mu \\ \mu \to \nu_\mu e \nu_e \end{pmatrix}$ | $10^3$ | $\sim 13000$ | $\nu_{\mu/e}\, Nucleus \to \mu/e + \cdots$ | Cherenkov |
| Nuclear reactor | $n \to \bar{\nu}_e e^- p$ | 1 | $\sim 1$ | $\bar{\nu}_e p \to e^+ n$ | scint. |
| Accelerator (short base) | $\pi/K \to \mu\nu_\mu$ | $10^{3-5}$ | $0.1-1$ | $\nu_\mu(\bar{\nu}_\mu)\, Nucleus \to l^\mp + \cdots$ | imag. |
| Accelerator (long base) | $\pi/K \to \mu\nu_\mu$ | $10^{3-4}$ | $300-900$ | $\nu_\mu(\bar{\nu}_\mu)\, Nucleus \to l^\mp + \cdots$ | imag. |

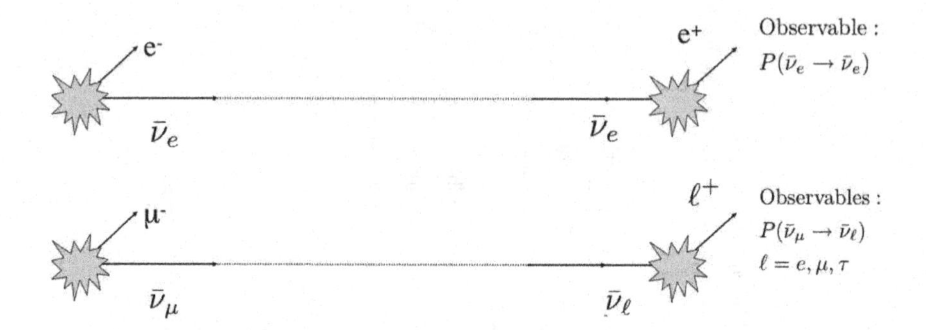


Figure 16.1 Production and detection of neutrinos via direct and inverse $\beta$ processes.

In 1962, Lederman, Schwarz and Steinberger and collaborators [33] at Brookhaven observed that neutrinos associated with muons in the decay of $\pi$ and $K$ mesons do not give rise to reactions in which an electron appears, but are invariably associated with a muon. This experiment shows that the electron and muon are associated with two different types of neutrino, subsequently denoted respectively as $\nu_e$ and $\nu_\mu$. The third known charged lepton, the $\tau$, is associated with a third neutrino, $\nu_\tau$, as shown by a recent experiment carried out at Fermilab by the *DONUT* collaboration [34].

## 16.1 OSCILLATIONS IN VACUUM

We can interpret the experiment of the two neutrinos by assuming the existence of two lepton numbers which are individually conserved: an *electron number*, which characterises $\nu_e$ and $e$, and a *muon number*, associated with $\nu_\mu$ and $\mu$. In modern terminology, these quantum numbers are known as *lepton flavours*.

If neutrinos are massless, different flavour neutrino states are degenerate and we can always find a basis in which the Hamiltonian and the lepton flavours are simultaneously diagonal. This is equivalent to *aligning* the neutrino states with the basis of the charged lepton states.

If neutrinos acquire a mass, it is no longer automatic that the mass matrices of the charged leptons and neutrinos are simultaneously diagonalisable and the phenomenon of *neutrino mixing* can occur, with violation of lepton flavour [35, 36].

Mixing of Two Flavours. We denote with $\nu_e$ and $\nu_\mu$ the neutrino states with the same lepton flavour as $e$ and $\mu$. In general the two states differ from the eigenstates of the mass matrix, which we denote as $\nu'$ and $\nu''$, by a unitary transformation represented by a $2 \times 2$ matrix:

$$|\nu_e\rangle = \cos\theta|\nu'\rangle + \sin\theta|\nu''\rangle\ ,$$
$$|\nu_\mu\rangle = -\sin\theta|\nu'\rangle + \cos\theta|\nu''\rangle\ . \tag{16.3}$$

The Hamiltonian of the system of neutrinos with momentum $\mathbf{p}$ is:

$$H = \sqrt{\mathbf{p}^2 + m^2} \simeq |\mathbf{p}| + \frac{m^2}{2|\mathbf{p}|}\ , \tag{16.4}$$

where $m$ is the neutrino mass matrix and we have used the ultra-relativistic approximation.

Referring to Fig. 16.1, the amplitude to observe in the detector a neutrino of flavour $j$, originating from the source as a neutrino of flavour $i$ ($i, j = e, \mu$) is calculated from the evolution matrix, $e^{-iHt}$. We approximate:

$$t = L\ \ ,\ \ |\mathbf{p}| = E_\nu\ ,$$

$$A(i \to j) = \langle j|e^{-iHL}|i\rangle\ . \tag{16.5}$$

If we denote the mass eigenstates with $|\nu_a\rangle$ ($a = ', ''$), we also have:

$$A(i \to j) = \sum_{a,b} \langle j|a\rangle\langle a|e^{-iHL}|b\rangle\langle b|i\rangle = e^{-iE_\nu L} \sum_a \left(\langle j|a\rangle e^{-i\frac{m_a^2}{2E_\nu}L}\langle a|i\rangle\right)\ , \tag{16.6}$$

from which we obtain the probability of appearance of a new flavour, for example $\nu_e \to \nu_\mu$:

$$\begin{aligned} P(\nu_e \to \nu_\mu; E, L) &= |A(\nu_e \to \nu_\mu)|^2 \\ &= \cos^2\theta\sin^2\theta|1 - e^{-i\frac{\Delta m^2 L}{2E_\nu}}|^2 \\ &= \sin^2(2\theta)\sin^2\left(\frac{\Delta m^2 L}{4E_\nu}\right)\ , \end{aligned} \tag{16.7}$$

where $\Delta m^2 = m_{\nu''}^2 - m_{\nu'}^2$. As expected, at short distances, depending on the values of $\Delta m^2$ and $E_\nu$, the identity of the detected neutrino agrees with that of the neutrino produced. At longer distances *neutrino oscillations* between the initial flavour and the other flavour develop.

We recall that the limit on the mass of $\nu_e$ from the $\beta$ spectrum is of order eV. Assuming that the neutrino masses are all of this order, we can see more clearly the spatial scale of the oscillations by expressing the argument of the sine in more convenient units[1]:

$$\phi = \frac{\Delta m^2 L}{4E_\nu} = \frac{\Delta m^2 L}{4\hbar c E_\nu}$$
$$= 1.27\,\frac{\Delta m^2(\mathrm{eV}^2)L(\mathrm{km})}{E_\nu(\mathrm{GeV})} = 1.27\,\frac{\Delta m^2(\mathrm{eV}^2)L(\mathrm{m})}{E_\nu(\mathrm{MeV})}\ . \tag{16.8}$$

Clearly the interesting effects occur when $\phi$ approaches $\pi/2$.

The $\nu_e \to \nu_\mu$ oscillation is not observable with neutrinos from reactors or the Sun, which do not have sufficient energy to produce a $\mu$ in the detector. The $\nu_e$ in this case oscillates into a non-observable, *sterile*, neutrino but we can observe a neutrino deficit, described by the *non-oscillation probability*:

$$P(\nu_e \to \nu_e; E, L) = 1 - P(\nu_e \to \nu_\mu; E, L)\ . \tag{16.9}$$

The interactions of neutrinos with matter include elastic scattering, in which the neutrino remains intact instead of transforming into the corresponding charged lepton. In this case (*neutral current* processes, cf. [13]) all the neutrinos into which $\nu_e$ oscillate are active and there should be no observed variation in the number of reactions with distance (allowing for the geometrical reduction in the flux). The simultaneous observation of a deficit in charged current processes, $P(\nu_e \to \nu_e) < 1$, with unchanged neutral current processes constitutes a very convincing proof of the oscillation phenomenon (cf. the *SNO* experiment results discussed later).

Three Flavours. The formulae in (16.3) and (16.7) are easily generalised to the case of three lepton flavours. We write:

$$|\nu_i\rangle = \sum_a U_{ia}|a\rangle \qquad (i = e, \mu, \tau; \quad a = 1, 2, 3)\ . \tag{16.10}$$

The index $a$ characterises the mass eigenstates, and $U$ is a unitary matrix which is, in general, complex:

$$U_{ia} = \langle a|i\rangle\ . \tag{16.11}$$

The non-diagonal oscillation amplitude is written:

$$A(\nu_e \to \nu_\mu) = \sum_a e^{-i\frac{m_a^2 L}{2E_\nu}}\, U^*_{\mu a} U_{ea}\ , \tag{16.12}$$

and the probability:

$$P(\nu_e \to \nu_\mu; E, L) = \sum_{a,b} e^{i\Delta_{ab}L}(U^*_{\mu a}U_{ea})(U_{\mu b}U^*_{eb})\ , \tag{16.13}$$

$$\Delta_{ab} = \frac{(m_b^2 - m_a^2)}{2E_\nu}\ , \quad \Delta_{12} + \Delta_{23} = \Delta_{13}\ . \tag{16.14}$$

[1]We recall that $\hbar c \sim 197$ MeV·fm and that $E(\mathrm{GeV}) = 10^9 E(\mathrm{eV})$.

In this case there are two differences between independent masses and therefore two different spatial scales over which the oscillations evolve.

In general terms, the $3 \times 3$ unitary matrix, $U$, is parameterised by three angles and a complex phase, which implies a violation of $CP$ symmetry (of which more later). The parameterisation commonly used in the literature is obtained in the following way:

The most general form of a complex vector $\nu_e$ in terms of three complex vectors $\nu_{1,2,3}$ is written:

$$\nu_e = \cos\theta_{13}\left[\cos\theta_{12}\nu_1 + \sin\theta_{12}\nu_2\right] + e^{i\delta}\sin\theta_{13}\nu_3 \ . \tag{16.15}$$

Proof. In general, we expect a phase factor in front of each of the three terms on the right-hand side. A redefinition of the phase of the left-hand side $\nu_e \to e^{i\alpha}\nu_e$ redefines the three phases for a constant value of the phase, which we choose so that the coefficients of $\nu_1$ are real. Now we can redefine the phases of $\nu_2$ and $\nu_3$ only, in a way so that the phase of $\nu_e$ does not change, therefore $\nu_2 \to e^{i\beta}\nu_2$ and $\nu_3 \to e^{-i\beta}\nu_3$. By this we can also reduce the phase of $\nu_2$ to zero and we are left with one phase for the coefficient of $\nu_3$, as in (16.15).

The definition of $\nu_e$ allows us to define two other orthogonal vectors (in the complex field). We choose:

$$\nu' = -\sin\theta_{12}\nu_1 + \cos\theta_{12}\nu_2 \ , \tag{16.16}$$

$$\nu'' = e^{-i\delta}\sin\theta_{13}\left[\cos\theta_{12}\nu_1 + \sin\theta_{12}\nu_2\right] - \cos\theta_{13}\nu_3 \ . \tag{16.17}$$

The most general form of vectors $\nu_\mu$ and $\nu_\tau$ will be that of two orthogonal combinations of $\nu'$ and $\nu''$, which we can choose in terms of a third angle without having to introduce new phases (the reasoning is the same as that used to have all real coefficients in equation (16.3)):

$$\begin{aligned} \nu_\mu &= \cos\theta_{23}\nu' + \sin\theta_{23}\nu'' \ , \\ \nu_\tau &= -\sin\theta_{23}\nu' + \cos\theta_{23}\nu'' \ . \end{aligned} \tag{16.18}$$

Overall, the expression for the unitary matrix $U$, for three flavours, is:

$$U = \begin{pmatrix} c_{13}c_{12} & c_{13}s_{12} & e^{i\delta}s_{13} \\ -c_{23}s_{12} + s_{23}s_{13}c_{12}e^{-i\delta} & c_{23}c_{12} + s_{23}s_{13}s_{12}e^{-i\delta} & -s_{23}c_{13} \\ s_{23}s_{12} + c_{23}s_{13}c_{12}e^{-i\delta} & -s_{23}c_{12} + c_{23}s_{13}s_{12}e^{-i\delta} & -c_{23}c_{13} \end{pmatrix} . \tag{16.19}$$

## 16.2 NATURAL AND ARTIFICIAL NEUTRINOS

The relevant characteristics of the sources and methods of detection of naturally occurring and artificially produced neutrinos are summarised in Table 16.1. The Sun is a source of *electron neutrinos*, $\nu_e$, produced in nuclear fusion reactions by several reaction cycles and with a complex energy spectrum which extends up to ~10 MeV, Figs. 16.2 and 16.3.

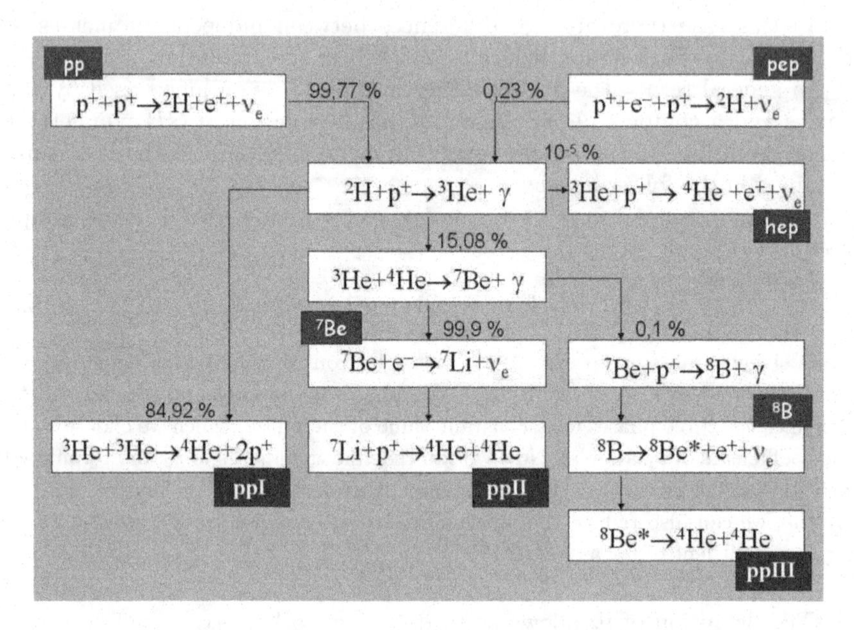


Figure 16.2 The fusion reactions in the Sun (Bethe cycles) [37].

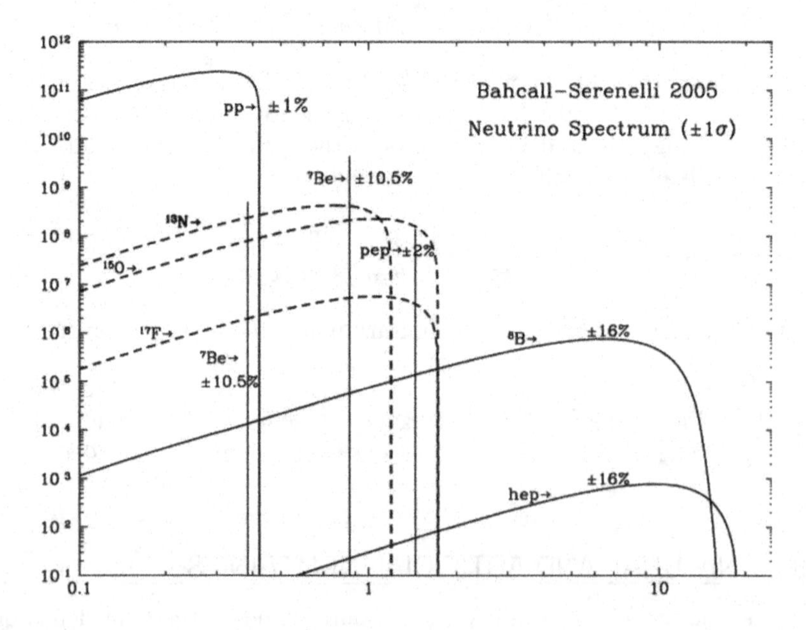


Figure 16.3 Spectrum of solar neutrinos in the *Solar Standard Model* [38]. Energy in MeV. Reprinted from Ref. [12] with permission. © APS (1998).

Neutrinos originating from *Be* and from *B* were detected in 1967 by Davies and collaborators with a detector deep underground, to shield it from cosmic rays, located in the Homestake mine[2]. The solar neutrinos were detected by the reaction $^{37}\mathrm{Cl} \to {}^{37}\mathrm{Ar}$ by means of radiochemical measurements of the radioactive isotope $^{37}\mathrm{Ar}$. These observations showed a deficit compared to estimates from solar models, and interpreted as the effect of neutrino oscillations by Pontecorvo and Gribov [39]. Neutrinos from the *pp* cycle were observed in 1992 with the *GALLEX*[3] and *SAGE*[4] experiments through the reaction $^{71}\mathrm{Ga} \to {}^{71}\mathrm{Ge}$ with radiochemical measurements of $^{71}\mathrm{Ge}$. These observations confirmed the deficit compared to the predicted flux, a very reliable prediction since the flux of *pp* neutrinos is directly related to the energy flux produced by the Sun (cf. the problem at the end of Section 2.4).

The inverse $\beta$ reaction

$$\nu_e + n \to e + p \,, \tag{16.20}$$

was observed with the *SuperKamiokande* detector[5] by the detection of the electron by means of Cherenkov light. The correlation between the direction of the electron and the instantaneous position of the Sun allowed discrimination against background events. More recently, the *SNO* detector[6] detected dissociation reactions of deuterium by solar neutrinos, followed by recombination of the neutron with emission of a photon which, in its turn, by Compton scattering, produces a fast electron, observed by the Cherenkov effect:

$$\begin{aligned} &\nu + d \to \nu + p + n \;\; (\mathrm{NC}) \,, \\ &n + p \to d + \gamma, \;\; \gamma + e \to \gamma + e \,. \end{aligned} \tag{16.21}$$

In reaction (16.21) all the flavours into which the original $\nu_e$ can be transformed are active. *SNO* also measures charged current reactions and scattering on electrons:

$$\nu + d \to e^- + p + p \quad (\mathrm{CC}) \,, \tag{16.22}$$

$$\nu + e^- \to \nu + e^- \qquad (\mathrm{ES}) \,. \tag{16.23}$$

The *SNO* results prove that there is no solar deficit for the neutrino elastic scattering processes, confirming the oscillatory nature of the deficit observed in the charged current reactions.

On 23 February 1987, at 7:35 am Universal Time, neutrino events were observed in two large underground detectors: *Kamiokande* in Japan and *IMB*[7]

---

[2]Homestake Gold Mine, Lead, South Dakota, USA.

[3]*GALLEX*, INFN Gran Sasso National Laboratory, Italy.

[4]*SAGE*, Baksan Neutrino Observatory, Caucasus, Russia.

[5]Kamioka Observatory, Hida-city, Toyama, Japan.

[6]Sudbury Neutrino Observatory, Creighton Mine, Sudbury, Ontario, Canada.

[7]Irvine-Brookhaven-Michigan experiment for detection of proton decays, Fairport mine, Lake Erie, USA.

Table 16.2 Observed deficit in solar neutrino experiments. For SNO, cf. Fig. 16.5.

| Experiment | Observed/expected | Years of operation |
|---|---|---|
| *Homestake* | $0.33 \pm 0.03 \pm 0.05$ | 1970–1995 |
| *Kamiokande* | $0.54 \pm 0.08^{+0.10}_{-0.07}$ | 1986–1995 |
| *SAGE* | $0.58 \pm 0.06 \pm 0.03$ | 1990–2006 |
| *GALLEX* | $0.60 \pm 0.06 \pm 0.04$ | 1991–1996 |
| *SuperKamiokande* | $0.465 \pm 0.005^{+0.016}_{-0.015}$ | 1996– |

in the USA. A few hours later optical signals from a supernova located in our galaxy, at a distance of 170,000 light years, were detected. The correlation between the neutrino events and the light signals allowed a significant limit to be set on the difference between the speed of neutrinos of ~10 MeV energy and the speed of light.

Cosmic rays produce neutrinos of energy around 1 GeV starting from $\pi$ and $K$ mesons produced in high layers of the atmosphere, via the decay chain

$$\pi^+/K^+ \to \mu^+\nu_\mu, \quad \mu^+ \to \bar{\nu}_\mu e^+ \bar{\nu}_e \ . \tag{16.24}$$

The chain with charge conjugate particles is obtained starting from $\pi^-/K^-$. Overall the expected fluxes of the two neutrino flavours are in the ratio $\nu_\mu : \nu_e = 2 : 1$.

Very large underground detectors can observe atmospheric neutrino events produced (i) in the region directly above the detector (zenith), with path lengths of order 20 km, (ii) in the atmosphere on the opposite side of the Earth (nadir) with path lengths of order 13,000 km, corresponding to the diameter of the Earth. *SuperKamiokande* distinguishes $\nu_\mu \to \mu$ from $\nu_e \to e$ events by the distribution of Cherenkov light emitted by the charged leptons and observed a ratio of around 2:1 between events classified as '*muons*' and '*electrons*', for the neutrinos from the zenith, while observing a ratio of about 1:1 for those from the nadir. The observation, now strengthened, is interpreted as due to oscillations of $\nu_\mu$ into the neutrino associated with a lepton which, in the Standard Model with three flavours, can only be the $\nu_\tau$. The limited energy spectrum of atmospheric neutrinos does not allow direct production of $\tau$s but only permits the deficit of atmospheric neutrinos to be established.

Concerning neutrinos from artificial sources, accurate measurements of the non-oscillation probability of $\bar{\nu}_e$ over distances of the order of kilometres have been carried out at nuclear reactors, so far with negative results. The *KAMLAND*[8] experiment adopted an innovative approach, detecting, with a large volume of scintillator placed inside *Kamiokande*, the antineutrinos ($\bar{\nu}_e$)

[8]Kamioka Liquid Scintillator Anti-Neutrino Detector. With 1000 tons of liquid scintillator, it is the largest scintillation detector so far constructed.

emitted by a group of nuclear reactors located in Japan and South Korea. The flux is dominated by a few of the reactors, situated at an average distance of 180 km. *KAMLAND* succeeded in detecting a reduction of about 40% compared to the expected flux, and a distortion of the spectrum in agreement with the oscillation hypothesis.

Finally, starting from the *Kamiokande* observations, beams of neutrinos produced by accelerators have been developed so that they can be detected in underground experiments placed at long distances from the source (long-baseline neutrino experiments).

At present there are three active beams of this type:

- *K2K*[9]: baseline $\sim 250$ km, $E_\nu = 1.3$ GeV ,
- *NuMi*[10]: baseline $\sim 700$ km, $E_\nu \simeq 3$ GeV ,
- *CNGS*[11]: baseline $\sim 700$ km, $E_\nu \sim 20$ GeV.

The objective of these beams is to study the atmospheric neutrino anomaly in controlled and reproducible conditions. *CNGS* also has the goal of observing $\tau$ leptons produced by possible $\nu_\mu$–$\nu_\tau$ oscillation. The observation of a $\tau$ event was reported by the *OPERA* collaboration at Gran Sasso in 2010[12].

Both *MINOS* and *K2K* announced the observation of electron events from the $\nu_\mu \to \nu_e$ oscillation in 2011.

## 16.3 INTERACTION WITH MATTER: THE MSW EFFECT

To correctly describe the propagation of neutrinos inside the Sun it is necessary to take account of their interactions with solar matter. To this end, to the expression for the energy of the neutrino in vacuum, equation (16.4), we must add the effect of the weak interaction with particles of the surrounding medium, i.e. electrons, protons and neutrons. We have to deal with two types of interaction:

- the $\nu_e$–$e$ interaction, produced by the exchange of the charged intermediate boson, cf. Section 15.4,
- the interaction due to the exchange of the neutral intermediate boson, $Z^0$, between a neutrino of any flavour and matter particles.

The second interaction is independent of flavour [13] and therefore contributes to the Hamiltonian a term proportional to the identity matrix that has no effect on the mixing or on the oscillations. We can therefore limit

[9]KEK to Kamiokande, Japan.

[10]Neutrinos to *MINOS*: from Fermilab to the *MINOS* experiment, Soudan mine, Minnesota, USA.

[11]CERN Neutrinos to Gran Sasso, Italy

[12]At the end of 2014, *OPERA* had identified a total of 4, $\nu_\tau$ initiated, events.

our considerations to the first interaction, which arises from the Hamiltonian density:

$$H_{\nu_e-e}(x) = -\mathcal{L}_{Fermi}(x) = +\frac{G}{\sqrt{2}}\left[\bar{e}\gamma_\mu(1-\gamma_5)\nu_e\right]\left[\bar{\nu}_e\gamma^\mu(1-\gamma_5)e\right] \ . \quad (16.25)$$

The sign is consistent with the sign found in Section 15.4, when we derived the Fermi Lagrangian starting from the exchange of the charged intermediate boson.

The calculation is greatly simplified if we exchange the fields $\bar{e}$ and $\bar{\nu}_e$, by expressing the Hamiltonian as the product of diagonal bilinears, relative to the field of the electron and $\nu_e$. This transformation, known as the *Fierz transformation*, has the property of leaving invariant the $V$–$A$ interaction:

$$\left[\bar{e}\gamma_\mu(1-\gamma_5)\nu_e\right]\left[\bar{\nu}_e\gamma_\mu(1-\gamma_5)e\right] = +\left[\bar{\nu}_e\gamma^\mu(1-\gamma_5)\nu_e\right]\left[\bar{e}\gamma_\mu(1-\gamma_5)e\right] \ . (16.26)$$

We will therefore use the Hamiltonian in the form:

$$H_{\nu_e-e}(x) = +\frac{G}{\sqrt{2}}\left[\bar{e}\gamma^\mu(1-\gamma_5)e\right]\left[\bar{\nu}_e\gamma_\mu(1-\gamma_5)\nu_e\right] \ . \quad (16.27)$$

Proof. If we write equation 16.25) in terms of left-handed fields, Chapter 13, we obtain the combination:

$$(\bar{e}_L\gamma_\mu\nu_L)(\bar{\nu}_L\gamma^\nu e_L) \ ,$$

which can be expressed in terms of bilinears:

$$(\bar{\nu}_L\Gamma^{(i)}\nu_L)(\bar{e}_L\Gamma^{(i)}e_L) \ ,$$

by the completeness of the Dirac matrices. On the other hand, from the presence of the projectors $1-\gamma_5$, the $\Gamma^{(i)}$ matrices must anticommute with $\gamma_5$ (cf. Chapter 13) or $\Gamma^{(i)} = V, A$. Taken between the left-handed fields, $\gamma_\mu$ and $\gamma_\mu\gamma_5$ give the same result, except for a sign. Therefore, if we take account of the sign changes from the exchange of $e$ and $\nu_e$ due to Fermi statistics, the relation to be proven is:

$$\left[\gamma_\mu(1-\gamma_5)\right]_{\alpha\beta}\left[\gamma^\mu(1-\gamma_5)\right]_{\delta\epsilon} = -\left[\gamma_\mu(1-\gamma_5)\right]_{\delta\beta}\left[\gamma^\mu(1-\gamma_5)\right]_{\alpha\epsilon} \ .$$

We multiply both sides by $\gamma^\rho_{\beta\alpha}$ and sum over $\beta$ and $\alpha$. The left-hand side gives:

$$Tr[\gamma^\rho\gamma_\mu(1-\gamma_5)]\left[\gamma^\mu(1-\gamma_5)\right]_{\delta\epsilon} = 4\gamma^\rho(1-\gamma_5),$$

and the right-hand side:

$$-\left[\gamma_\mu(1-\gamma_5)\gamma^\rho\gamma^\mu(1-\gamma_5)\right]_{\delta\epsilon} = -2\gamma_\mu\gamma^\rho\gamma^\mu(1-\gamma_5) = +4\gamma^\rho(1-\gamma_5) \ . \quad (16.28)$$

Q. E. D.

Matrix Element. The energy density of the neutrino is found by taking the matrix element of equation (16.27) between the states with the initial and final electron at rest, and the initial and final neutrino with equal momentum, **p**:

$$\mathcal{E}_W = +\frac{G}{\sqrt{2}} \langle e(\boldsymbol{p}=0)|\bar{e}\gamma^\mu(1-\gamma_5)e|e(\boldsymbol{p}=0)\rangle \times \langle \nu_e(\boldsymbol{p})|\bar{\nu}_e\gamma_\mu(1-\gamma_5)\nu_e|\nu_e(\boldsymbol{p})\rangle \ . \quad (16.29)$$

With the electron at rest, only $\mu = 0$ counts. We can approximate the mass of the neutrino to zero and consider it a Weyl field (in this limit, the same result holds for a Majorana neutrino, Chapter 13). We use the expansion (13.10) and we find for the energy density:

$$\mathcal{E}_W = +\frac{G}{\sqrt{2}}\frac{1}{V^2}\, u_e(0)^\dagger u_e(0) \times 2 \times u_{\nu_e}{}_L^\dagger(\boldsymbol{p}) u_{\nu_e}{}_L(\boldsymbol{p}) \ , \quad (16.30)$$

with the factor 2 arising from the action of $1-\gamma_5$ on the left-handed neutrino.

The normalisation chosen corresponds to a neutrino and an electron in a volume $V$. The neutrino energy is obtained by multiplying by $V$ and by the number of electrons present, $N_e$. With $\rho_e = N_e/V$, the number of electrons per unit volume, we obtain:

$$E_W = +\sqrt{2}G\rho_e = \mathcal{V} \ , \quad (16.31)$$

and the total energy of the neutrino can now be written:

$$E = |\boldsymbol{p}| + \frac{m^2}{2p} + \mathcal{V} = |\boldsymbol{p}| + \Delta E \ . \quad (16.32)$$

The estimated mass density at the centre of the Sun is equal to:

$$\rho_{Sun}(r=0) \simeq 162 \ \text{g/cm}^3 \ . \quad (16.33)$$

To a good approximation in the Sun there is one proton for every electron, therefore the density of 1 g/cm$^3$ corresponds to $N_A = 6.6 \cdot 10^{23}$ electrons/cm$^3$, from which:

$$(\rho_e)_{Sun}(r=0) \simeq \frac{\rho_{Sun}(r=0)}{162 \ \text{g/cm}^3} N_A \cdot 162(\hbar c)^3 = \frac{\rho_{Sun}(r=0)}{162 \ \text{g/cm}^3}\, 8.18 \cdot 10^{11} \ \text{eV}^3 \ , \quad (16.34)$$

from which

$$\mathcal{V} = 1.35 \cdot 10^{-5} \ \frac{\rho_{Sun}(r=0)}{162 \ \text{g/cm}^3} \ \frac{\text{eV}^2}{\text{MeV}} \ . \quad (16.35)$$

with $\rho_{Sun}$ in g/cm$^3$.

MSW Effect. In the $(\nu_e, \nu_\mu)$ basis, we can express the neutrino mass matrix in terms of $\Delta m^2_{12} = m^2_2 - m^2_1$ and the vacuum mixing angle:

$$\begin{aligned}\langle i|m^2|j\rangle &= \sum_a \langle i|a\rangle m^2_a \langle a|j\rangle = [U(m^2)_{diag}U^\dagger]_{ij} = \\ &= -\frac{\Delta m^2_{12}}{2}\begin{pmatrix} \cos(2\theta_{12}) & \sin(2\theta_{12}) \\ \sin(2\theta_{12}) & -\cos(2\theta_{12})\end{pmatrix} + \cdots ,\end{aligned} \tag{16.36}$$

where we have omitted a term proportional to the unity matrix. Similarly, we write the potential as:

$$\begin{aligned}\mathcal{V} &= +\sqrt{2}G\rho_e \begin{pmatrix} 1 & 0 \\ 0 & 0 \end{pmatrix} \\ &= +\frac{\sqrt{2}G}{2}\rho_e \begin{pmatrix} 1 & 0 \\ 0 & -1 \end{pmatrix} + \cdots ,\end{aligned} \tag{16.37}$$

from which:

$$\begin{aligned}\Delta E &= \frac{m^2}{2E_\nu} + \mathcal{V} \\ &= \left[-\frac{\Delta m^2_{12}\cos(2\theta_{12})}{4E_\nu} + \frac{\sqrt{2}G}{2}\rho_e\right]\sigma_3 + \frac{\Delta m^2_{12}\sin(2\theta_{12})}{4E_\nu}\sigma_1 + \cdots ,\end{aligned} \tag{16.38}$$

where $\sigma_{1,3}$ are the familiar Pauli matrices and we have omitted terms proportional to the identity matrix.

If $\Delta m^2_{12}\cos(2\theta_{12}) > 0$, equation (16.38) shows that *a critical value of the density exists for which the coefficient of* $\sigma_3$ *vanishes* and, correspondingly, the mixing angle becomes maximal, even if the vacuum mixing angle is very small. In these conditions a $\nu_e \to \nu_\mu$ resonant conversion, the *Mikheyev–Smirnov–Wolfenstein effect* [40, 41], can occur. The critical density is:

$$\bar{\rho}_e = \frac{\Delta m^2_{12}\cos(2\theta_{12})}{2\sqrt{2}GE_\nu} . \tag{16.39}$$

The reactions which generate the neutrinos take place in the central region of the Sun ($r = 0$) where the density is maximum; in their journey towards the surface, the neutrinos cross regions whose density decreases until it vanishes at the surface of the Sun, $r = R_{Sun}$. The mixing matrix at the exit from the Sun is obtained by solving the Schrödinger equation along the trajectory. We can consider two extreme cases [16]:

- the potential term is negligible:

$$\frac{\Delta m^2_{12}\cos(2\theta_{12})}{4E_\nu} >> \frac{\sqrt{2}G}{2}\rho_e(r=0) \quad \text{or} \quad \rho_e(r=0) << \bar{\rho}_e . \tag{16.40}$$

  The neutrino deviates further from the critical condition, the matter effect is negligible and the formulae for oscillations in vacuum apply.

- the potential term is dominant:

$$\frac{\Delta m_{12}^2 \cos(2\theta_{12})}{4E_\nu} << \frac{\sqrt{2}G}{2}\rho_e(r=0) \quad \text{or} \quad \rho_e(r=0) >> \bar{\rho}_e \ . \tag{16.41}$$

The neutrino at the origin is found in the eigenstate with maximum eigenvalue of $\mathcal{V} = +\frac{\sqrt{2}G}{2}\rho_e(r=0)$. If, as happens in the Sun, the variation in density along the trajectory is sufficiently slow (adiabatic conditions), the neutrino is found at all times in an eigenstate of the mixing matrix, corresponding to the density at that point. In addition, it can be shown that the levels do not cross, from which, at the exit from the Sun, the neutrino is found in the eigenstate of the mass matrix with maximum eigenvalue, or:

$$\nu_e(r = R_{Sun}) \sim \nu_2 \ . \tag{16.42}$$

After this, the eigenstate propagates to the detector without oscillation, from which:

$$P(\nu_e \to \nu_e) = |U_{e2}|^2 = \sin^2\theta_{12} \ \text{(MSW adiabatic)} \ . \tag{16.43}$$

## 16.4 ANALYSIS OF THE EXPERIMENTS

The results of recent decades have led to a determination of the parameters which describe neutrino oscillations of the three known flavours. We summarise in Table 16.4 the values obtained in two recent measurements. These numbers result from a rather complex global fit, but it is possible to provide a few simple considerations for guidance, starting from the fact that $\theta_{13}$ is much smaller than the other two angles and that $\Delta m_{12}^2 << \Delta m_{23}^2$.

In the limit $\theta_{13} = 0$, equations (16.15) and (16.18) show that $\nu_e$ is mixed only with the superposition of $\nu_\mu$ and $\nu_\tau$ which we called $\nu'$. Therefore the oscillations of $\nu_e$ and $\bar{\nu}_e$, including solar neutrinos and the effects of their passage through the interior of the Sun, are, to an excellent approximation, those of a simple system of two neutrinos, $\nu_e$–$\nu'$.

*KAMLAND*. Measurement of the reduction in the flux of $\nu_e$ compared to what is observed in experiments in close proximity to reactors, and the distortion of the energy spectrum, allows a direct measurement of the oscillation parameters, Table 16.3.

Solar Neutrinos. The first proposal to explain the deficit observed at Homestake, advanced by Pontecorvo and Gribov, was the *"just so"* solution, in which the oscillation phase is of order $\pi/2$ just at the orbit of the Earth. Referring to the values from *GALLEX* (deficit $R \sim 0.58\%$, $E \sim 0.5$ MeV) we

Table 16.3 Oscillation parameters for $\nu_e$–$\nu'$ determined by the $KAMLAND$ experiment [42].

| $\Delta m^2_{12}$ ($10^{-5}$ eV$^2$) | $\tan^2(\theta_{12})$ |
|---|---|
| $7.58 \pm 0.14$(stat)$\pm 0.15$ (syst) | $0.56 \pm 0.10$ (stat) $\pm 0.10$ (syst) |

find:

$$\text{solution } \textbf{just so} : \tag{16.44}$$

$$\Delta m^2_{12} \sim \frac{\pi}{2} \frac{E}{1.27 L_{Sun}} \sim 4.3 \cdot 10^{-12} ,$$

$$\sin^2 \theta_{12} \sim 0.12 . \tag{16.45}$$

The *just so* solution appears in the lower part of Fig. 16.4, but is not consistent with all the other information, in particular with the $KAMLAND$ observations, which require an oscillation length of terrestrial scale. For $\Delta m^2_{12}$ much larger than the *just so* value, the oscillation in vacuum averages to a value independent of distance:

$$P(\nu_e \to \nu_e) \to 1 - \frac{1}{2} \sin^2(2\theta_{12}) \qquad \text{(vacuum oscillation)} . \tag{16.46}$$

Referring to the value of $GALLEX$ we obtain:

$$\sin^2(2\theta_{12}) = 2(1 - R) = 0.84 \quad \to \sin^2 \theta_{12} = 0.30 . \tag{16.47}$$

This solution corresponds to a vertical line with $\tan^2 \theta_{12} = 0.43$, from the *just so* solution upwards, roughly the dark region of Fig. 16.4.

For a more precise analysis, it is necessary to take account of the MSW effect in the Sun, equation (16.38). Using the $KAMLAND$ parameters, Table 16.3, and the values of the energy, $E_\nu = 0.5, E_\nu = 7$ MeV for gallium and boron, respectively, we find:

$$\frac{\Delta m^2_{12} \cos(2\theta_{12})}{4E_\nu} = 2.4 \cdot 10^{-5} \, \frac{\text{eV}^2}{\text{MeV}} \quad \text{(Ga)} ,$$

$$= 0.17 \cdot 10^{-5} \, \frac{\text{eV}^2}{\text{MeV}} \quad \text{(B)} ,$$

$$\frac{\sqrt{2}G}{2} \rho_e(r = 0) = 0.63 \cdot 10^{-5} \, \frac{\text{eV}^2}{\text{MeV}} .$$

We are, therefore, in the conditions:

$$\bar{\rho}_e(Ga) > \rho_{Sun}(r = 0) > \bar{\rho}_e(B) , \tag{16.48}$$

from which, to a sufficiently good approximation, we can conclude that:

- for the deficit from $Ga$, equation (16.47) applies, which, with $R(Ga) = 0.58$, gives $\sin^2\theta_{12}(Ga) = 0.30$;
- for the deficit from $B$, equation (16.43) applies, which, with $R(B) = 0.3$ gives $\sin^2\theta_{12}(B) = 0.33$.

The values of the deficit from Homestake and $GALLEX/SAGE$, while different from each other, are explained by the same oscillation parameters, consistent; moreover, with those observed by $KAMLAND$ and other experiments, as shown in Table 16.4.

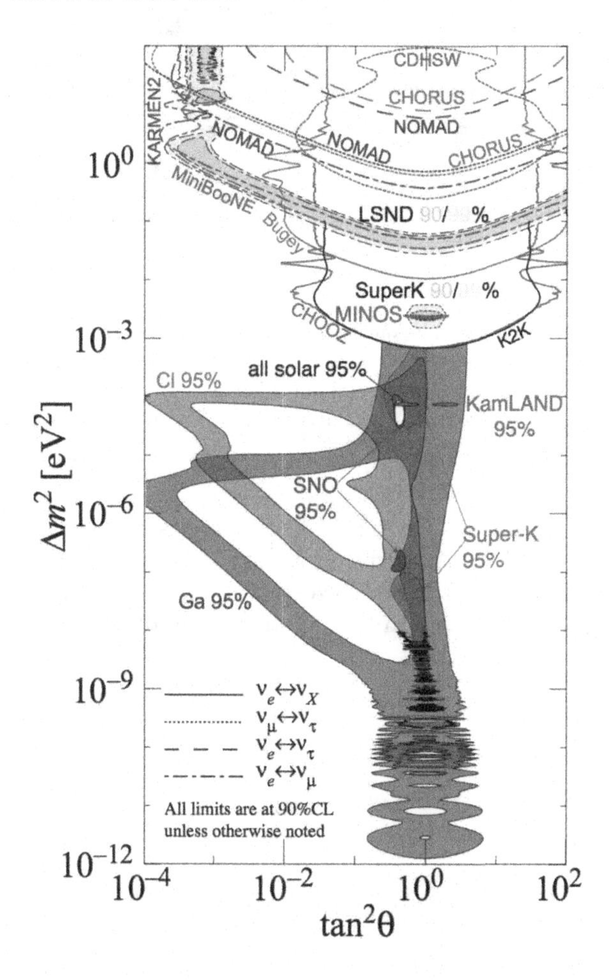


Figure 16.4 Allowed regions from the measurement of the probability of non-oscillation; $\nu_e$ of solar origin and $\bar{\nu}_e$ produced on Earth [16]. Reprinted from Ref. [12] with permission. © APS (1998).

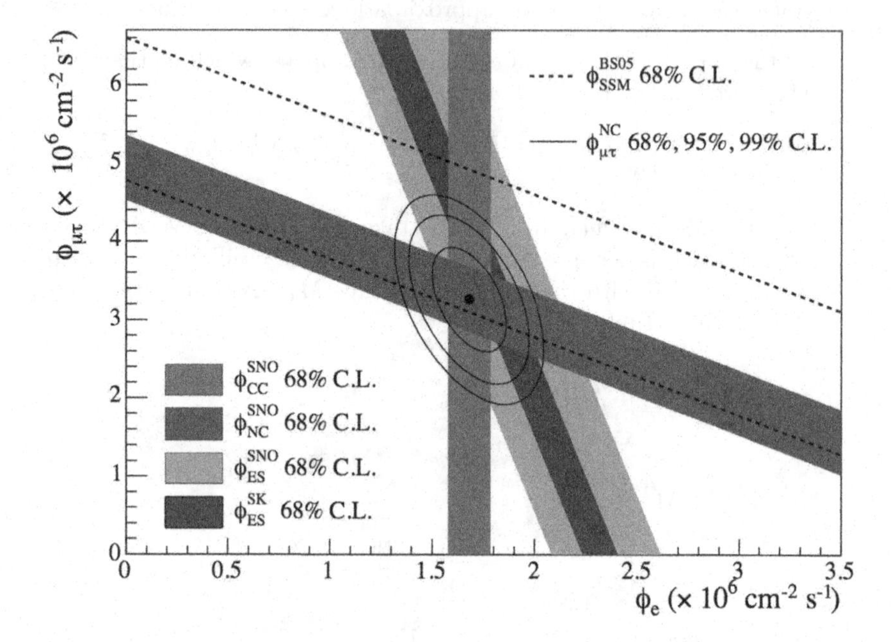


Figure 16.5 SNO results [43]. The measurement of charged current (CC) and neutral current (NC, ES) reactions allows separation of the contributions of $\nu_e$ and the superposition $\nu_\mu$–$\nu_\tau$, and to compare them with the predictions of the Standard Model and the predicted spectrum for neutrinos from boron. Reprinted from Ref. [12] with permission. © APS (1998).

Atmospheric Neutrinos. On the scale of energies and distances of atmospheric neutrinos, cf. Table 16.1, the oscillation associated with $\Delta m^2_{12}$ is negligible. If we set $\theta_{13}$ and $\Delta m^2_{12}$ equal to zero, the formula for the non-oscillation amplitude of $\nu_\mu$, cf. equation (16.12), simplifies considerably:

$$\begin{aligned} |A(\nu_\mu \to \nu_\mu; E, L)| &= |\sum_a e^{-i\frac{m_a^2 L}{2E_\nu}} U^*_{\mu a} U_{\mu a}| \\ &= ||U_{\mu 1}|^2 + |U_{\mu 2}|^2 + e^{i\Delta_{23}L}|U_{\mu 3}|^2| \\ &= |1 - |U_{\mu 3}|^2 + e^{i\Delta_{23}L}|U_{\mu 3}|^2| \; . \end{aligned} \tag{16.49}$$

In addition:

$$|U_{\mu 3}|^2 = \sin^2\theta_{23}\cos^2\theta_{13} \sim \sin^2\theta_{23} \; , \tag{16.50}$$

from which:

$$\begin{aligned} P(\nu_\mu \to \nu_\mu; E, L) &= \cos^4\theta_{23} + \sin^4\theta_{23} + 2\cos^2\theta_{23}\sin^2\theta_{23}\cos\Delta_{23} \\ &= 1 - \sin^2 2\theta_{23}\sin^2\frac{\Delta m^2_{23} L}{4E_\nu} \; . \end{aligned} \tag{16.51}$$

Table 16.4 Recent determinations of the oscillation parameters of neutrinos of three flavours from a global fit to all available data.

| Parameter | Fogli *et al.* [44] | Schwetz *et al.* [45] |
|---|---|---|
| $\Delta m^2_{12}(10^{-5}\ \mathrm{eV}^2)$ | $7.58^{+0.22}_{-0.26}$ | $7.59^{+0.20}_{-0.18}$ |
| $\lvert\Delta m^2_{23}\rvert(10^{-3}\ \mathrm{eV}^2)$ | $2.35^{+0.12}_{-0.09}$ | $2.50^{+0.09}_{-0.16}$ |
| $\sin^2\theta_{12}$ | $0.312^{+0.017}_{-0.016}$ | $0.312^{+0.017}_{-0.015}$ |
| $\sin^2\theta_{23}$ | $0.42^{+0.08}_{-0.03}$ | $0.52^{+0.06}_{-0.07}$ |
| $\sin^2\theta_{13}$ | $0.025\pm0.007$ | $0.013^{+0.007}_{-0.005}$ |

Using the parameters from Tables 16.1 and 16.4 with $E_\nu \sim 1$ GeV, we find:

$$P(\nu_\mu \to \nu_\mu; 1\ \mathrm{GeV}, 13000\ \mathrm{km}) = 0.25\ , \tag{16.52}$$

which is a good approximation to the result of *SuperKamiokande*, $R$(nadir)= observed flux/calculated flux $\sim 0.57$.

The Transformation $\nu_\mu \to \nu_e$. The probability of appearance of $\nu_e$ in a long-baseline beam composed initially of $\nu_\mu$ is easily calculated in the approximation $\Delta m^2_{12} = 0$. Exchanging $e$ with $\mu$ in equation (16.12), gives:

$$\begin{aligned} |A(\nu_\mu \to \nu_e)| &= |U_{e1}U^*_{\mu 1} + U_{e2}U^*_{\mu 2} + e^{-i\Delta_{13}L}U_{e3}U^*_{\mu 3}| \\ &= |U_{e3}U^*_{\mu 3}(1 - e^{-i\Delta_{13}L})|\ , \end{aligned} \tag{16.53}$$

therefore:

$$\begin{aligned} P(\nu_\mu \to \nu_e) &= |U_{e3}U^*_{\mu 3}|^2\ 2[1 - \ cos(\Delta_{31}L)] \\ &= \sin^2\theta_{23}\sin^2(2\theta_{13})\sin^2\left(\frac{\Delta m^2_{23}L}{4E_\nu}\right)\ . \end{aligned} \tag{16.54}$$

The observation by *MINOS* and *K2K* of electrons produced by a $\nu_\mu$ beam, if confirmed, allows the determination of the third angle, $\theta_{13}$, cf. Table 16.4.

Matter–Antimatter Symmetry. If we move from the states, equation (16.10), to fields, along the lines of Chapter 13, we can connect the oscillations of neutrinos and antineutrinos in a precise way. The neutrino field is written schematically as:

$$\psi \sim ae^{-ikx} + b^\dagger e^{+ikx}\ , \tag{16.55}$$

and neutrino and antineutrino states are obtained by applying the relevant creation operators, $a^\dagger$ and $b^\dagger$, to the vacuum.

Since the mixing of the fields is described by a *unique* matrix which we denote as $U^*$, it follows that:

- the mixing of the neutrino states is described by the matrix $U$,
- the mixing of the antineutrinos is described by $U^*$.

If $U$ is real, oscillations of neutrinos (from the Sun or from $\pi^+$ decays) are identical to those of antineutrinos (from reactors or from $\pi^-$ decays), as we have implicitly assumed so far. The presence or absence of a phase in the mixing matrix is connected to the violation of matter–antimatter symmetry[13]. This connection appears explicitly by comparing the formula (16.12) with the analogous one for antineutrinos

$$A(\nu_e \to \nu_\mu; E, L) = \sum_a e^{-i\frac{m_a^2 L}{2E_\nu}}\, U^*_{\mu a} U_{ea} \,,$$
$$A(\bar{\nu}_e \to \bar{\nu}_\mu; E, L) = \sum_a e^{-i\frac{m_a^2 L}{2E_\nu}}\, U_{\mu a} U^*_{ea} \,. \tag{16.56}$$

We see that:

- $P(\bar{\nu}_e \to \bar{\nu}_\mu; E, L) = P(\nu_e \to \nu_\mu; E, L)$ *if, and only if, $U$ is real.*

Even for complex $U$, equation (16.56) implies a connection between the oscillations of neutrinos and antineutrinos, i.e.:

- $P(\bar{\nu}_e \to \bar{\nu}_\mu; E, L) = P(\nu_e \to \nu_\mu; E, -L) = P(\nu_\mu \to \nu_e; E, L)$ *for any unitary $U$.*

If we recall that $L \equiv t$, we see that the previous equation implies an exact symmetry for matter–antimatter exchange united to time reversal. It is the consequence of the $CPT$ transformation symmetry which every relativistic quantum field theory obeys.

As we saw in equation (16.19), the theory with three flavours allows a mixing matrix with a complex phase. We can see explicitly how this leads to a violation of matter-antimatter symmetry, by calculating the difference:

$$\Delta P = \frac{1}{2}[P(\nu_\mu \to \nu_e; E, L) - P(\bar{\nu}_\mu \to \bar{\nu}_e; E, L)]$$
$$= \frac{1}{2}[P(\nu_\mu \to \nu_e; E, L) - P(\nu_\mu \to \nu_e; E, -L)] \,. \tag{16.57}$$

[13]This is described by the $CP$ product of the charge conjugation $C$ and parity $P$ transformations. For a field theory discussion of the $C$ and $P$ symmetries and of time reversal, $T$ see Chapter 12.

Using the definition of equation (16.14) and the antisymmetry of $\Delta_{ab}$, we find:

$$\begin{aligned}
\Delta P &= \frac{-i}{2}\sum_{ab} \sin(\Delta_{ab}L)\ (U^*_{\mu a}U_{ea})(U_{\mu b}U^*_{eb}) \\
&= \sum_{ab} \sin(\Delta_{ab}L)\ \mathrm{Im}\left[(U^*_{\mu a}U_{ea})(U_{\mu b}U^*_{eb})\right] \\
&= 2\sum_{a<b} \sin(\Delta_{ab}L)\ \mathrm{Im}\left[(U^*_{\mu a}U_{ea})(U_{\mu b}U^*_{eb})\right] \\
&= 2\{\sin(\Delta_{12}L)\ \mathrm{Im}\left[(U^*_{\mu 1}U_{e1})(U_{\mu 2}U^*_{e2})\right]+ \\
&\qquad + \sin(\Delta_{13}L)\ \mathrm{Im}\left[(U^*_{\mu 1}U_{e1})(U_{\mu 3}U^*_{e3})\right]+ \\
&\qquad\qquad + \sin(\Delta_{23}L)\ \mathrm{Im}\left[(U^*_{\mu 2}U_{e2})(U_{\mu 3}U^*_{e3})\right]\}\ .
\end{aligned} \tag{16.58}$$

We use the orthogonality relation on the rows of $U$:

$$U^*_{\mu 1}U_{e1} + U^*_{\mu 2}U_{e2} + U^*_{\mu 3}U_{e3} = 0\ , \tag{16.59}$$

to eliminate $(U_{\mu 2}U^*_{e2})$, for example:

$$\mathrm{Im}\left[(U^*_{\mu 1}U_{e1})(U_{\mu 2}U^*_{e2})\right] = -\mathrm{Im}\left[(U^*_{\mu 1}U_{e1})(U_{\mu 3}U^*_{e3})\right]\ , \tag{16.60}$$

and the second condition from equation (16.14) to eliminate $\Delta_{13}$. We find:

$$\begin{aligned}
\Delta P &= -2\mathrm{Im}\left[(U^*_{\mu 1}U_{e1})(U_{\mu 3}U^*e3)\right] \\
&\qquad \times \{\sin(\Delta_{12}L) - \sin[(\Delta_{12}+\Delta_{23})L] + \sin(\Delta_{23}L)\} \\
&= -2\mathrm{Im}\left[(U^*_{\mu 1}U_{e1})(U_{\mu 3}U^*_{e3})\right]\times \\
&\qquad \times \{\sin(\Delta_{12}L)[1-\cos(\Delta_{23}L)] + \sin(\Delta_{23}L)[1-\cos(\Delta_{12}L)]\} \\
&= F \times G\ ,
\end{aligned} \tag{16.61}$$

and the two factors are, explicitly; cf. equation (16.19):

$$\begin{aligned}
F &= \cos\theta_{13}\sin(2\theta_{12})\sin(2\theta_{23})\sin(2\theta_{13})\sin\delta\ , \\
G &= 2\left[\sin(\Delta_{12}L)\sin^2(2\Delta_{23}L) + \sin(\Delta_{23}L)\sin^2(2\Delta_{12}L)\right]\ .
\end{aligned} \tag{16.62}$$

From these formulae we learn that:

- the phase in the matrix $U$ produces a violation of $CP$ symmetry,
- the violation only occurs if *all* the mass differences and *all* the angles are different from zero; in the converse case, as we saw when setting $\theta_{13}$ or $\Delta m^2_{12}$ to zero, the system reduces to a mixing between two states, which is described by a real matrix.

## 16.5 OPEN PROBLEMS

The observation of the phenomena just illustrated has shown beyond any reasonable doubt that neutrinos have a mass and that their flavour quantum numbers are not conserved. Several problems remain open, which make neutrinos still a frontier area of astrophysics, cosmology and the physics of fundamental forces. Without any attempt at completeness, we illustrate a few of these issues.

$CP$ Violation in the Lepton Sector. The recent evidence for a non-zero value of $\theta_{13}$, Table 16.4, opens the way to the search for $CP$ violation among neutrinos, which would match the already observed violation in weak interactions of hadrons, cf. [13]. To find evidence of an asymmetry in the oscillations of neutrinos and antineutrinos, even for large values of the phase $\delta$ which appears in the mixing matrix, (16.19), requires the construction of accelerators providing extremely high intensities of muons: *muon factories, muon colliders,* etc.

Majorana Neutrino. Everything which has been said so far holds for both Dirac or Majorana neutrinos since, for masses which are anyway $<< E$, the $V$–$A$ interaction inhibits transitions between particles and antiparticles, of the type $\nu_e \to e^+$, cf. Chapter 13. The observation of neutrinoless double–$\beta$ decay is the characteristic signal of a Majorana neutrino with mass, and is actively being sought in several laboratories, for example Gran Sasso. The present situtation is illustrated in the next Section.

Mass Hierarchy. As is seen from equation (16.51) and in Table 16.1, the data do not define the sign of $\Delta m^2_{13}$; therefore, we have two possible orderings of the neutrino masses:

$$\begin{aligned} \text{normal order}: \quad & m_1^2 < m_2^2 < m_3^2 \ , \\ \text{inverted order}: \quad & m_3^2 < m_1^2 < m_2^2 \ . \end{aligned} \tag{16.63}$$

In the case of quark masses, normal ordering holds, and in addition each mass squared is much less than the squared mass which follows. If this scheme were to be repeated for neutrinos, we would have:

$$0 \sim m_1^2 << \Delta m^2_{12} \sim m_2^2 << \Delta m^2_{23} \sim m_3^2 \ , \tag{16.64}$$

but they could also have squared masses roughly equal to each other and much larger than the differences:

$$m_1^2 \sim m_2^2 \sim m_3^2 >> \Delta m^2 \ . \tag{16.65}$$

The absolute values of the masses can be obtained with accurate measurements of the electron spectrum in $\beta$ decay of tritium, Chapter 15, or from the observation of neutrinoless double-$\beta$ decay.

Sterile Neutrinos. An experiment carried out at Los Alamos laboratory [47] with the *LSND* detector[14] has observed positrons (around 60 events) and electrons (around 18 events) possibly originating in the decay chain:

$$\begin{aligned} &\pi^+ \to \mu^+ + \nu_\mu && (\pi^+ \text{ in flight}) , \\ &\mu^+ \to \bar{\nu}_\mu + e^+ + \nu_e && (\mu^+ \text{ at rest}) . \end{aligned} \tag{16.66}$$

From the energy and distance scales involved, the interpretation of the result as a $\nu_\mu \to \nu_e$ or $\bar{\nu}_\mu \to \bar{\nu}_e$ oscillation requires a value of $\Delta m^2$ very different from the values observed with atmospheric neutrinos:

$$(\Delta m^2)_{LSND} \sim 1 \text{ eV} . \tag{16.67}$$

Three different values of $\Delta m^2$ require a fourth neutrino. However, the decay of the neutral vector boson is compatible only with three neutrino types [13]. Therefore the fourth neutrino, if confirmed, must be a sterile neutrino, not subject to the usual weak interactions.

Searches to confirm or disprove the *LSND* effect are at present underway with beams of muon neutrinos (*MiniBooNE*[15] experiment) and at several nuclear reactors (disappearance of $\nu_e$ at short distances).

## 16.6 PROBLEM FOR CHAPTER 16

### Sect. 16.1

1. Show that the phase arbitrariness of quantum states can be used to make the two neutrino mixing matrix to be real, as in Eq. (16.3).

[14] Liquid Scintillator Neutrino Detector, Los Alamos Meson Facility, USA.
[15] Fermilab, Chicago, USA.

CHAPTER 17

# NEUTRINOLESS DOUBLE-BETA DECAY

As pointed out in Section 16.5, clear-cut evidence of massive Majorana neutrinos can only be obtained from the observation of neutrinoless double beta decay, wherein a parent nucleus emits a pair of virtual $W$ bosons which exchange a Majorana neutrino and produce the two emitted electrons. This chapter provides an outline of the derivation of the double beta decay rate, as well as a summary of the status of the experiments aimed at establishing the existence of the Majorana neutrino and determining its mass.

## 17.1 DOUBLE BETA DECAY

Double beta decay is a rare nuclear transition between two nuclei having the same mass number $A$, in which the nuclear charge number $Z$ changes by two units. This process was first considered by Maria Goeppert-Mayer shortly after the development of Fermi's theory of neutron beta decay [48][1], which provided a remarkably accurate description of the reactions

$$\mathcal{N}(A,Z) \to \mathcal{N}'(A,Z+1) + e^- + \overline{\nu}_e \ , \tag{17.1}$$

$$\mathcal{N}(A,Z) \to \mathcal{N}'(A,Z-1) + e^+ + \nu_e \ . \tag{17.2}$$

where $\mathcal{N}$ and $\mathcal{N}'$ denote the parent and daughter nucleus, respectively; see Section 15.1.

The stability of a nucleus of mass number $A$ against beta decays is driven by the dependence of its mass, $M_A$, on the charge number $Z$. In the vicinity of the minimum, the available experimental data turns out to be accurately

[1] In her paper, Goeppel-Mayer actually reports that the possible occurrence of double-beta nuclear disintegrations had been pointed out to her by E. Wigner.


DOI: 10.1201/9781003436263-17

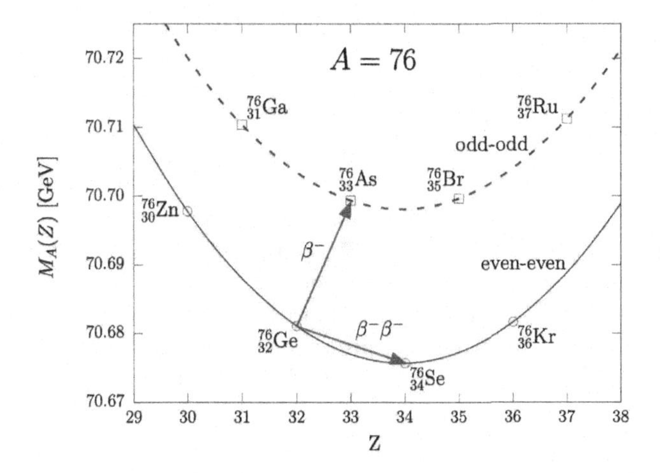


Figure 17.1 $Z$-dependence of the masses of $A = 76$ isobars, as described by Eq. (17.3). The upper and lower parabolae correspond to odd-odd and even-even nuclei, respectively. It is apparent that the $^{76}_{32}$Ge $\rightarrow$ $^{76}_{33}$As beta decay is forbidden, while the $^{76}_{32}$Ge $\rightarrow$ $^{76}_{34}$Se double beta decay is energetically allowed.

described by the quadratic parametrisation

$$M_A(Z) = Zm_p + (A - Z)m_n \tag{17.3}$$
$$+ \alpha - \beta A^{-1/3} - \gamma \frac{Z^2}{A^{1/3}} - \delta \frac{(A - 2Z)^2}{4A} - \epsilon(A, Z) \ .$$

with $m_p$ and $m_n$ being the proton and neutron mass, respectively. The last contribution to the right-hand side of the above equation accounts for the experimental evidence that nuclei with even number of both protons and neutrons tend to be more stable than those with odd $Z$ and odd $N = A - Z$, as well as those with odd $A$. This property originates from the pairing of like nucleons, and can be described setting $\epsilon = \pm 11.2 \ A^{-1/2}$ MeV for even $A$, even-even or odd-odd nuclei respectively, and $\epsilon = 0$ for odd $A$.

Figure 17.1, illustrates the $Z$-dependence of the masses of isobars with $A = 76$. The upper and lower parabolæ correspond to odd-odd and even-even nuclei, respectively. It is apparent that beta decays of even-even nuclei, leading to the appearance of an odd-odd isobar—for example, the transition $^{76}_{32}$Ge $\rightarrow$ $^{76}_{33}$As—is energetically forbidden. On the other hand, $^{76}_{32}$Ge can decay to $^{76}_{34}$Se emitting two electrons.

Double beta decays can occur through two different reaction modes, characterised by different final states. These are the two-neutrino double beta ($2\beta_{2\nu}$) decays

$$\mathcal{N}(A, Z) \rightarrow \mathcal{N}'(A, Z + 2) + e_1^- + e_2^- + \overline{\nu}_1 + \overline{\nu}_2 \ , \tag{17.4}$$

$$\mathcal{N}(A, Z) \rightarrow \mathcal{N}'(A, Z - 2) + e_1^+ + e_2^+ + \nu_1 + \nu_2 \ , \tag{17.5}$$

analysed in the pioneering work of Goeppert-Mayer, and the neutrinoless double beta ($2\beta_{0\nu}$) decays

$$\mathcal{N}(A,Z) \to \mathcal{N}'(A,Z+2) + e_1^- + e_2^- \ , \tag{17.6}$$
$$\mathcal{N}(A,Z) \to \mathcal{N}'(A,Z-2) + e_1^+ + e_2^+ \ , \tag{17.7}$$

first discussed by Furry in 1939 [49]. The decays proceed mainly from the ground state of a parent nucleus with spin-parity $0^+$ to the $0^+$ ground state of the daughter nucleus, although in some instances the transition to excited $0^+$ or $2^+$ states is also energetically allowed.

The occurrence of $2\beta_{0\nu}$ decay processes—which is forbidden in the Standard Model of weak interactions outlined in Chapter 15—is only possible if neutrinos behave as massive Majorana particles, the properties of which are discussed in Chapter 13.

Neglecting the recoil energy of the daughter nucleus, the $Q$-value of double beta decays reduces to

$$Q_{2\beta} = M_{\mathcal{N}} - M_{\mathcal{N}'} - 2m_e \ , \tag{17.8}$$

where $M_{\mathcal{N}}$ and $M_{\mathcal{N}'}$ are the masses of the initial and final nuclei, respectively, and $m_e$ is the electron mass. Note that, because $2\beta_{2\nu}$ decays lead to the appearance of a final state comprising four leptons, the sum of the kinetic energies of the two charged leptons, $E_{e_1} + E_{e_2}$, exhibits a continuous spectrum extending from zero to $Q$. In the case of $2\beta_{0\nu}$ decays, on the other hand, there are only two final-state leptons, and $E_{e_1} + E_{e_2} = Q$.

### 17.1.1 Two-neutrino Double Beta Decay

The $2\beta_{2\nu}$ decay is a second order process in the weak interaction described by the Fermi Lagrangian introduced in Chapter 9, which can be conveniently cast in the form

$$\mathcal{L}_F = -\frac{G}{\sqrt{2}} L_\rho H^\rho + \text{h.c.} \ , \tag{17.9}$$

with

$$L_\rho = \overline{\psi}_e \gamma_\rho (1 - \gamma_5) \psi_\nu \ , \tag{17.10}$$

and

$$H^\rho = \overline{\psi}_N \gamma^\rho (1 - \frac{g_A}{g_V} \gamma_5) \tau^+ \psi_N \ . \tag{17.11}$$

In the above equation, the nucleon is described by the doublet

$$N = \begin{pmatrix} \psi_p \\ \psi_n \end{pmatrix} \ , \tag{17.12}$$

where $\psi_p$ and $\psi_n$ denote the proton and neutron fields, respectively, while $\tau^+ = (\tau_1 + i\tau_2)/2$—with the $\tau_i$ being Pauli matrices—is the operator raising the lower component of the doublet.

The $S$-matrix element describing the decay (17.4) can be written in the form

$$S_{2\nu} = \int d^4x_1 d^4x_2 \, \langle e_1, e_2, \overline{\nu}_1, \overline{\nu}_2, \mathcal{N}' | T \left\{ \mathcal{L}_F(x_1) \mathcal{L}_F(x_2) \right\} | \mathcal{N} \rangle \qquad (17.13)$$
$$= (2\pi)^4 \delta^{(4)}(P_f - P_i) \, M_{2\nu} \, ,$$

where $P_f$ and $P_i$ are the four-momenta of the initial and final states, respectively, and $M_{2\nu}$ denotes the transition ampliude. The expression of $S_{2\nu}$ involves the lepton tensor

$$L_{\rho\sigma} = J_\rho(x_1) J_\sigma(x_2) \, ,$$

with[2],

$$J_\rho(x_i) = \langle e_i, \overline{\nu}_i | \, L_\rho(x) \, | 0 \rangle = \frac{1}{2} \sqrt{\frac{1}{E_{e_i} E_{\nu_i}}} \overline{e}_i \gamma_\rho (1 - \gamma_5) \nu_i e^{i(p_{e_i} + p_{\nu_i}) x_i} \, , \qquad (17.14)$$

and the nuclear tensor

$$H^{\rho\sigma} = \langle \Psi_{\mathcal{N}'} | \, H^\rho_A(x_1) H^\sigma_A(x_2) \, | \Psi_{\mathcal{N}} \rangle \, , \qquad (17.15)$$

with $|\Psi_{\mathcal{N}}\rangle$ and $|\Psi_{\mathcal{N}'}\rangle$ being the ground states of the parent and daughter nuclei, respectively. The nuclear current is defined as

$$H^\rho_A = \sum_{i=1}^{A} H^\rho_i \, . \qquad (17.16)$$

with $H^\rho_i$ being the current of Eq. (17.11) associated with the $i$-th nucleon.

Figure 17.2 provides a diagrammatic representation of the reaction (17.4), in which beta decays are associated with the exchange of the positively-charged $W$ bosons, as described by the IVB Lagrangian $\mathcal{L}_{IVB}$ of Eq. (15.38). The Fermi Lagrangian $\mathcal{L}_F$ can be obtained from $\mathcal{L}_{IVB}$ in the $M_W \to \infty$ limit, safely applicable to the description of nuclear beta decays; see Section 15.4.

The calculation of $H^{\rho\sigma}$ is carried out by rewriting the nuclear tensor in the form

$$H^{\rho\sigma} = \sum_m \langle \Psi_{\mathcal{N}'} | \, H^\rho_A(x_1) \, | \Psi_m \rangle \langle \Psi_m | \, H^\sigma_A(x_2) \, | \Psi_{\mathcal{N}} \rangle , \qquad (17.17)$$

[2] Here we use the spinor normalisation of Eq. (13.8), which turns out to be more convenient in the $m \to 0$ limit.

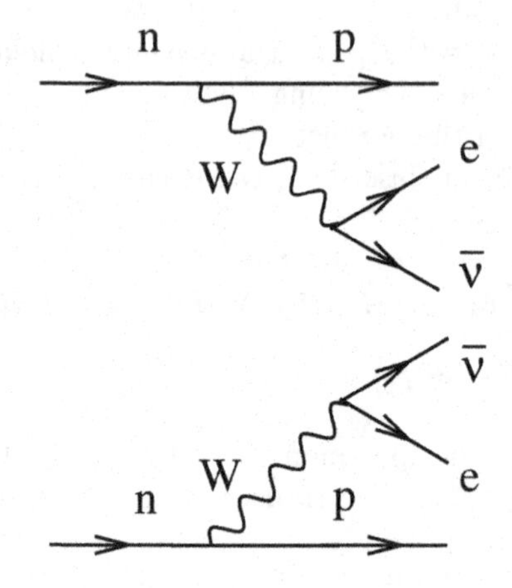


Figure 17.2 Diagrammatic representation of the two-neutrino double beta decay reaction (17.4).

with $\{\Psi_m\}$ being a complete set of eigenstates of the unobserved intermediate-state nucleus, having mass number $A$ and charge number $Z+1$. The time integration in Eq. (17.13) leads to the familiar result of second order perturbation theory

$$(2\pi)\delta\left(E_{\mathcal{N}'} + E_{e_1} + E_{\nu_1} + E_{e_2} + E_{\nu_2} - E_{\mathcal{N}}\right) \times \sum_m \left[ \frac{\langle \Psi_{\mathcal{N}'} | H_A^\rho(\mathbf{x}_1) | \Psi_m \rangle \langle \Psi_m | H_A^\sigma(\mathbf{x}_2) | \Psi_{\mathcal{N}} \rangle}{E_m + E_{e_2} + E_{\nu_2} - E_{\mathcal{N}}} J_\rho(\mathbf{x}_1) J_\sigma(\mathbf{x}_2) + \ldots \right], \tag{17.18}$$

where the ellipses refer to presence of similar additional contributions, arising from different time ordering and exchange terms. Numerical calculations are often performed using the approximation

$$E_{e_1} + E_{\nu_1} \approx E_{e_2} + E_{\nu_2} \approx \frac{M_{\mathcal{N}} - M_{\mathcal{N}'}}{2}, \tag{17.19}$$

with $M_{\mathcal{N}}$ and $M_{\mathcal{N}'}$ being the masses of initial and final nuclei, respectively, which allows to decouple the lepton and nuclear parts of the amplitude.

The half-time of the parent nucleus, $T_{1/2}^{2\nu}$, trivially related to the decay rate $\Gamma$ through $[T_{1/2}^{2\nu}]^{-1} = \Gamma/\ln 2$, can be cast in the form

$$[T_{1/2}^{2\nu}]^{-1} = |\mathcal{M}_{2\nu}|^2 G_{2\nu}, \tag{17.20}$$

where

$$\mathcal{M}_{2\nu} = M_{2\nu}^{GT} - \left(\frac{g_V}{g_A}\right)^2 M_{2\nu}^F. \tag{17.21}$$

In the above equation, $M_{2\nu}^F$ and $M_{2\nu}^{GT}$—referred to as Fermi and Gamow-Telller matrix elements—describe the nuclear transition amplitudes involving the vector and axial-vector weak currents, respectively. Using the approximation of Eq. (17.19) and treating the nucleons as non relativistic particles they can be written in the form

$$M^F_{2\nu} = \sum_m \frac{\langle \Psi_{\mathcal{N}'} | \sum_{j=1}^{A} \tau^+_j | \Psi_m \rangle \langle \Psi_m | \sum_{k=1}^{A} \tau^+_k | \Psi_{\mathcal{N}} \rangle}{E_m - (M_{\mathcal{N}} + M_{\mathcal{N}'}/2)} , \tag{17.22}$$

$$M^{GT}_{2\nu} = \sum_m \frac{\langle \Psi_{\mathcal{N}'} | \sum_{j=1}^{A} \boldsymbol{\sigma}_j \tau^+_j | \Psi_m \rangle \cdot \langle \Psi_m | \sum_{k=1}^{A} \boldsymbol{\sigma}_k \tau^+_k | \Psi_{\mathcal{N}} \rangle}{E_m - (M_{\mathcal{N}} + M_{\mathcal{N}'})/2} , \tag{17.23}$$

with $\boldsymbol{\sigma}_j$ being the matrix associated with the spin of the $j$-th nucleon.

The phase-space factor is obtained from

$$G_{2\nu} = \frac{1}{\ln 2} \frac{G^4 g_A^4}{64\pi^7 m_e^2} \int d\Omega_{2\nu} F_0(Z+2, E_{e_1}) F_0(Z+2, E_{e_2}) , \tag{17.24}$$

with

$$\begin{aligned} d\Omega_{2\nu} &= E^2_{\nu_1} dE_{\nu_1} E^2_{\nu_2} dE_{\nu_2} |\mathbf{p}_{e_1}| E_{e_1} dE_{e_1} |\mathbf{p}_{e_2}| E_{e_2} dE_{e_2} d\cos\theta \\ &\times \delta(E_{\mathcal{N}'} + E_{e_1} + E_{\nu_1} + E_{e_2} + E_{\nu_2} - E_{\mathcal{N}}) , \end{aligned}$$

where $\theta$ is the angle between the momenta of the emitted electrons. The Fermi factor $F_0(Z, E_i)$ appearing in Eq. (17.24) takes into account the effect of the Coulomb attraction between the emitted electron and the daughter nucleus, having charge number $Z + 2$. The corresponding expression is

$$F_0(Z+2, E) = \frac{E}{|\boldsymbol{p}|} \frac{2\pi\alpha(Z+2)}{1 - e^{-2\pi\alpha(Z+2)}} , \tag{17.25}$$

with $\alpha$ being the fine-structure constant. Note that nuclear matrix elements defined according to the above equations turn out to be dimensionless.

The occurrence of $2\beta_{2\nu}$ decay has been unambiguously observed in several nuclei—ranging from $^{48}$Ca to $^{150}$Nd—using different detection techniques. The measured values of the half-life lie in the range $10^{18} \lesssim T^{2\nu}_{1/2} \lesssim 10^{23}$ years [50]. Precision measurements of $2\beta_{2\nu}$ reactions can provide information useful to constrain the models employed to perform calculations of the nuclear matrix elements, which are also needed for the determination of the $2\beta_{0\nu}$ decay rates to be discussed in the next section.

### 17.1.2 Neutrinoless Double Beta Decay

The observation of the hypothetical $2\beta_{0\nu}$ decay reaction would provide evidence of matter creation, associated with a two-unit violation of lepton number conservation. Such a departure from the predictions of the Standard Model of weak interactions may be explained considering the process illustrated in Fig. 17.3. Since no antineutrino is emitted, the two antineutrino lines attached to the weak interaction vertices in Fig. 17.2 should be joined to represent a neutrino propagator. If neutrinos behaved as Dirac particles, however, this interpretation would not be allowed, because:

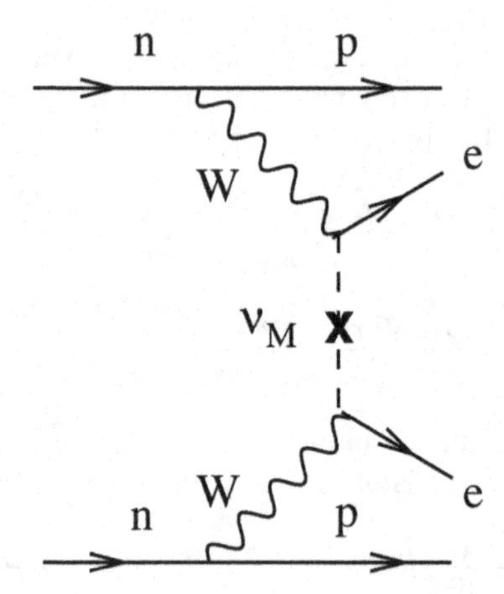


Figure 17.3 Diagrammatic representation of the neutrinoless double beta decay reaction (17.6).

1) The antineutrino emitted from the upper leptonic vertex could not be absorbed by the lower vertex, which can only absorb a neutrino;

2) The helicity of the antineutrino emitted from the upper leptonic vertex would be positive, while the lower leptonic vertex can only absorb a neutrino with negative helicity.

As a consequence, the occurrence of $2\beta_{0\nu}$ decay entails two necessary conditions:

1) The neutrino must be a Majorana fermion, in which case $\overline{\nu}_e = \nu_e$ and the total lepton number is not conserved;

2) The neutrino mass $m_{\nu_e}$ must be non zero, in which case a neutrino of negative helicity can be emitted from the upper leptonic vertex with amplitude $m_{\nu_e}/E_{\nu_e}$, and absorbed by the lower leptonic vertex with unit amplitude.

For convenience, we recall here the expression of the field describing massive Majorana fermions, discussed in Chapter 13

$$\psi_\nu(x) = \sum_{\mathbf{p},r} \frac{1}{\sqrt{2VE_\nu}} \left[a_r(\mathbf{p})u_r(\mathbf{p})e^{-ipx} + a_r^\dagger(\mathbf{p})u_r^\star(\mathbf{p})e^{ipx}\right] , \qquad (17.26)$$

showing that a Majorana particle is identical to its antiparticle and inherently charge neutral.

The expression of the $S$-matrix element of $2\beta_{0\nu}$ decay, to be contrasted with the corresponding expression for $2\beta_{2\nu}$, Eq. (17.13), reads

$$S_{0\nu} = -\int d^4x_1 d^4x_2 \, \langle e_1, e_2, \mathcal{N}'| \, T\left\{\mathcal{L}_F(x_1)\mathcal{L}_F(x_2)\right\} |\mathcal{N}\rangle \, . \qquad (17.27)$$

Comparison between the diagrams of Figs. 17.2 and 17.3 shows that the nuclear parts of the $2\beta_{0\nu}$ and $2\beta_{2\nu}$ amplitudes are identical. The lepton part of Eq. (17.27)

$$L_{\rho\sigma} = \langle e_1 e_2 | T \left\{ L_\rho(x_1) L_\sigma(x_2) \right\} |0\rangle \tag{17.28}$$
$$= \langle e_1 e_2 | T \left\{ \overline{\psi}_e(x_1) \gamma_\rho \left(1 - \gamma_5\right) \psi_{\nu_e}(x_1) \overline{\psi}_e(x_2) \gamma_\sigma \left(1 - \gamma_5\right) \psi_{\nu_e}(x_2) \right\} |0\rangle \ .$$

can be rewritten substituting the expression of the $\nu_e$ field with a superposition of mass eigenstates $\nu_i$ acording to Eq. (16.10). The resulting expression is

$$L_{\rho\sigma} = \sum_{j,k=1}^{3} U_{ej} U_{ek} \tag{17.29}$$
$$\times \langle e_1 e_2 | T \left\{ \overline{\psi}_e(x_1) \gamma_\rho \left(1 - \gamma_5\right) \psi_{\nu_j}(x_1) \overline{\psi}_e(x_2) \gamma_\sigma \left(1 - \gamma_5\right) \psi_{\nu_k}(x_2) \right\} |0\rangle \ ,$$

whith the $U_{ej}$ being elements of the neutrino mixing matrix. Let us now assume that the neutrino be a Majorana particle, so that

$$\psi^T_{\nu_k} = -\overline{\psi}_{\nu_k} \mathcal{C} \ , \tag{17.30}$$

where $\mathcal{C}$ is the charge-conjugation matrix. Substituting the above relation in $L_{\rho\sigma}$ we obtain the result

$$L_{\rho\sigma} = \sum_{j,k=1}^{3} U_{ej} U_{ek} \tag{17.31}$$
$$\times \langle e_1 e_2 | T \left\{ \overline{\psi}_e(x_1) \gamma_\rho \left(1 - \gamma_5\right) \psi_{\nu_j}(x_1) \psi^T_{\nu_k}(x_2) \left(1 - \gamma_5\right) \gamma_\sigma \psi^C_e(x_2) \right\} |0\rangle \ ,$$

showing that implementation of the Majorana condition, Eq. (17.30), leads to the appearance of the propagator describing the internal neutrino line appearing in Fig. 17.3

$$\sum_{k=1}^{3} U^2_{ek} \langle 0 | T \left\{ \psi_{\nu_k}(x_1) \psi^T_{\nu_k}(x_2) \right\} |0\rangle \tag{17.32}$$
$$= -i \sum_{k=1}^{3} U^2_{ek} \int \frac{d^4 q}{(2\pi)^4} \frac{\not{q} + m_k}{q^2 - m_k^2} e^{-iq(x_1 - x_2)} \ .$$

Using the result

$$(1 - \gamma_5)(\not{q} + m_j)(1 - \gamma_5) = -2m_j(1 - \gamma_5) \ , \tag{17.33}$$

we find that the lepton part of the $2\beta_{0\nu}$ amplitude is proportional to the effective Majorana mass, defined as

$$m_{ee} = \sum_{k=1}^{3} U^2_{ek} m_k \ . \tag{17.34}$$

In addition, performing the $q_0$ integration in the right-hand side of Eq. (17.32), we obtain the result

$$\int \frac{d^4q}{(2\pi)^4} \frac{e^{-iq(x_1-x_2)}}{q^2 - m_j^2} = \int \frac{d^3q}{(2\pi)^3} \frac{e^{-i[\omega_q(x_{1_0}-x_{2_0})-\boldsymbol{q}\cdot(\boldsymbol{x}_1-\boldsymbol{x}_2)]}}{2\omega_q} , \tag{17.35}$$

with $\omega_q = \sqrt{\mathbf{q}^2 + m_j^2}$.

As noted above, the nuclear part of the amplitude is the same as in the case of $2\beta_{2\nu}$ decays. In neutrinoless decays, however, its time dependence is combined with a different time dependence of the leptonic part. Carrying out the time integration, we find the result

$$(2\pi)\delta\left(E_{\mathcal{N}'} + E_{e_1} + E_{e_2} - E_{\mathcal{N}}\right) \tag{17.36}$$
$$\times \sum_m \left[ \frac{\langle \Psi_{\mathcal{N}'}|H_A^\rho(\mathbf{x})|\Psi_m\rangle\langle\Psi_m|H_A^\sigma(\mathbf{y})|\Psi_{\mathcal{N}}\rangle}{E_m + \omega_q + E_{e_2} - E_{\mathcal{N}}} + \dots \right] .$$

The sum over intermediate nuclear states is performed using the closure approximation, which amounts to replacing the energy $E = E_m + E_{e_2} - E_{\mathcal{N}}$, appearing in the denominator of the above equation, with an average value $\langle E \rangle \sim 10$ MeV. The sum can then be performed exploiting completeness of the set $\{\Psi_m\}$, implying $\sum_m |\Psi_m\rangle\langle\Psi_m| = \mathbf{1}$. This procedure drastically simplifies the calculation of the nuclear matrix elements, which reduce to the form

$$\mathcal{M}^F_{0\nu} = \langle \Psi_{\mathcal{N}'}| \sum_{j,k=1}^{A} \tau_j^+ \tau_k^+ H(r_{jk}) |\Psi_{\mathcal{N}}\rangle , \tag{17.37}$$

$$\mathcal{M}^{GT}_{0\nu} = \langle \Psi_{\mathcal{N}'}| \sum_{j,k=1}^{A} (\boldsymbol{\sigma}_j \cdot \boldsymbol{\sigma}_k)\, \tau_j^+ \tau_k^+ H(r_{jk}) |\Psi_{\mathcal{N}}\rangle , \tag{17.38}$$

where $r_{jk}$ is the distance between the two nucleons undergoing beta decay. The function $H(r_{jk})$, called neutrino potential, originates from the integration of Eq. (17.35) over the space components of the virtual neutrino momentum, which can be performed using the approximation $\omega_q \approx |\boldsymbol{q}|$. The resulting expression is

$$H(r) = \frac{2R_A}{\pi r} \int_0^{+\infty} dq\, \frac{\sin(qr)}{q + \langle E\rangle} , \tag{17.39}$$

with $R_A = 1.2\ A^{1/3}$ being the radius of the parent nucleus.

Collecting the nuclear and leptonic parts together we can cast the expression of the $2\beta_{0\nu}$ half-life in the form

$$\left[T^{0\nu}_{1/2}\right]^{-1} = |\mathcal{M}_{0\nu}|^2\, G_{0\nu}\, \frac{|m_{ee}|^2}{m_e^2} \tag{17.40}$$

where, $m_{ee}$ is the effective Majorana mass, defined in Eq. (17.34), and $\mathcal{M}_{0\nu}$ is the dimensionless nuclear transition matrix element, including the contributions of Fermi and Gamow-Teller transitions

$$\mathcal{M}_{0\nu} = M_{0\nu}^{GT} - \left(\frac{g_V}{g_A}\right)^2 M_{0\nu}^{F} \ . \tag{17.41}$$

Finally, the phase-space factor, $G_{0\nu}$, is given by

$$G_{0\nu} = \frac{2}{\ln 2} \frac{G^4 g_A^4 m_e^2}{16\pi^5} \int d\Omega_{0\nu} F_0(Z+2, E_{e_1}) F_0(Z+2, E_{e_2}) \ , \tag{17.42}$$

where

$$d\Omega_{0\nu} = |\mathbf{p}_{e_2}|\ E_{e_2} |\mathbf{p}_{e_1}| E_{e_1} dE_{e_1} \ , \tag{17.43}$$

with $E_{e_2} = E_{\mathcal{N}} - E_{\mathcal{N}'} - E_{e_1}$.

## 17.2 EXPERIMENTAL STUDIES OF DOUBLE BETA DECAY

As pointed out above, the observation of $2\beta_{0\nu}$ decays would unambiguously establish the existence of Majorana neutrinos and provide a measurement of the effective Majorana mass $m_{ee}$, the definition of which involves both the elements of the neutrino mixing matrix, $U_{ei}$, and the mass eigenvalues $m_i$; see Eq. (17.34). Complementary information on neutrino masses are inferred from cosmological observations. The 2018 analysis of the data collected by the Planck observatory provides the constraint $\sum_i m_i < 0.12$ meV [51].

As discussed in Section 15.1, a limit on the neutrino mass can also be obtained from the observation of the endpoint of the electron energy distribution in Tritium beta decay experiments. The latest measurement of the effective electron neutrino mass, defined as $m_\nu^2 = \sum_i |U_{ei}|^2 m_i^2$, reported by the KATRIN Collaboration, yields the upper limit $m_\nu < 0.8$ eV [52].

The detection of $2\beta_{0\nu}$ decays is based on the measurement of the total energy of the two emitted electrons, which is expected to exhibit a sharp peak at $E_{e_1} + E_{e_2} = Q_{2\beta}$, with $Q_{2\beta}$ defined by Eq. (17.8).

The data obtained from the oscillation analyses, discussed in Chapter 16, provide differences between the squared neutrino mass eigenvalues, $\Delta_{ij}$, but do not allow to pin down which of the three mass eigenstates is the heaviest. Experimental studies of $2\beta_{0\nu}$ decay have the potential to resolve this issue. To see this, consider that, by using the available experimental information on mixing angles and mass splittings, $m_{ee}$ can be written as a function of the lightest neutrino mass, $m_{\rm light}$, assuming either normal order (NO), corresponding to $\Delta_{23} = m_3^2 - m_2^2 > 0$, or inverted order (IO), corresponding to $\Delta_{23} = m_3^2 - m_2^2 < 0$. As a consequence, experimental searches of $2\beta_{0\nu}$ decay turn out to be sensitive to differences between the predictions of the two scenarios, and may shed new light on the neutrino mass spectrum.

Table 17.1 Current limits on the $2\beta_{0\nu}$ decay half-life and the effective Majorana mass, reported by experiments performed using different nuclei.

| Nucleus | Experiment | $T^{0\nu}_{1/2}$ [years] | $m_{ee}$ [meV] |
|---|---|---|---|
| $^{76}$Ge | GERDA [53] | $> 1.8 \times 10^{26}$ | $< 79 - 180$ |
| $^{76}$Ge | MAJORANA [54] | $> 8.3 \times 10^{25}$ | $< 113 - 269$ |
| $^{136}$Xe | KamLAND-Zen [55] | $> 2.3 \times 10^{26}$ | $< 36 - 156$ |
| $^{136}$Xe | EXO-200 [56] | $> 3.5 \times 10^{25}$ | $< 93 - 286$ |
| $^{130}$Te | CUORE [57] | $> 2.2 \times 10^{25}$ | $< 90 - 305$ |
| $^{100}$Mo | CUPID-Mo [58] | $> 1.8 \times 10^{24}$ | $< 280 - 490$ |
| $^{82}$Se | CUPID-0 [59] | $> 4.6 \times 10^{24}$ | $< 263 - 545$ |
| $^{48}$Ca | CANDLES-III [60] | $> 5.6 \times 10^{22}$ | $< 1600 - 2900$ |

The results of recent experimental determinations of both the lower bound of $T^{0\nu}_{1/2}$ and the upper bound of $m_{ee}$ are listed in Table 17.1. The limits on the half-life are mainly driven by exposure, defined as the product between the amount of active isotope employed by the experiment and the duration of data taking. According to the estimates of Engels and Menendéz [61], assuming $m_{ee} \approx 50$ meV, the detection of a $2\beta_{0\nu}$ decay will be possible with an experimental sensitivity of $5-8\times 10^{26}$ years, to be compared to the best limits on $T^{0\nu}_{1/2}$ obtained so far. If, on the other hand, one assumes $m_{ee} \approx 10$ meV, a sensitivity as high as of $1 - 2 \times 10^{28}$ years will be needed. The results of Table 17.1 suggest that in the best case scenario the required sensitivity may be achieved by some of the current experiments.

The largest source of uncertainty in the determination of the effective Majorana mass from Eq. (17.41)—resulting in the spread of values of $m_{ee}$ reported in Table 17.1—is the calculation of the nuclear matrix element, generally performed using the formalism of the nuclear shell model, which unavoidably involve approximations. Furthermore, the effects of nucleon-nucleon correlations—which is neglected altogether in the mean-field approximation underlying the shell model—have been found to be large [62]. Overall, the uncertainty arising from the discrepancy between the results of different calculations of $\mathcal{M}_{2\nu}$ turns out to be as large as 50$-$75 %

The values of $m_{ee}$ obtained from the measured half-lives can be compared with the predictions based on the results of oscillation experiments. Following the procedure outlined above one finds

$$m_{ee} = \left| \sum_{i=1}^{3} U_{ei} m_i \right| \tag{17.44}$$

$$= \begin{cases} \left| m_{\text{light}} c_{12}^2 c_{13}^2 + s_{12}^2 c_{13}^2 e^{2i(\eta_2 - \eta_1)} \sqrt{\Delta_{12}^2 + m_{\text{light}}^2} \right. \\ \qquad \left. + s_{13}^2 e^{-2i(\delta+\eta_1)} \sqrt{\Delta_{23}^2 + \Delta_{12}^2 + m_{\text{light}}^2} \right| & \text{(NO)} \\ \left| m_{\text{light}} s_{13}^2 + s_{12}^2 c_{13}^2 e^{2i(\delta+\eta_2)} \sqrt{m_{\text{light}}^2 - \Delta_{23}^2} \right. \\ \qquad \left. + s_{13}^2 e^{2i(\delta+\eta_1)} \sqrt{m_{\text{light}}^2 - \Delta_{23}^2 + \Delta_{12}^2} \right| & \text{(IO)} \end{cases}$$

Here, $c_{ij} = \cos\theta_{ij}$, $s_{ij} = \sin\theta_{ij}$, with $\theta_{ij}$ being a neutrino mixing angle, and $\delta$ is the phase associated with the violation of the CP symmetry discussed in Section 16.1. The additional phases $\eta_1$ and $\eta_2$ which also describe CP violation, only appear in the case of Majorana neutrinos.

The effective Majorana masses obtained from the above equations using the results of the global analysis of oscillation data of Esteban *et al.* [63] are shown in Fig. 17.4.

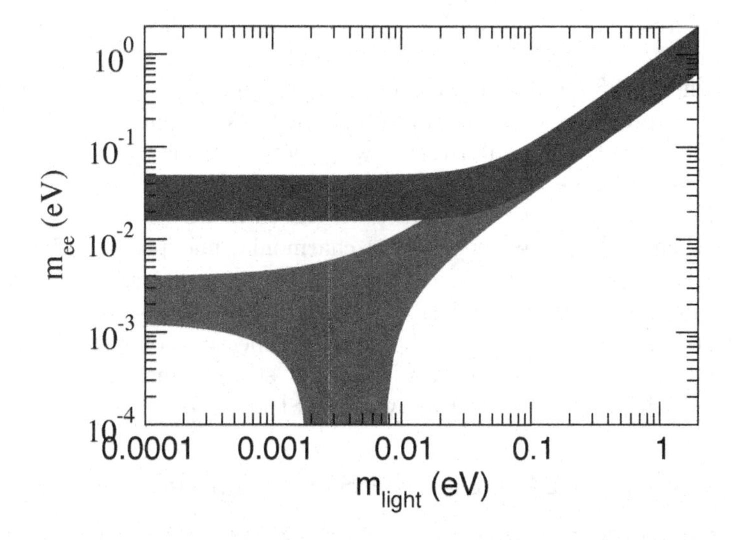


Figure 17.4 The effective Majorana neutrino mass $m_{ee}$ as a function of the lightest neutrino mass, $m_{\text{light}}$. The lower and upper bands correspond to normal and inverted order; that is, $m_{\text{light}} = m_1$, and $m_{\text{light}} = m_3$, respectively. Reprinted from Ref. [12] with permission. © APS (1998).

CHAPTER 18

# A LEAP FORWARD: CHARMONIUM

In Chapter 9 we have introduced the Electromagnetic, Weak and Strong Interactions, the three forces that act at the elementary particle level.

The correct theory of the Strong Interactions, the very intense forces that bind protons and neutrons in the atomic nuclei, was found in 1973. It took indeed almost three decades to decipher the puzzle, following the discovery in 1947 of the mediator of nuclear forces, the $\pi$-meson, and the exploration of the spectrum of a wealth of particles subject to the Strong Interactions, similar to protons, neutrons and $\pi$-mesons and called collectively *hadrons*; see Section 9.4.

The theory of strong interaction, named Quantum Chromo Dynamics (QCD), was prepared by important steps partly anticipated in Sections 9.3 and 9.4; here, QCD will be illustrated with more detail in Section 18.1 below.

In Section 18.2 we introduce charm-anticharm mesons, called *charmonia* in analogy with the positronium states discussed in Chapter 12.

The calculation of the spectrum of charmonia, unexpectedly, has turned out to be an interesting exercise in Non Relativistic Quantum Mechanics that deserves to appear in this Volume. All the more so because the discovery of unexpected lines intermixed with the predicted spectrum of charmonia has led to the discovery of a family of *exotic hadrons*, still waiting for a quantitative description within the present theory of the Strong Interactions.

## 18.1 A PRIMER: BARYONS, MESONS, QUARKS AND QCD

Hadrons divide into two great families: baryons and mesons. Baryons are particles with half-integer spin, with proton and neutron being the lightest ones; mesons have integer spin, the lightest mesons being the $\pi$ meson, in its three charged states $\pi^+, \pi^0, \pi^-$.


DOI: 10.1201/9781003436263-18

### 18.1.1 Conserved Quantum Numbers

Besides electric charge, light hadrons are characterised by four quantum numbers that are conserved by the Strong Interactions, *Baryon number*, $B$, *Isotopic Spin* and its projection, $I$ and $I_3$, and *Strangeness*, $S$.

**Baryon number.** Baryons have $B = 1$ (antibaryons $B = -1$). Mesons have all $B = 0$. Proton and neutron are the lightest baryons and $B$ conservation implies that higher mass baryons (called hyperons) will eventually decay into a final state containing one proton or one neutron and other $B = 0$ particles (mesons, leptons and photons). Charged $\pi^\pm$ decay into leptons, $\pi^0$ decays into pure radiation (two photons), higher mass mesons decay eventually into states containing $\pi$ mesons, leptons, photons and, if heavy enough, baryon-antibaryon pairs.

**Isotopic Spin.** The close mass of proton and neutron has prompted very early the suggestion that the Strong Interaction hamiltonian is symmetric under unitary transformations of $p$ and $n$

$$\begin{pmatrix} p \\ n \end{pmatrix} \to U \begin{pmatrix} p \\ n \end{pmatrix} \quad \text{with} : U = \text{unitary matrix}, \ \det U = 1. \tag{18.1}$$

These transformations are analogous to spin rotations, hence the name *Isotopic Spin* Symmetry. The symmetry can be extended to the other hadrons. Proton and neutron (Nucleons) have isotopic spin $I = 1/2$ with $I_3 = +1/2(p), \ -1/2(n)$, $\pi$ mesons have $I = 1$, with $I_3 = +1(\pi^+), \ 0(\pi^0), \ -1(\pi^-)$.

**Strangeness.** The $\Lambda^0$ hyperon and the $K^+$ meson, the next heavier particles than proton and $\pi$, have very long lifetimes, typical of the Weak rather than Strong interactions. This unexpected fact led to introduce a new quantum number called *Strangeness*, conserved by Strong and Electromagnetic Interactions and violated by Weak Interactions. Conventionally $S = -1$ for $\Lambda^0$, $S = +1$ for $K^+$ and $S = 0$ for proton, neutron and $\pi$ mesons.

Strangeness conservation implies that particles with $S \neq 0$ are produced in pairs in hadronic collisions initiated by $S = 0$ particles (associate production), e.g. : $\pi + p \to \Lambda^0 + K^+ + (S = 0)$ hadrons, as it is in fact observed. We have identified hyperons with $S = -2$ ($\Xi^{0,-}$) and $S = -3$ ($\Omega^-$), and mesons with $S = \pm 1, \ 0$[1].

### 18.1.2 Quarks and QCD

The idea that some hadrons may be elementary (proton and neutron) and be the constituents of the other hadrons was considered in 1949 by E. Fermi and C. N. Yang [64]. The scheme was extended in 1956 by S. Sakata [65], to include strange particles. At the beginning of the 1960s, however, it was recognised that the most natural principle to understand the hadrons was Nuclear Democracy: *all hadrons are to be treated on the same footing* (G. Chew and S. Frautschi).

[1]Systematic classification of hadron resonances and related conventions are given in [12].

Table 18.1 Baryon number B, Isospin $I_3$, Electric charge $Q$, Strangeness $S$, of $u, d$ and $s$ quarks introduced by Gell-Mann and Zweig. We have listed also the heavy quarks (charm, beauty, top) each with its conserved quantum numbers. Quark masses (in GeV) are derived from the mass spectrum of the corresponding mesons (see Sect.18.2 for charm quark masses).

| Quark name | $B$ | $I_3$ | $Q$ | $S$ | Charm | Beauty | Top | mass |
|---|---|---|---|---|---|---|---|---|
| u (up) | 1/3 | +1/2 | 2/3 | 0 | 0 | 0 | 0 | 0.31 |
| d (down) | 1/3 | -1/2 | -1/3 | 0 | 0 | 0 | 0 | 0.31 |
| s (strange) | 1/3 | 0 | -1/3 | -1 | 0 | 0 | 0 | 0.48 |
| c (charm) | 1/3 | 0 | 2/3 | 0 | 1 | 0 | 0 | 1.32 |
| b(beauty) | 1/3 | 0 | -1/3 | 0 | 0 | 1 | 0 | 4.58 |
| t(top) | 1/3 | 0 | 2/3 | 0 | 0 | 0 | 1 | 173 |

**Quarks.** A decisive step was made in 1964 by M. Gell-Mann [66] and by G. Zweig [67] independently: Nuclear Democracy holds, subnuclear particles are *all* composite, the elementary constituents being three kinds of spin 1/2 particles, unobserved until then and called *quarks* by Gell-Mann. Mesons and Baryons have the following quark compositions

$$\begin{aligned} \text{mesons} &= q\bar{q} \ , \\ \text{baryons} &= qqq \ . \end{aligned} \tag{18.2}$$

It is customary to say that Gell-Mann's quarks come with three different qualities, or *flavours*, named according to $q = u, d, s$ (*up, down* and *strange*).

Quark quantum numbers are reported in Table 18.1. It is easy to check the matching of quantum numbers in the simplest cases:

$$\begin{aligned} &\pi^+ = (u\bar{d}) \text{ spin0} \ ; \ K^+ = (u\bar{s}) \text{ spin } 0 \ ; \\ &p = (uud), \text{ spin } 1/2 \ ; \ n = (udd), \text{ spin } 1/2 \ ; \\ &\Lambda^0 = (uds), \text{ spin } 1/2 \ ; \ \Xi^0 = (uss), \text{ spin } 1/2 \ ; \ \Omega^- = (sss), \text{ spin } 3/2 \ . \end{aligned}$$

Unitary transformations of the 3-dimensional vector $q = (u, d, s)$ correspond to a group called $SU(3)_{flavour}$, which extends the Isotopic Spin symmetry and is the natural approximate symmetry of the Strong Interactions, as proposed earlier by Gell-Mann and by Y. Ne'eman [68], with consequences well supported by the observed properties of the hadrons known at the time.

**Heavy Quarks.** In 1970, Glashow, Iliopoulos and Maiani [28] have proposed a fourth quark, considerably heavier than the other three, to explain the observed suppression of weak interaction processes involving strangeness-changing neutral weak currents. The fourth quark removed the obstacles that until then had precluded a unified Electro-Weak theory of quarks. Like the strange quark, the new quark and the particles composed by it, carries a quantum number $C$, or Charm, similar to Strangeness and conserved in Strong and Electromagnetic Interactions. New particles containing a $\bar{c}c$ pair (hidden charm) or one unpaired $c$ quark (charmed particles) have been observed from 1974 onwards, with the properties anticipated in [28].

In 1973, Kobayashi and Maskawa [29] have proposed the existence of a further pair of quarks, to explain the CP-violation observed in $K$ decays. Hadrons with *beauty* and *top* quarks have been first observed in 1976 and 1994, respectively.

Heavy quarks decay by Weak Interactions into a lighter quark plus particles with total baryon number $B = 0$. Thus they all have the same baryon number (1/3) of the lightest ones, which is determined by the fact that proton and neutron are made by three quarks. Indeed charmed and beauty baryons, with composition $(qq'c)$ and $(qq'b)$ have been observed, as well as doubly charmed baryons $(qcc)$ ($q$ and $q'$ denote light quarks).

**QCD.** The first baryon with $S = 0$ heavier than the proton is the so called 3-3 resonance $\Delta^{++}(1232)$[2] (in parenthesis the mass in MeV). The quark model composition of $\Delta^{++}$ in its $s_3 = +3/2$ state is:

$$\Delta^{++}_{s_3=+3/2} = (u^{\uparrow}u^{\uparrow}u^{\uparrow}) \ . \tag{18.3}$$

Assuming all $u$ quarks to be in the fundamental state, Eq. (18.3) would be a fully symmetric configuration of three quarks, in conflict with the Fermi statistics obeyed by spin 1/2 particles.

At the end of the years 1960s, the view prevailed that the $u$ quark, as well as all other quark flavors, had to have a hidden quantum number, to restore the required antisymmetry. First ideas were advanced in a seminal paper by Han and Nambu [69], the modern view was proposed by Bardeen, Fritzsch and Gell-Mann in 1972 [70].

It is assumed that quark fields have an additional (three valued) index associated to unitary transformations of a new symmetry called $SU(3)_{colour}$[3]:

$$q \to q^a : q = u, d, s \text{ and } a = 1, 2, 3 \ . \tag{18.4}$$

The further hypothesis is done that the observed hadrons are invariant under the new symmetry, i.e. they are *colour singlets*. In the case of (18.3), this implies:

$$\Delta^{++} \to (u^{\uparrow a}u^{\uparrow b}u^{\uparrow c})\epsilon_{abc} \ . \tag{18.5}$$

with $\epsilon_{abc}$ the completely antisymmetric tensor in three dimensions (sum over repeated indices is understood). The full antisymmetry of $\epsilon$ restores the antisymmetry of the state of three quarks with equal flavour and spin up.

To describe the strong interactions, Han-Nambu and Bardeen *et al.* proposed quark interactions to be invariant under gauge transformations based

[2] $\Delta$ is the spin 3/2 resonance discovered at the Chicago Cyclotron by E. Fermi in 1951–1952 in $\pi^+ + p$ collisions.

[3] The name *colour* is given in analogy to the *flavor* quantum numbers exhibited by quarks, colour symmetry is assumed for heavy quarks as well.

on $SU(3)_{colour}$, analogous to the gauge transformations of QED described in Section 9.1, see Eq. (9.4):

$$u^a(x) \rightarrow \sum_{A=1,\dots,8} \left[e^{-i\frac{\lambda^A}{2}\cdot\alpha^A(x)}\right]^a_b u^b(x) \,. \tag{18.6}$$

where the $\lambda^A$ are eigth $3 \times 3$ hermitian and traceless matrices (the generators of $SU(3)_{colour}$).

### 18.1.3 Infrared Confinement and Asymptotic Freedom

The one-dimensional gauge invariance of QED is associated with the massless photon. Similarly, the eight-dimensional gauge transformations (18.6) are associated with eight massless vector particles called *gluons*, which are supposed to be the mediators of the basic strong interactions which glue together quarks inside the hadrons[4]. The resulting quark-gluon interaction has been named *Quantum Chromo Dynamics* (QCD) in [70].

As indicated by Eq. (18.4) trasformations of *colour* are independent from and therefore *commute* with the Weak and Electromagnetic interactions. In correspondence, *gluons are electrically neutral* and hadron spectroscopy remains the same as in the colourless theory, only with quark colours arranged so as to produce mesons and baryons in strictly colourless states (i.e. $SU(3)_{colour}$ singlets) as in (18.5).

Invariance under the gauge transformations (18.6) determines the quark gluon interaction lagrangian. Using the same argument that led to the electron-photon interaction in Chapter 9, Eq. (9.7), one finds:

$$\begin{aligned}
&\mathcal{L}_q(q, D^\mu q) = \bar{q}(i\not{D} - m)q \,; \\
&D^\mu q^a(x) = \partial^\mu q^a(x) - ig_s\, g^{A\mu}(x)\left[\frac{\lambda^A}{2}\right]^a_b q^b(x) \,; \\
&q^a = (u, d, s, c)^a \,,\ m = (m_u, m_d, m_s, m_c) \,.
\end{aligned} \tag{18.7}$$

with $g_s$ the strong quark gluon coupling, $g^{A\mu}(x)$, $A = 1, \dots, 8$ the gluon fields, and $m$ the quark masses. Unlike photons in QED, gluons interact among themselves, with 3-gluon and 4-gluons interactions of order $g_s$ and $g_s^2$, respectively.

**Quark Confinement.** Gluons self-interaction, in the presence of a coupling constant $g_s = \mathcal{O}(1)$ is supposed to be at the origin of *quark confinement*, the fact that quarks are permanently confined inside colour singlet hadrons. Indeed, quarks have never been observed in isolation[5].

To visualise this phenomenon, consider a (colour singlet) meson made by a quark-antiquark pair bound by colour forces. For weak coupling, we have colour lines of force that start from the quark, like in electrostatics, and have

[4]A gauge theory based on a non-commutative symmetry like $SU(3)_{colour}$ is called a Yang-Mills theory, after the names of the authors who first worked out these theories in 1954 [71].

[5]The lightest quark must be stable due to its fractional electric charge. Isolated quarks produced in the collisions of cosmic rays or at particle accelerators have been intensively and unsuccessfully searched in the years nineteen sixties.

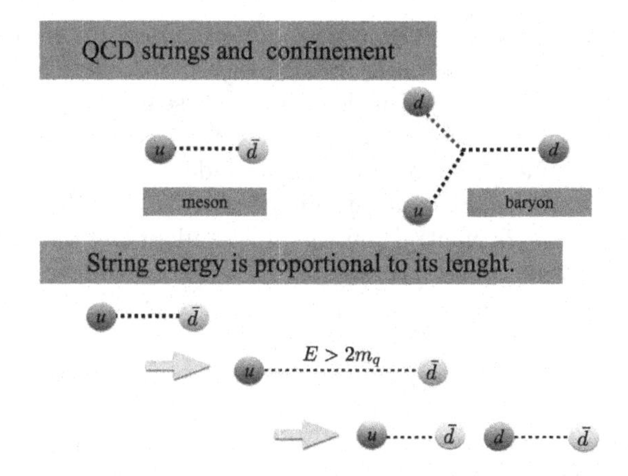


Figure 18.1 Inside hadrons, colour fields condense in "strings" that go from the quark to the antiquark (mesons) or to a three-string vertex (baryons). The energy of the string is proportional to the lenght. If we apply a force to the quark, e.g. from a collision with an external electron, the string is stretched until string energy reaches the $q\bar{q}$ threshold and it gives rise to two mesons (baryon plus meson), still with confined quarks.

to land all in the antiquark, since the meson is colour neutral. It has been supposed that, going to a strong coupling, gluon self-attraction makes the force-lines to condense into a *string* that goes from the quark to the antiquark, with an associated energy which increases (e.g. linearly) with the length. In the case of baryons, the strings originating from the three quarks merge into a colour invariant triple vertex with their colour indices saturated by the antisymmetric tensor $\epsilon_{abc}$ as in Eq. (18.5).

If we give energy and momentum to one quark, e.g. from a collision with an external electron, the string is stretched until string energy reaches the $q\bar{q}$ threshold. At this point an additional quark pair is created, see Fig. 18.2, and a state with two mesons (or a baryon plus meson) is produced, still with confined quarks.

**Asymptotic Freedom.** In 1973 David Gross and Franck Wilczek [72] and, indipendently, David Politzer [73], computed the asymptotic behaviour of the $SU(3)_{colour}$ coupling, Eq. (18.7), for large values of the momentum transfer. Unlike what happens in all cases previously studied (QED, Yukawa and $\lambda\phi^4$ interaction) they found that the colour coupling $g_s$, defined to be very large in the confinement region $q < 1$ GeV, decreases logarithmically for large momentum transfer:

$$\alpha_s(q) = \frac{g_s^2(q)}{4\pi} \approx \frac{C}{\ln q^2} \to 0 \ \ (q^2 \to \infty) \ . \tag{18.8}$$

Asymptotically, quark behave *as free particles*!

The unexpected result agreed with the scaling relations observed in the experimental $e-p$ deep inelastic cross-sections since the first data in 1968. The scaling behaviour had been anticipated by J. Bjorken, on the basis of quark current commutators, and was explained by R. Feynman in 1971 as indicating that the proton, seen at very large momentum transfer, behaves like a cloud of free, point-like particles that Feynman called *partons.*

Experimental investigations of deep inelastic scattering of electrons and neutrinos off protons and neutrons, in the years 1970s and 1980s, have shown that the proton seen at large momentum can indeed be described as an incoherent mixture of free quarks and antiquarks with different flavours, with an additional component of neutral partons, to be identified with gluons. Each parton is characterised by *structure functions* that describe the probability to find that particular constituent (i.e. $u$ or $d$ quarks or gluons) with a fraction $x$ of proton's momentum.

Owing to the logarithmic approach to asymptotic freedom, Eq. (18.8), at large but finite energies there are deviations from free-parton behaviour, which have been accurately computed in QCD, see Ref. [74], and compared to experimental data at increasingly large energies.

The $q$ dependence of the coupling, the so-called *running coupling,* has been computed in QCD to leading logarithmic (LLO) and next-to-leading logarithmic order (NLLO) to be[6]:

---

[6]The running coupling is determined by the *renormalization group* equation

$$\frac{\partial\alpha(t)}{\partial t} = \beta(\alpha),\ t = \ln\, q^2\ , \tag{18.9}$$

where $\beta(\alpha) = \beta_1\alpha^2 + \beta_2\alpha^3$, $\beta_1$ and $\beta_2$ are constants computed in [75] to leading logarithmic order (LLO) and next-to-leading logarithmic order (NLLO), respectively, and reported in (18.16). The solution of (18.9) is:

$$\int^{\alpha_s(q^2)} \frac{d\alpha}{\beta_1\alpha^2 + \beta_2\alpha^3} = D(\alpha_s(q^2)) = \int^{q^2} dt = \ln\frac{q^2}{\Lambda^2_{QCD}}\ , \tag{18.10}$$

where $-\ln\Lambda^2_{QCD}$ is an integration constant and $D(\alpha)$ is the primitive of the integrand in (18.10): $D'(\alpha) = (\beta_1\alpha^2 + \beta_2\alpha^3)^{-1}$:

$$D(\alpha) = -\frac{1}{\alpha\beta_1} - \frac{\beta_2}{\beta_1^2}\ln(\frac{\alpha\beta_2}{\beta_1 + \alpha\beta_2})\ . \tag{18.11}$$

To LLO (and $\beta_1 < 0$):

$$\frac{1}{\alpha_{s1}(q^2)} = (-\beta_1)\ln\frac{q^2}{\Lambda^2_{QCD}},\ \text{i.e.}\ \alpha_{s1}(q^2) = \frac{1}{(-\beta_1)\ln\frac{q^2}{\Lambda^2_{QCD}}}\ . \tag{18.12}$$

To find $\alpha_s$ to NLLO we use (18.11) inserting $\alpha_{s1}$ in the NLLO correction, to find

$$\alpha_s(q^2) = \frac{1}{(-\beta_1)\ln\frac{q^2}{\Lambda^2_{QCD}}}\left\{1 + \frac{\beta_2}{\beta_1}\alpha_{s1}(q^2)\ln\left[\frac{\beta_2}{\beta_1}\alpha_{s1}(q^2)\right]\right\}, \tag{18.13}$$

which gives Eq. (18.15). To find the value of $\Lambda_{QCD}$ given $M_Z$, we use Eq. (18.11) yielding

$$\Lambda_{QCD} = M_Z\ Exp\left\{-\frac{1 + \frac{\beta_2}{\beta_1}\alpha_{sMZ}\ln(\frac{\beta_2}{\beta_1}\alpha_{sMZ})}{2(-\beta_1)\ \alpha_{sMZ}}\right\} = 0.244\ \text{GeV}\ , \tag{18.14}$$

for 5 quark flavours with masses $m << M_Z$.

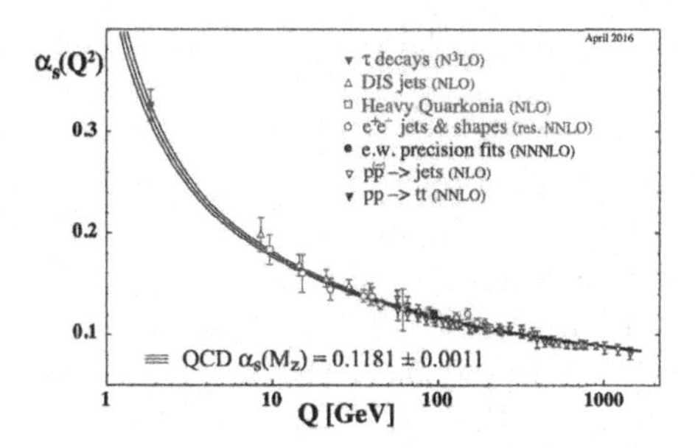


Figure 18.2 Taken from Ref. [76], the figure shows the strong coupling $\alpha_s = g_s^2/4\pi$ as a function of momentum transfer, compared to recent data from LHC and other sources.

$$\alpha_s(q^2) = \frac{1}{(-\beta_1)\ \ln\frac{q^2}{\Lambda^2_{QCD}}} + \frac{\beta_2}{\beta_1}\frac{1}{[(-\beta_1)\ln\frac{q^2}{\Lambda^2_{QCD}}]^2}\ln[\frac{\beta_2}{\beta_1}\frac{1}{(-\beta_1)\ln\frac{q^2}{\Lambda^2_{QCD}}}]\ ,$$
$$\Lambda_{QCD} \simeq 0.25\ \text{GeV}\ , \tag{18.15}$$

with $\beta_1$ $\beta_2$ the LLO and NLLO coefficients of the beta-function, see [75]:

$$\beta_1(f) = -\frac{11 - 2/3f}{4\pi}\ ;\ \beta_2(f) = \frac{-102 + 38/3f}{(4\pi)^2}\ , \tag{18.16}$$
$$f = 5 = \text{number quarks with } m << M_Z\ .$$

The value of $\Lambda_{QCD}$ is determined from the condition: $\alpha_s(M_Z) = 0.1181$, which is the result of the fit to the data in [76]. The behaviour implied by (18.15) is compared with recent experimental data from LHC experiments and other sources in Fig. 18.2, also taken from Ref. [76].

The extraordinary agreement between theoretical predictions and observations is a solid confirmation of QCD.

## 18.2 CHARMONIA

The critical parameter that regulates the passage from strong to weak regime, $\Lambda_{QCD} = 0.25$ GeV, Eq. (18.15), can be compared with the quark masses given in Table 18.1. Quarks $u$ and $d$ fall in the strong coupling region, $s$ marginally out, but heavy quarks are definitely in the weak coupling region. The result suggests the possibility to *compute* the mass spectrum of $\bar{c}c$ and $\bar{b}b$ mesons with methods similar to those employed for atomic spectra.

After the discovery of asymptotic freedom, *T.* Applequist and *D.* Politzer observed that the Coulomb-like interactions associated to the exchange of gluons would produce a series of $\bar{c}c$ bound states [77], in all similar to the positronium states introduced in Section 12.5.2. By analogy, Applequist and Politzer proposed to name *charmonium* each of these states and observed that the smallness of $\alpha_s$ at the charm quark mass scale would make their spectrum calculable in perturbation theory. As it happens for positronium, the hyperfine (spin-spin) interaction is expected to split the ground state into "Paracharmonium" ($J = 0$) and "Orthocharmonium" ($J = 1$).

Due to charge-conjugation, Para- and Orthocharmonium must decay in states with two or three gluons respectively [77], which would then evolve into final states of light mesons. The smallness of $\alpha_s$ would make the decay width of Ortopositronium (proportional to $\alpha_s^3$) much smaller than the Parapositronium width (proportional to $\alpha_s^2$) estimated in [77] to be of about ~6 MeV (see Sect. 12.5.2 to compare with positronium).

After $J/\Psi$ discovery in 1974, A. De Rujula and S. L. Glashow [78] identified this, very narrow width particle ($\Gamma \sim 100$ keV) with the Orthocharmonium proposed by Appelquist and Politzer and opened the road to the investigation of the spectrum of charmed particles and charmonia[7]. It was also clarified that the threshold below which charmonia had only gluon (or photon) decays had to be identified with the threshold for the production of charmed meson pairs, $(\bar{c}q) + (c\bar{q})$, see Fig. 18.4 below.

In 1974, the Cornell potential [80–82] was introduced to supplement the Coulomb-like potential envisaged in [77], which dominates at short distances, with a term linear in the radius to simulate the confining forces that dominate at larger distances, the latter are determined by a new, phenomenological constant, the string tension $k$.

### 18.2.1 The Cornell Potential and Its Relativistic Corrections

The Cornell potential[8]:

$$V = -\frac{4}{3}\frac{\alpha_s}{r} + kr + 2M_c = V_V + V_S \ , \tag{18.17}$$

provides the basic charmonium wave functions via the spin-independent Schrödinger equation

$$(-\frac{1}{2\mu_c}\nabla + V)\psi = E\psi \ , \tag{18.18}$$

$\mu_c = M_c/2$ is the reduced mass of the charm quark-antiquark pair.

Following the treatment of the non-relativistic hydrogen atom (Appendix B), we separate angular and radial wave functions for each value of the orbital angular momentum $L$, according to

$$\psi(n, L, m)(x) = R(n, L)(r) \cdot Y(\theta, \phi)^L_m \ , \tag{18.19}$$

with $n = 1, 2, \ldots$ the principal quantum number, $R$ the radial wave function and $Y^L_m$ the spherical harmonics (we shall use the standard notation: $L = S, P, D, \cdots = 0, 1, 2,$). Further, we put:

$$R(r) = \frac{\chi(r)}{r} \ , \tag{18.20}$$

[7]The association of $J/\Psi$ with the opening of the $c\bar{c}$ threshold had been made previously in [79].

[8]$-4/3$ is a group theoretical factor reflecting the fact that the $c\bar{c}$ pair, like all hadrons, is in a colour singlet state.

with $\chi(r)$, the reduced radial wave function, satisfying the boundary condition

$$\lim_{r\to 0^+} \chi(r) = 0 \ . \tag{18.21}$$

Similarly to atomic physics, hyperfine, spin-orbit and tensor interactions (discussed e.g. in [83]) arise as part of the expansion in powers of $v^2/c^2$ of the relativistic quark-antiquark potential and are to be treated as first-order perturbations to the quark-antiquark interaction (18.17) (see also [84,85]).

**Hyperfine Interaction.** The spin-spin interaction, taken from atomic physics, is:

$$V_{SStot} = \frac{2}{3M_c^2}\nabla(V_V)(\mathbf{S}\cdot\mathbf{S}) = \frac{16\pi\alpha_s}{9M_c^2}\delta^{(3)}(r)\ [S(S+1) - \frac{3}{2}] \ . \tag{18.22}$$

The hyperfine correction applies to $L = 0$ states only, for which:

$$|\psi(0)|^2 = \frac{1}{4\pi}\lim_{r\to 0^+}(\frac{\chi(r)}{r})^2 \neq 0 \ . \tag{18.23}$$

Defining

$$\bar{V}_{SS} = \frac{16\pi\alpha_s}{9M_c^2}|\psi(0)|^{2\prime} \tag{18.24}$$

one has:

$$< V_{SStot} >_{S-wave} = \bar{V}_{SS} \times \begin{cases} +1/2, & (S=1) \\ -3/2, & (S=0) \end{cases} \ . \tag{18.25}$$

**Spin Orbit Interaction.**

We define

$$\begin{aligned} V_{LS} &= \frac{1}{2M_c^2 r}(3\frac{dV_V}{dr} - \frac{1}{r}\frac{dV_S}{dr}) = \frac{1}{2M_c^2}(\frac{4\alpha_s}{r^3} - \frac{k}{r}) \ , \\ V_{LStot} &= V_{LS}(\boldsymbol{L}\cdot\boldsymbol{S}) \ , \end{aligned} \tag{18.26}$$

with $\mathbf{S}$ being the total $\bar{c}c$ spin. To first order in this potential, we compute the integral

$$\bar{V}_{LS} = \frac{1}{2M_c^2}\int_0^{+\infty}(\frac{4\alpha_s}{r^3} - \frac{k}{r})r^2(\frac{\chi(r)}{r})^2 dr \ . \tag{18.27}$$

The integral is convergent at the origin for $L \geq 1$ since $(\chi(r)/r)^2 \sim r^{2L}$ for $r \to 0$. We restrict to $P$-waves and find

$$\delta m_{LS} =< V_{LStot} >_{P-wave} = \bar{V}_{LS}(\boldsymbol{L}\cdot\boldsymbol{S}) = \bar{V}_{LS} \times \begin{cases} -2 \ (J=0) \\ -1 \ (J=1) \\ +1 \ (J=2) \end{cases} . \tag{18.28}$$

**Tensor Interaction.**
One defines the tensor interaction starting from (see [83, 84])

$$V_T = \frac{1}{6M_c^2}(3\frac{d^2V_V}{dr^2} - \frac{1}{r}\frac{dV_S}{dr}) ,$$

$$V_{Ttot} = -3V_T[N_{ij}S_iS_j] = \frac{1}{2M_c^2}(\frac{8\alpha_s}{r^3} + \frac{k}{r})[N_{ij}S_iS_j] , \quad (18.29)$$

with:

$$N_{ij} = (\hat{n}_i\hat{n}_j - \frac{1}{3}\delta_{ij}), \ \hat{n}_i = \frac{r_i}{r} . \quad (18.30)$$

The average value of $N_{ij}$ in a state with definite orbital angular momentum $L$ is[9]:

$$< (\hat{n}_i\hat{n}_j - \frac{1}{3}\delta_{ij}) >_L = a(L)\big[L_iL_j + L_jL_i - \frac{2}{3}\delta_{ij}L(L+1)\big] , (18.31)$$

$$a(L) = -\frac{1}{4L(L+1)-3} , \quad (18.32)$$

so that:

$$< V_{Ttot} >_L = < \frac{1}{2M_c^2}(\frac{8\alpha_s}{r^3} + \frac{k}{r}) > \times$$

$$\times 2a(L)\big[(\mathbf{S}{\cdot}\mathbf{L})^2 - \frac{1}{3}S(S+1)L(L+1)\big] . \quad (18.33)$$

The effect, of course, vanishes for $L = 0$ or $S = 0$. For $L = S = 1$, we compute the integral

$$V_1 = < \frac{1}{2M_c^2}(\frac{8\alpha_s}{r^3} + \frac{k}{r}) >_{P-wave} =$$

$$= \int_0^\infty \frac{1}{2M_c^2}(\frac{8\alpha_s}{r^3} + \frac{k}{r})r^2(\frac{\chi(r)}{r})^2 dr , \quad (18.34)$$

and find

$$\delta m_T = < V_{Ttot} >_{P-wave} = \frac{2}{15}V_1 \times \begin{cases} -8 \ (J=0) \\ +1 \ (J=1,2) \end{cases} = \bar{V}_T \begin{cases} -8 \ (J=0) \\ +1 \ (J=1,2) \end{cases} ;$$

$$\bar{V}_T = \frac{1}{15}\frac{1}{M_c^2} < (\frac{8\alpha_s}{r^3} + \frac{k}{r}) >_{P-wave} . \quad (18.35)$$

---

[9]This result is reported in the book of Landau and Lifshitz [86]. The r.h.s of (18.31) is the only symmetric and traceless tensor with two indices, which can be formed with $L_i$, the only vector remaining after integration over $r$. To obtain (18.32), one multiplies both sides of (18.31) by $L_i$ (on the left) and $L_j$ (on the right), summing over $i$ and $j$. The result follows from the fact that $\hat{n}$ is orthogonal to $L = \vec{r} \times \vec{p}$ and from angular momentum commutators.

**Charmonium masses.** We denote by $E(n, L)$ the eigenvalues of (18.18). Recalling that $V$ includes the charm quark pair rest mass, charmonium masses are given by:

$$M(n, L, S, J) = E(n, L) + \\ +\delta m_{SStot}(n, L, J) + \delta m_{LStot}(n, L, J) + \delta m_{Ttot}(n, L, J) \quad (18.36)$$

The usual spectroscopic notation for the states with quantum numbers $n, L, S, J$ is (see [84])

$$n\ ^{2S+1}L_J :\ 1\ ^1S_0,\ 1\ ^3S_1,\ \dots\ . \quad (18.37)$$

It is also usual to denote the $S$-wave, spin singlet and triplet states with the symbols $\eta_c$ and $\psi$ respectively, $P$-wave states with $\chi_{cJ}$, $(S = 1, J = 0, 1, 2)$ and $h_c$, $(S = 0, J = 1)$.

Parity and charge conjugation properties of charmonia are the same as those of positronium (Eqs. (12.84) and (12.86)); that is:

$$P = (-1)^{L+1},\ C = (-1)^{(L+S)}\ . \quad (18.38)$$

### 18.2.2 Strategy and Numerical Results

The formulae given in the previous paragraph identify an ambitious model, in which the spectrum of charmonia is determined by only three parameters: $M_c$, the string tension $k$ and $\alpha_s(M_c)$.

Due to quark confinement, however, and *unlike atomic and nuclear physics*, we cannot determine the mass of the elementary constituents (i.e. the charm quark) and their interactions in isolation, independent of the spectrum of bound states.

In mathematical terms, confinement implies that we cannot determine *a priori* the zero of the Cornell potential and we should add to the Cornell potential an *a priori* undetermined constant $V_0$. We can include $V_0$ in the definition of the charm quark mass, and consider the mass a free constant *to be determined from charmonium spectrum.* The same holds true for the string tension. The third parameter, the running coupling at the charm mass scale $\alpha_s(M_c)$, with only a logarithmic dependence upon $M_c$, can be obtained from Eq. (18.15) via the well determined value of $\alpha_s(M_Z) = 0.1185$. All this amounts to take the masses of two $\bar{c}c$ states as input and predict the masses of the other charmonia.

Following Ref. [84], we take:

$$\alpha_s = 0.331,\ M_c = 1.317\ \text{GeV},\ k = 0.18\ \text{GeV}^2\ , \quad (18.39)$$

which reproduce well the Orto and Para-charmonium masses.

We solve numerically the Schrödinger equation, Eq. (18.18), restricting to the lowest levels: $n = 1, 2;\ L = S, P$, and use the wave functions to compute hyperfine, spin-orbit and tensor interaction corrections.

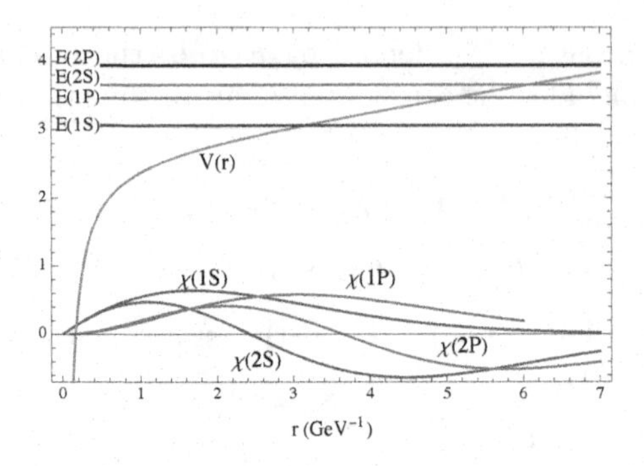


Figure 18.3 Radial wave functions and eigenvalues of the states: $n = 1, 2$; $L = S, P$. In ordinates E in GeV. Note: $1\text{GeV}^{-1} \sim 0.2$ fm.

To solve the Scrödinger equation, we use the program *schroedinger.nb* [87]. Eigenfunctions and eigenvalues are displayed in Fig. 18.3. The numerical values of the constants to determine hyperfine, spin-orbit and tensor corrections are reported in Table 18.2.

The resulting spectrum is displayed in Fig. 18.4, together with the masses obtained numerically, labeled "th".

Arrows in Fig. 18.4 indicate cascade decays between charmonia levels, with emission of $\gamma$ rays or light hadrons. Radiative decays, typical of transitions with $\Delta L = 1$ or $\Delta S = 1$, have been an importante guide to discover and identify charmonium levels.

The agreement of masses and transitions predicted from the simple, QCD motivated, Cornell potential and the observed spectrum of charmonia is remarkable. For the extension of the spectrum to higher masses, further developments and improvements, the reader may resort to Refs. [83, 84, 93].

## 18.3 CHARMONIA END EXOTICS

In November 1974, $J/\Psi = \psi(3097)$ has been discovered in proton-Be collisions at Brookhaven, as a narrow peak in $e^+e^-$ invariant mass distribution, and as a peak in the total cross section of $e^+e^- \to$ hadrons, at SLAC. One week after $J/\Psi$ discovery at SLAC, a second peak has been observed at SLAC, corresponding to $\psi'(3686) = \psi(2S)$.

Table 18.2 Values of the constants needed for hyperfine, spin-orbit and tensor corrections, in GeV, see Eqs. (18.24), (18.27), (18.35).

| – | $\bar{V}_{SS}$ | $\bar{V}_{LS}$ | $\bar{V}_T$ |
|---|---|---|---|
| 1S | 0.055 | 0 | 0 |
| 1PS | 0 | 0.013 | 0.011 |
| 2S | 0.041 | 0 | 0 |

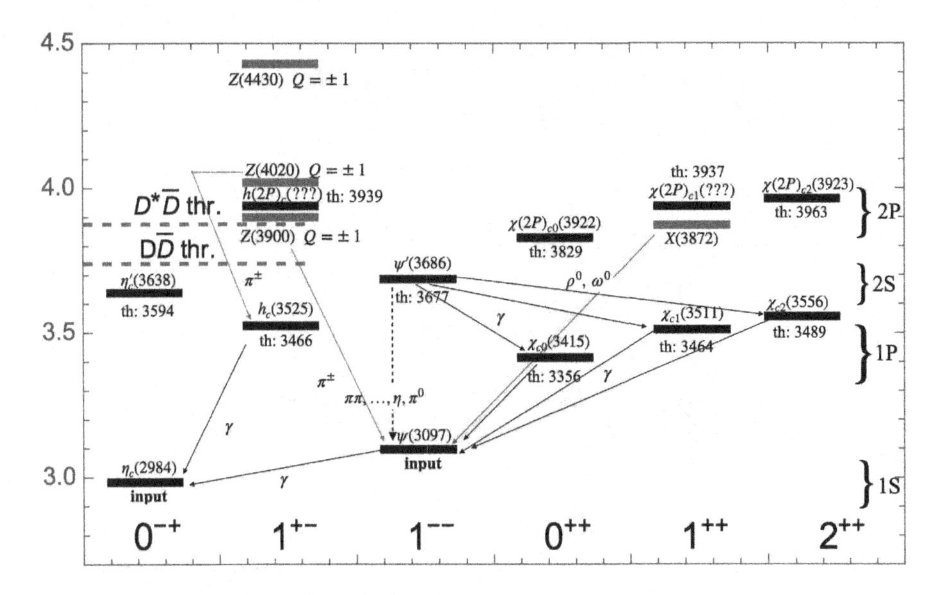


Figure 18.4 *Black lines.* Expected and observed ground and excited levels of Charmonia ($\bar{c}c$ mesons) up to masses of 4 GeV. $J^{PC}$ quantum numbers are indicated at the bottom of the plot and particles are identified by the denominations introduced in the text. The observed masses are reported in parenthesis, next to the particle name, and are taken from [12]. Charmonia $h(2P)_c$ and $\chi(2P)_{c1}$ have not been identified yet. The mass values obtained with the Cornell potential plus hyperfine, spin-orbit and tensor corrections are reported for each particle, with the "th" label. *Gray lines.* The particle named $X(3872)$, $J^{PC} = 1^{++}$, was the first example of "exotic hadron" found among hidden charm particles, followed by the electrically charged $Z(3900)$, $Z(4020)$ and $Z(4430)$, $J^{PC} = 1^{+-}$.

The first evidence[10] of intermediate P-wave states was found in 1975 by the spectrometer DASP [89] at DESY (Hamburg), in the two-photon cascade: $\psi'(3686) \to \gamma + \chi_{c0}(3415) \to 2\gamma + \psi(3097)$. Later, this cascade and cascades to other intermediate $\chi$ states were seen at SPEAR [90, 91] and at DESY by the Desy-Hamburg and the PLUTO Collaborations [88].

The $\eta_c(2984)$ has been identified in 1977 in the radiative decay $\psi(3097) \to \gamma + \eta_c(2984)$ by the DASP Collaboration [92] in DESY.

**Exotics.** In a word made of gluons and charmed quarks/antiquarks, light hadrons can be created, to a great extent, only in isospin zero states, i.e. $\eta$ $(I = 0)$ and not $\pi^0$ $(I = 1)$ mesons, or $\omega$ $(I = 0)$ but not $\rho^0$ $(I = 1)$ mesons. This is well illustrated by the small ratio [12]

$$\frac{\Gamma(\psi'(3686) \to \pi^0 \psi(3097)}{\Gamma(\psi'(3686) \to \eta\psi(3097))} = (4.5 \pm 0.1)10^{-2} \ . \tag{18.40}$$

The result is contradicted by the much larger $\rho^0$ vs $\omega$ production in $X(3872)$

[10]See H. Schopper in Ref. [88] for details.

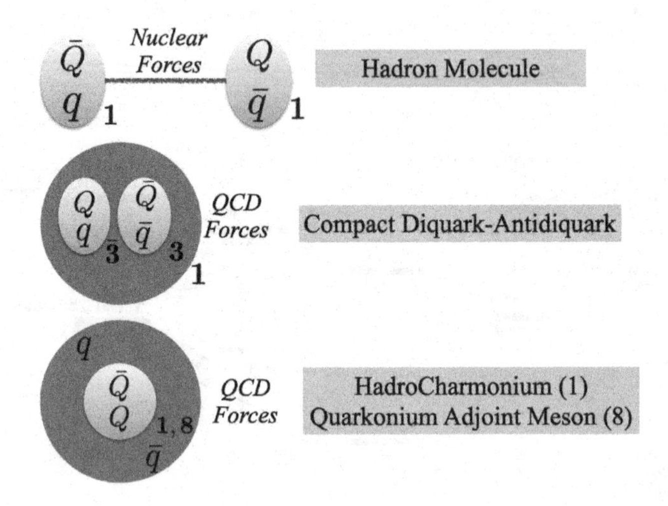


Figure 18.5 Schematic representations of different models of exotic tetraquark systems.

decays [94]

$$\frac{\Gamma(X(3872) \to \rho^0\psi(3097)}{\Gamma(X(3872) \to \omega\psi(3097))} = (2.9 \pm 0.4)10^{-1} \ , \tag{18.41}$$

which suggests that a pair of light quarks may be "intrinsically" present in the constitution of $X(3872)$, together with the $c\bar{c}$ pair.

A first proposal, based on the fact that $X(3872)$ is very close to the $D^0\bar{D}^{*0}$ threshold, was the $X(3872)$ to be a *composite meson-antimeson state*, bound by the same nuclear force that binds proton and neutron into the deuteron: a bound state that, by analogy, has been called "deuson" [95, 96].

In alternative, based on previous investigations on the constitution of light spin 0 mesons, the hypothesis has been made of $X(3872)$ as a *compact tetraquark*: a $[cq][\bar{c}\bar{q}]$ state (with $q$ a light quark) bound by the fundamental QCD forces [97].

The observation of hadrons that are: (i) electrically charged and (ii) decay into a final state containing a true charmonium has confirmed the possible coexistence of one light and one heavy quark-antiquark pair in the same hadron. This is realised in the three levels reported in Fig. 18.4: $Z(3900)^\pm \to \pi^\pm + \psi'(3986)$, $Z(4020)^\pm \to \pi^\pm + h_c(3525)$ and $Z(4430)^\pm \to \pi^\pm + \psi'(3986)$.

The experimental study of exotic hadrons is in full development. Recent acquisitions are the discovery of "pentaquarks" [98], with composition $(c\bar{c}qq'q'')$ and "tetra-charm" states [99] observed to decay into pairs of $\psi$ mesons, with composition $(\bar{c}c\bar{c}c)$.

Theoretical studies are also in evolution, to clarify the internal dynamics of exotic hadrons, its relations to QCD and to the well known dynamics of $(q\bar{q})$ mesons and $(qqq)$ baryons, see Fig. 18.5 for illustration and Ref. [100] for a theoretical introduction.

## 18.4 PROBLEMS FOR CHAPTER 18

### Sect. 17.1

1. Show that the dimension of the algebra of the group with $N$ colours, i.e. the number of $N \times N$ hermitian and traceless matrices, is $D = N^2 - 1$.

2. Starting from Eq. (18.6), explain why there are 8 gluons in QCD.

### Sect. 17.2

1. Using the radial equation for $\chi(r)$, Eq. (B.16), prove that:

$$\frac{\chi(r)}{r} \sim r^L, \text{ for } r \to 0^+$$

   where $L$ is the orbital angular momentum.

2. Using the computer program *schroedinger.nb* and the data given in (18.39), reconstruct the eigenvalues and wave functions reported in Fig. 18.3.

3. Compute the masses of $\psi(3S)$ and $\psi(4S)$ and compare to the computed and observed values reported in [84].

CHAPTER 19

# THE BORN-OPPENHEIMER APPROXIMATION FOR THE DOUBLY CHARMED BARYON

Molecules and crystals are made of two kinds of particles with very different masses, light electrons and heavy nuclei, both moving in the fields generated by Coulomb forces. Different masses entail different scales for the space variation of the wave functions and this is the basis of an approximation, the Born-Oppenheimer approximation, widely used in the theory of complex atomic systems. A recent illustration of the Born-Oppenheimer approximation in QED is found in Weinberg's book on Quantum Mechanics [101].

The similarity with QED, has led several authors to apply the Born-Oppenheimer approximation to hadronic systems which contain both heavy and light quarks, and treat them as molecule-like systems bound by QCD forces [104–107]. We illustrate here an application of the Born-Oppenheimer approximation (BOA), to estimate the mass of the recently observed double charm baryon, $\Xi_{cc}^{++} = (ccu)$.


DOI: 10.1201/9781003436263-19

## 19.1 BORN-OPPENHEIMER APPROXIMATION IN BRIEF

To be definite, consider a system with two heavy and two light particles with Hamiltonian (see [101]):

$$H = H_{heavy} + H_{light} = \frac{1}{2M} \sum_{heavy} P_i^2 + V(\boldsymbol{x}_A,\ \boldsymbol{x}_B) +$$

$$+\frac{1}{2m} \sum_{light} p_i^2 + V_l(\boldsymbol{x}_A,\ \boldsymbol{x}_B,\ \boldsymbol{x}_1,\ \boldsymbol{x}_2)\ . \tag{19.1}$$

First, consider the heavy particles as classical sources with fixed coordinates and quantum numbers, and find the ground state of the light particles, solving the eigenvalue equation:

$$\begin{aligned} &H_{light}\ f_0(\boldsymbol{x}_A,\ \boldsymbol{x}_B,\ \boldsymbol{x}_1,\ \boldsymbol{x}_2) = \mathcal{E} f_0\ , \\ &\mathcal{E} = \mathcal{E}(\boldsymbol{x}_A,\ \boldsymbol{x}_B)\ . \end{aligned} \tag{19.2}$$

Then search for solutions of the complete Schrödinger equation for wave functions of the form:

$$\Phi = \Psi(\mathbf{x}_A, \mathbf{x}_B)\ f_0(\boldsymbol{x}_A,\ \boldsymbol{x}_B,\ \boldsymbol{x}_1,\ \boldsymbol{x}_2)\ .$$

In the equation:

$$(H_{heavy} + H_{light})\Psi(\boldsymbol{x}_A, \boldsymbol{x}_B)\ f_0(\boldsymbol{x}_A,\ \boldsymbol{x}_B,\ \boldsymbol{x}_1,\ \boldsymbol{x}_2) = E\Psi f_0\ ,$$

we may replace $H_{light}$ by its eigenvalue and rewrite the equation as

$$(\frac{1}{2M} \sum_{heavy} P_i^2 + V(\boldsymbol{x}_A,\ \boldsymbol{x}_B) + \mathcal{E})\Psi f_0 = E\Psi f_0\ . \tag{19.3}$$

Applying $H_{heavy}$ to $\Phi$ we encounter terms of the kind:

$$-i\mathbf{P}\Phi = \frac{\partial}{\partial x_A}\Phi = (\frac{1}{f_0}\frac{\partial f_0}{\partial x_A} + \frac{1}{\Psi}\frac{\partial \Psi}{\partial x_A})\Psi f_0\ . \tag{19.4}$$

The Born-Oppenheimer approximation consists in neglecting systematically the first with respect to the second term. If we call $a$ and $b$ the lengths over which $f_0$ or $\Psi$ show an appreciable variation with the source coordinates, the BO approximation is valid for $a >> b$, namely for small values of the ratio

$$\Lambda = \frac{1/a}{1/b} << 1\ . \tag{19.5}$$

In the hydrogen molecule, $a$ and $b$ are, respectively, the electron and proton Bohr radius ($1/a = \alpha m$, $1/b = \alpha M$) and the error is of order $m/M$ (see [101] for the more complicated case of crystals). We will discuss the error in QCD later.

The upshot is the Born-Oppenheimer (BO) equation:

$$\left( \sum_{\text{heavy}} \frac{P_i^2}{2M} + V(\boldsymbol{x}_A, \boldsymbol{x}_B) + \mathcal{E}(\boldsymbol{x}_A, \boldsymbol{x}_B) \right) \Psi = E\Psi \; , \tag{19.6}$$

and in the following, we will denote

$$V(\boldsymbol{x}_A, \boldsymbol{x}_B) + \mathcal{E}(\boldsymbol{x}_A, \boldsymbol{x}_B) = V_{BO}(\boldsymbol{x}_A, \boldsymbol{x}_B) \; . \tag{19.7}$$

The remarkable fact about Eq. (19.6) is that the wave function $f_0$ has dropped out: what matters is only the energy of the light quarks in presence of the heavy sources, which makes the BO approximation amenable to numerical calculations (see e.g. [102] for a review of recent Lattice QCD calculations).

The BO approximation requires all quantum numbers of the heavy particles to be fixed when computing the light quarks energy, $\mathcal{E}$. Besides position and flavour, we have to fix spin and colour quantum numbers. In the two cases of interest, the $c$ quarks are colour triplets with spin 1/2, with colours combined to produce a colour $\bar{\mathbf{3}}$ representation and total spin 1[1].

## 19.2 COLOUR GYMNASTIC FOR QUARK-QUARK POTENTIALS

The Coulomb potential part of the quark-quark Cornell potential introduced in (18.17), can be written as

$$V_V(r) = \frac{\alpha_s}{r} \cdot \langle T_1^A T_2^A \rangle_R \; , \tag{19.8}$$

where $T_{1,2}^A$ are the matrices $\frac{\lambda^A}{2}$ representing the generators of $SU(3)_c$ acting on quarks 1 or 2 (sum over the repeated indices A understood) and R indicates the $SU(3)_{colour}$ irreducible multiplet to which the product of the quark and antiquark fields belong.

The $T_{1,2}^A$ act on different spaces, and obviously commute. To be precise, one should write

$$T_1^A = (\frac{\lambda^A}{2})_1 \otimes 1_2 \; , \; T_2^A = 1_1 \otimes (\frac{\lambda^A}{2})_2 \; . \tag{19.9}$$

One can write the generator of the whole $1+2$ system according to:

$$T_{1+2}^A = T_1^A + T_2^A = (\frac{\lambda^A}{2})_1 \otimes 1_2 + 1_1 \otimes (\frac{\lambda^A}{2})_2 \; , \tag{19.10}$$

[1]For the $(c_1 c_2 q)$ baryon, color $\bar{\mathbf{3}}$ is required to make a colour singlet with the colour triplet $q$ and spin 1 is required by Fermi statistics, since $c_1$ and $c_2$ are already antisymmetrised by colour as in Eq. (18.5). For the tetraquark $(c_1 c_2 \bar{u}\bar{d})$, not consider here, we would chose the charm pair to be a colour anti-triplet, because it is the only attractive channel, see below, and spin 1 follows from Fermi statistics as before.

and (sum over $A$ understood):

$$
\begin{aligned}
& T^A_{1+2} T^A_{1+2} = \boldsymbol{T}^2 = \\
& = (\frac{\lambda^A \lambda^A}{4})_1 \otimes 1_2 + 1_1 \otimes (\frac{\lambda^A \lambda^A}{4})_2 + 2(\frac{\lambda^A}{2})_1 \otimes (\frac{\lambda^A}{2})_2 = \\
& = \boldsymbol{T}_1^2 + \boldsymbol{T}_2^2 + 2\boldsymbol{T}_1 \cdot \boldsymbol{T}_2 \ . \qquad (19.11)
\end{aligned}
$$

For each representation, $\boldsymbol{R}_1$, $\boldsymbol{R}_2$ or $\boldsymbol{R}_{12}$ to which $q_1$, $q_2$ or their product $q_1 q_2$ belong, the operator $\boldsymbol{T}^2 = C_2(\boldsymbol{R})$ is called the *quadratic Casimir operator* and is a constant multiple of the identity over the representation.

The last term in Eq. (19.11) is just the coefficient in front of the potential (19.8). Expressing it in terms of the Casimir operators, we write the potential (19.8) as[2]:

$$
\begin{aligned}
& V_V = \lambda(R)\frac{\alpha_s}{r} \ ; \\
& 2\,\lambda(R) = C_2(\boldsymbol{R}_{12}) - C_2(\boldsymbol{R}_1) - C_2(\boldsymbol{R}_2) \ . \qquad (19.12)
\end{aligned}
$$

We note the results:

$$
\begin{aligned}
C_2(\mathbf{1}) &= 0 \ , \ C_2(\boldsymbol{R}) = C_2(\bar{\boldsymbol{R}}) \ , \\
C_2(\mathbf{3}) &= 4/3 \ ; \ C_2(\mathbf{6}) = 10/3 \ ; \ C_2(\mathbf{8}) = 3 \ . \qquad (19.13)
\end{aligned}
$$

The dependence of $\lambda$ on Casimir operators shows an interesting pattern of forces vs $\boldsymbol{R}$:

1. quark-antiquark:

$$
\boldsymbol{R} = (\mathbf{1}): \text{attractive } (\lambda = -4/3); \ \boldsymbol{R} = (8): \text{repulsive } (\lambda = +1/6) \ , \qquad (19.14)
$$

2. quark-quark

$$
\boldsymbol{R} = (\bar{\mathbf{3}}): \text{attractive } (\lambda = -2/3); \ \boldsymbol{R} = (\mathbf{6}): \text{repulsive } (\lambda = +1/3) \ . \qquad (19.15)
$$

In conclusion: quark-antiquark pairs may bind in colour singlet mesons, while diquarks may bind: (i) to another quark, to make a colour-singlet baryon, or (ii) to an antidiquark, to make a colour-singlet tetraquark.

Non perturbatively, colour lines of force are supposed to condense in strings going from quarks to antiquarks, see Section 18.1.3, or from quark to quark, when colour force is attractive. In this case, it has been argued that the

[2]The similarity with the familiar system of two particles with spins $s_1$ and $s_2$ and total spin $S = s_1 + s_2$ is evident. In this case, the quadratic Casimir operator of total spin is: $C_2(\mathbf{S}) = \mathbf{S}^2 = S(S+1)$ and $2(s_1 \cdot s_2) = [S(S+1) - s_1(s_1+1) - s_2(s_2+1)]$.

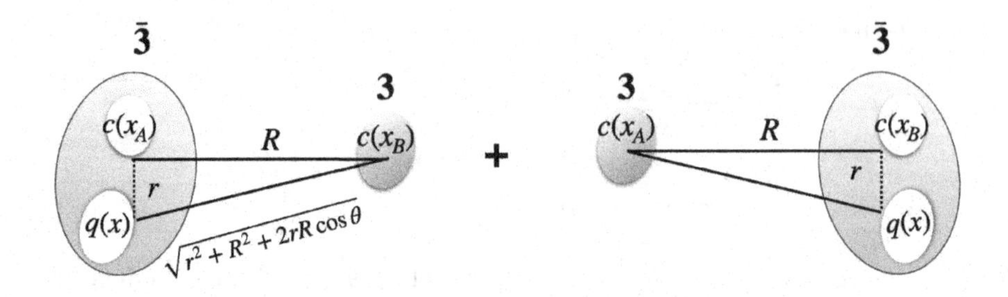


Figure 19.1 Schematic representation of the doubly charmed baryon within the BO approximation.

strenght of the attractive string (the constant $k$ in the Cornell potential) is proportional to $|\lambda|$, with $\lambda$ given in (19.12).

Residual quark-quark or quark-antiquark interactions are local chromomagnetic, spin-spin, interactions of the form:

$$H_{ij}(s_1, s_2) = -\frac{4\pi\alpha_s}{3m_i m_j}(T_1^A T_2^A)[2\ (\mathbf{s}_1 \cdot \mathbf{s}_2)]\ \delta^{(3)}(\mathbf{x}_1 - \mathbf{x}_2)\ . \qquad (19.16)$$

The formula given in Eq. (18.22) follows since $\lambda = -4/3$ for quark-antiquark in colour singlet, as indicated above.

## 19.3 THE DOUBLY CHARMED BARYON

In 2020 the LHCb Collaboration found convincing evidence for a doubly charmed, doubly charged baryon $\Xi_{cc}^{++}(3621)$ of mass $M(\Xi_{cc}) = 3620.6 \pm 1.5$ in the decay channel[3]

$$\Xi_{cc}^{++}(3621) \to \Xi_c^+ + \pi^+ \qquad (19.17)$$

Quark composition is $(ccu)$ and, given that the $cc$ pair must be in total spin $S_{cc} = 1$, as noted before, the baryon total angular momentum may be $J = 1/2, 3/2$. There is no convincing evidence of a lighter baryon of this kind[4] and we assume provisionally $J = 1/2$.

### 19.3.1 The BO Approximation for $\Xi_{cc}^{++}$

We assume the $c$ quarks at fixed positions $\mathbf{x}_A$ and $\mathbf{x}_B$ and in analogy with the BO approximation for the $H_2^+ = (ppe)$, see Ref. [103], consider the

[3]According to GIM [28] the charm quark decays weakly as $c \to s + u + \bar{d}$. In terms of quarks, the decay (19.17) reads $(ccu) \to (csu) + (u\bar{d})$.

[4]Previous evidence from the SELEX experiment of a $\Xi_{cc}^+(3520)$ has not been confirmed, see PdG 2023 [12].

state with the light quark bound to $c(\mathbf{x}_A)$ with a Cornell-like potential, see Eq. (19.15)

$$V_{\boldsymbol{A}} = -\frac{2}{3}\frac{\alpha_s}{r} + \frac{1}{2}kr + V_0;\ r = |\boldsymbol{x} - \boldsymbol{x}_A| \ , \tag{19.18}$$

where $V_0$ is the constant introduced in the Cornell potential in Sect. 18.2.2, to fix the zero of the energy.

Of course, one has to also consider the degenerate state, with $q$ bound to $c(\mathbf{x}_B)$. We assume the light quark ground level, in the presence of the two heavy sources, to be the superposition of the two states, see Fig. 19.1:

$$f_0(\boldsymbol{x}) = \frac{\psi(\boldsymbol{x}) + \phi(\boldsymbol{x})}{\sqrt{2(1+S)}} = \frac{1}{\sqrt{4\pi}}\,\frac{R(|\boldsymbol{x} - \boldsymbol{x}_A|) + R(|\boldsymbol{x} - \boldsymbol{x}_B|)}{\sqrt{2(1+S)}} \ , \tag{19.19}$$

where $\psi$ and $\phi$ are the wave functions of the two states. We indicate explicitly in Fig. 19.1 that the radial wave functions are *the same function* of two different variables: the distances of $q$ from the source in $\boldsymbol{x}_A$ or in $\boldsymbol{x}_B$. In correspondence, $\psi$ and $\phi$ are not orthogonal and we denote by $S$ the overlap integral:

$$S = \int d^3x\ \psi(x)\phi(x)\ \text{(real wave functions assumed)}\ , \tag{19.20}$$

a function of the distance between the two sources $R = |\boldsymbol{x}_A - \boldsymbol{x}_B|$, which appears in the normalization of the state (19.19).

***The cq orbital.*** In the jargon of molecular physics, the *cq* wave fuction is an *orbital.* It is determined by the Schrödinger equation

$$(-\frac{1}{2\mu_q}\nabla^2 + V_{\boldsymbol{A}} - V_0)\psi = E_0\psi \ , \tag{19.21}$$

where $\mu_q$ is the reduced light quark mass, $V_{\boldsymbol{A}}$ the potential in (19.18) and we have suppressed the constant $V_0$. The energy is then $E_0 + V_0$ plus the quark rest masses.

**First order correction to the light quark energy.** We consider now the effect of the interaction of the light quark with the other heavy source:

$$H_{\text{pert}} = -\frac{2}{3}\alpha_S\,\frac{1}{|\boldsymbol{x} - \boldsymbol{x}_B|} \ . \tag{19.22}$$

Treating (19.22) as a first order perturbation, the energy of the light quark is:

$$\mathcal{E}(R) = E_0 + V_0 + \Delta E(R) \ , \tag{19.23}$$

where

$$\begin{aligned}\Delta E(R) &= \langle f_0|H_{\text{pert}}|f_0\rangle = \\ &= -\frac{2\alpha_S}{3}\,2\,\frac{1}{2(1+S)}\,[I_1(R) + I_2(R)] \ .\end{aligned} \tag{19.24}$$

The factor 2 in the numerator of Eq. (19.24) arises because there are two equal contributions, from the souces $A$ and $B$, see Fig. 19.1.

The $I_{1,2}$ are functions of $R$ defined in terms of the orbital wave functions $\psi$ and $\phi$:

$$I_1(R) = \int d^3\xi\, |\psi(\xi)|^2 \frac{1}{|\boldsymbol{\xi} - \boldsymbol{x}_B|}\,, \tag{19.25}$$

$$I_2(R) = \int d^3\xi\, \psi(\xi)\phi(\xi)\, \frac{1}{|\boldsymbol{\xi} - \boldsymbol{x}_B|}\,, \tag{19.26}$$

where the vector $\boldsymbol{\xi}$ originates from $A$, taken in the origin, and $|\boldsymbol{x}_B| = R$. Analytic expressions for $S, I_{1,2}$ are given in [103] for the hydrogen wave functions. We shall evaluate them numerically.

The complete expression of $\mathcal{E}$, to be inserted in the BO potential, requires the addition of quark masses, so that:

$$\mathcal{E}(R) = \Delta E(R) + C\,, \tag{19.27}$$

with $C = E_0 + V_0 +$ quark masses.

**Boundary condition at $R = 0$.** For $R \to 0$ the two charm quarks become a single colour $\bar{\mathbf{3}}$ source and the system reproduces. to all effects, the colour fields configuration of a charmed meson, except for the larger mass, $2M_c$ of the source. This is in essence the doubly heavy quark-single heavy antiquark symmetry introduced by several authors in the study of doubly heavy baryons [108–110]. Charmed meson masses, with spin effects subtracted, are well reproduced by the sum $M_c^{mes} + M_q^{mes}$ with[5]

$$\begin{aligned} &M_c^{mes} = 1.667;\; M_q^{mes} = 0.308\,, \\ &2M_c^{mes} + M_q^{mes} = 3.642 \text{ (masses in GeV)}\,, \end{aligned} \tag{19.28}$$

and the boundary condition reads:

$$\Delta E(0) + C = (M_c^{mes} + M_q^{mes}) + M_c^{mes}\,, \tag{19.29}$$

that is

$$C = -\Delta E(0) + 2M_c^{mes} + M_q^{mes}\,, \tag{19.30}$$

and we obtain:

$$\mathcal{E}(R) = \Delta E(R) - \Delta E(0) + 2M_c^{mes} + M_q^{mes}\,. \tag{19.31}$$

**Boundary condition at $R = \infty$.** At distances much larger than the radius of the orbital $c(\boldsymbol{x}_A)q$, where forces due to gluons converge to zero like

[5]Charmed baryon masses require different charm and light quark masses, see e.g. Ref. [100], the difference being interpreted as due to the different colour field configurations in the two cases. For comparison, $2M_c^{bar} + M_q^{bar} = 3.782$ GeV [100].

$1/R$, the charm quark in $\mathbf{x}_B$ sees the other two particles as a $\bar{\mathbf{3}}$ source with which it combines to form a colour singlet. Therefore, the $cc$ interaction must contain at large distances a confining string potential, with the same strength as the Cornell potential introduced in (18.17).

To take this effect into account, we add to the BO potential a linearly rising term determined by the string tension $k$ of charmonium, see Sect. 18.2.1

$$V_{\rm conf}(R, R_0) = k \times (R - R_0) \times \theta(r - R_0) \; , \qquad (19.32)$$

and leave, for the moment, the onset point $R_0$ as a free parameter. The Born-Oppenheimer potential then reads

$$V_{\rm BO}(R) = -\frac{2}{3}\frac{\alpha_s}{R} + \Delta E(R) - \Delta E(0) + V_{\rm conf}(R, R_0) + \\ +2M_c^{mes} + M_q^{mes} \; . \qquad (19.33)$$

**Hyperfine interactions.** We consider first the interaction of the light quark spin, $s$, with the spin of the heavy sources, $S_A$ ans $S_B$, described by Eq. (19.16). With $T_q^A T_c^A = -2/3$, Eq. (19.15) we write

$$< f_0|H_{hf}(s, s_A) + H_{hf}(s, s_B)|f_0 >= \\ = +\frac{8\pi\alpha_s}{9M_qM_c}\frac{1}{2(1+S)}\left[\psi(\mathbf{x}_A) + \phi(\mathbf{x}_A)]^2\right]\; 2(\mathbf{s}\cdot\mathbf{s}_A)] + (A \to B) \; .$$

We set $|\boldsymbol{x}_A| = 0$, $|\boldsymbol{x}_B| = R$ and note that $\phi(\boldsymbol{x}_A) = \psi(\boldsymbol{x}_B) = \psi(R)$ and $\phi(\mathbf{x}_B) = \psi(\mathbf{x}_A) = \psi(0)$; thus, we find

$$< f_0|H_{hf}(s, s_A) + H_{hf}(s, s_B|f_0 >= \\ = +\frac{8\pi\alpha_s}{9M_qM_c}\frac{1}{2(1+S)}\left[\psi(0) + \psi(R)]^2\right]\; 2\mathbf{s}\cdot(\mathbf{s}_A + \mathbf{s}_B) \; . \qquad (19.34)$$

For a $\Xi_{cc}$ with spin $J = 1/2$, we have $2\boldsymbol{s}\cdot(\boldsymbol{s}_A + \boldsymbol{s}_B) = -2$ and the spin potential, to be added to the BO potential (19.33) is

$$V_{BOspin} = +\frac{16\pi\alpha_s}{9M_qM_c}\frac{1}{2(1+S)}\left[\psi(0) + \psi(R)]^2\right] \; . \qquad (19.35)$$

The complete BO potential becomes

$$V_{\rm BOs}(R) = -\frac{2}{3}\frac{\alpha_s}{R} + \Delta E(R) - \Delta E(0) + V_{\rm conf}(R, R_0) + V_{BOspin} + \\ +2M_c^{mes} + M_q^{mes} \; . \qquad (19.36)$$

Finally, Eq. (19.16) applied to the $cc$ hyperfine interaction, taking into account that $2(\boldsymbol{s}_A \cdot \boldsymbol{s}_B) = +1/2$ for total $cc$ spin $S_{cc} = 1$, gives a first order

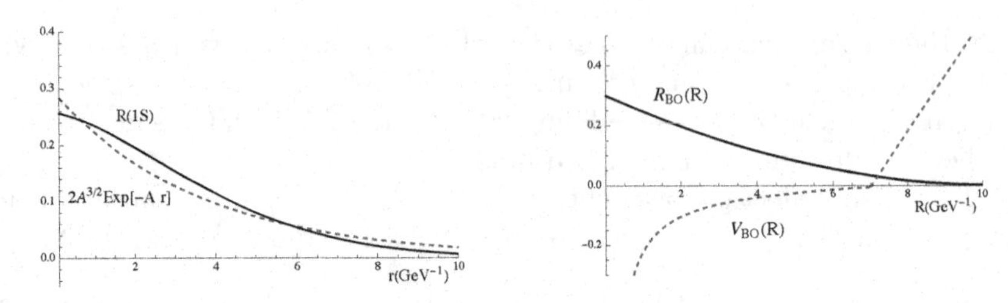


Figure 19.2 Left: Single exponential (dashed curve) fitted to the radial wave function of the $(qc)$ orbital (solid curve). Right: BO potential (no spin) and radial BO wave function.

correction to the eigenvalue $E_{BO}$ of the Schrödinger equation with the spinless potential (19.33):

$$E_{hf}(s_A, s_B) = +\frac{4\pi\alpha_s}{9M_c^2}|\Psi(0)|^2 , \tag{19.37}$$

to be computed from the corresponding BO eigenfunction $\Psi$ [6].

## 19.3.2 Numerical Results

The radial wave function $R(x)$ of the $cq$ orbital can be obtained by solving numerically Eq. (19.21). To reduce the time length of successive calculations it is useful to approximate $R(x)$ with a single exponential Exp$[-Ar]$, finding the value of $A$ from a best fit to the numerical determination (see Ref. [107]). Fig. 19.2 (left) shows the error made by the hydrogen like approximation (see Eq. (B.45))[7]:

$$R(x) = 2(A)^{3/2}e^{-Ax};\ A = 0.277\ \text{GeV}. \tag{19.39}$$

[6] The program Schrödinger.nb gives the reduced radial wave function $\chi(r)$ and

$$|\Psi(0)|^2 = \frac{1}{4\pi}\lim_{r\to 0}|\frac{\chi(r)}{r}|^2 . \tag{19.38}$$

[7] Correspondingly, the integrals to compute $\Delta E$ in the VBO potential become:

$$I_1(R) = 2A^3\int_0^\infty dr\int_{-\pi}^{+\pi} d\theta \sin\theta\big[r^2\frac{e^{-2Ar}}{\sqrt{r^2-2rR\cos\theta+R^2}}\big] ,$$

$$I_2(R) = 2A^3\int_0^\infty dr\int_{-\pi}^{+\pi} d\theta \sin\theta\big[r^2\frac{e^{-Ar}e^{-A\sqrt{r^2-2rR\cos\theta+R^2}}}{r}\big] ,$$

$$S(R) = 2A^3\int_0^\infty dr\int_{-\pi}^{+\pi} d\theta \sin\theta\big[r^2e^{-Ar}e^{-A\sqrt{r^2-2rR\cos\theta+R^2}}\big] ,$$

and the hyperfine $cq$ interaction potential is

$$< f_O|H_{hf}(s,s_A)+H_{hf}(s,s_B|f_O>= +\frac{8\pi\alpha_s}{9M_qM_c}\frac{2A^3(1+e^{-AR})^2}{1+S(R)}\ 2\mathbf{s}\cdot(\mathbf{s}_A+\mathbf{s}_B) .$$

Table 19.1 Eigenvalues of the Born-Oppenheimer equation without and with the hyperfine potential and the resulting $M(\Xi_{cc})$ mass, for $R_0 = 7 \pm 2$ GeV$^{-1}$ see text . Energy and mass in MeV.

| $R_0$ $(GeV^{-1})$ | $E_{BO}$, Eq. (19.33) | $E_{BOs}$, Eq. (19.36) | $E_{hf}(cc)$ | $2M_c^m + M_q^m$ | $M(\Xi_{cc})$ |
|---|---|---|---|---|---|
| 5 | 48 | 34 | 1.58 | 3642 | 3678 |
| 7 | 22 | 9 | 1.1 | " " | 3652 |
| 9 | 11 | 1.6 | 0.9 | " " | 3645 |

We report in Fig. 19.2 (right) the no-spin BO potential, Eq. (19.33), and the corresponding radial wave function, $\Psi(R)$.

Table 19.1 summarises the eigenvalues of the Born-Oppenheimer equation without, Eq. (19.33), and with, Eq. (19.36), the $qc$ hyperfine potential for $R_0 = 7 \pm 2$ GeV$^{-1}$. The hyperfine $cc$ interaction is computed with Eq. (19.37). To obtain $M(\Xi_{cc})$ the quark mass combination $2M_c^{mes} + M_q^{mes}$ has been added. In conclusion, we find:

$$M(\Xi_{cc}^{++}, J = 1/2) = 3652^{+26}_{-7} \text{ for } R_0 = 7 \pm 2 \text{ GeV}^{-1} , \tag{19.40}$$

to be compared to $M(\Xi_{cc}^{++})_{exp} = 3621$. The agreement with the observed $\Xi_{cc}^{++}$ mass is reasonable (given the error of the BO approximation estimated in the next Section) at the expense, however, of having introduced the ad-hoc parameter $R_0$. The real measure of the effectiveness of the Born-Oppenheimer approximation has to wait for the comparison of this formula with the different levels of doubly charmed baryons predicted by the quark model, as it was the case for the charmonium spectrum vs. the Cornell potential.

### 19.3.3 About the BO Approximation Error in QCD

Recall that we have characterised the BO approximation error by the ratio

$$\Lambda = \frac{1/a}{1/b} , \tag{19.41}$$

where $a$ and $b$ are the lengths over which $f_0$ and $\Psi$ show an appreciable variation.

The length $a$ is simply the radius of the $cq$ orbital: $1/a = A \sim 0.3$ GeV, i.e. $a \sim 0.7$ fm.

The length $b$ has to be formed from the dimensional quantities over which the Born-Oppenheimer equation (19.33) depends. In the case of double heavy baryons Eq. (19.33) depends on $M$, $A$ and on the string tension $k$, which has dimensions of GeV$^2$. The simplest possibility is $1/b = (AkM)^{+1/4}$; that is:

$$\Lambda = A^{3/4}(kM)^{-1/4} , \tag{19.42}$$

which is 0.53 for charm, using Eqs. (18.39) and (19.28). For convenience we have included quark masses in $V_{BO}$, but it is worth noticing that the error we are estimating is the error on binding energies, which turns out to be of the order of 50 MeV or smaller in absolute value. So, the BO approximation error corresponding to (19.42) is expected to be of the order of 25–30 MeV.

## 19.4 PROBLEMS FOR CHAPTER 19

### Sect. 18.3

1. Consider the decay $\Xi_{cc}^{++} \to \Xi_c^+ + \pi^+$, with both baryons with $J^P = 1/2^+$ and $\pi$ meson with $J^P = 0^-$. Find the values of the orbital angular momentum of $\pi^+$ allowed by total angular momentum conservation.

2. Same problem, assuming a $\Xi_{cc}^{++}$ with $J^P = 3/2^+$.

3. Which of the orbital angular momenta found in 1. and 2. correspond to parity conserving decays?

APPENDIX A

# Basic Elements of Quantum Mechanics

This appendix summarises the basic principles of quantum mechanics. The objective above all is to recall the most important ideas and to define the notation which will be used in what follows. For a deeper discussion of the physical basis of the theory, the reader is invited to consult the work of Dirac [111]. A concise and modern discussion of the fundamentals and the philosophical difficulties of quantum mechanics can be found in the book by Bell [112].

## A.1 THE PRINCIPLE OF SUPERPOSITION

At a given instant of time, the states of a quantum system are represented by the elements of an abstract space, $\mathcal{H}$. These elements will be denoted using Dirac notation:

$|A>$, $|B>$, $|C>\ldots$, known as *kets*, corresponding to the physical states A, B, C, etc. of the system.

The mathematical structure of $\mathcal{H}$ is fixed by the *principle of superposition*, according to which, if $|A>$ and $|B>$ represent two possible states of the system, other states can always be expressed in terms of an arbitrary linear combination of $|A>$ and $|B>$ with complex coefficients $\alpha$ and $\beta$:

$$|C>=\alpha|A>+\beta|B> \tag{A.1}$$

$\mathcal{H}$ is a complex vector (or linear) space, in general with an infinite number of dimensions.

Characterisation of the physical states necessarily implies the experimental determination of values of one or more observable quantities of the states themselves. Let us suppose that the kets $|A>$ and $|B>$ which appear in equation (A.1) correspond respectively to two distinct values which we call $a$ and $b$, of the same observable $X$ (for example the energy). The physical

DOI: 10.1201/9781003436263-A

interpretation of the state $|C>$ as a superposition of $|A>$ and $|B>$ is as follows:

- The superposition of $|A>$ with itself (the case $\beta = 0$) gives rise to the same physical state; $|A>$ and $\alpha|A>$ represent the same state for all values of $\alpha \neq 0$.
- In the case in which $\alpha$ and $\beta$ are both $\neq 0$, the result of a measurement of $X$ on the state $|C>$ can be $a$ or $b$; *only one of these two values can be the outcome of the measurement.*
- It is not possible to predict which of the two values will be the result of a given measurement. However, if the states $|A>$ and $|B>$ are *correctly normalised* (in the sense defined below), the frequencies with which the outcomes $a$ and $b$ occur are in the ratio $|\alpha|^2/|\beta|^2$.

To connect the superposition coefficients to the probability of different measurement results, normalisation of the vectors which characterise the states is necessary. This requires that the scalar product of $|A>$ and $|B>$ in $\mathcal{H}$ can be defined, which we denote as:

$$< B|A >=< A|B >^* \tag{A.2}$$

(the asterisk implies the operation of complex conjugation).

The scalar product $< B|A >$ must be **linear** in $A$ and hence **antilinear** in $B$

$$< C|\alpha A + \beta B >= \alpha < C|A > +\beta < C|B > \tag{A.3}$$

$$< \alpha A + \beta B|C >= \alpha^* < A|C > +\beta^* < B|C > \tag{A.4}$$

as well as positive definite:

$$\begin{aligned} &< A|A > \; > 0, \\ &< A|A >= 0 \quad \text{if, and only if,} \; |A >= 0. \end{aligned}$$

With this final constraint, the space $\mathcal{H}$ is a *Hilbert space*, and the quantity $< A|A >$ is the square of the norm of the vector $|A >$.

Besides $\mathcal{H}$, we can consider a new vector space, the *dual space* $\mathcal{H}^*$, defined as the space of linear functionals (with complex values) defined on $\mathcal{H}$. It is not difficult to be convinced that the elements of $\mathcal{H}^*$ have a one-to-one mapping onto those of $\mathcal{H}$. In effect, according to a well known theorem due to Riesz, every linear functional $f(|A >)$ can be written as:

$$f(|A >) =< f|A > \tag{A.5}$$

with $|f >$ a fixed ket. Therefore we can make the functional $f(|A >)$ in $\mathcal{H}^*$ correspond to $|f >$ in $\mathcal{H}$. (A.5) explains the Dirac notation for the elements of $\mathcal{H}^*$, according to which we denote with:

$$< f| \tag{A.6}$$

the element of $\mathcal{H}^*$ which corresponds to the ket $|f>$. The vectors $<f|$ are known by the name *bra*. As a result of (A.4), the bra $<f|$ depends antilinearly on the ket $|f>$ and the scalar product (A.2) can be interpreted as the product of the bra $<B|$ with the ket $|A>$ (bra* ket = bracket = product, from which the terms bra and ket are derived).

Turning to the probabilistic interpretation of the results of the measurement of $X$ on the state $|C>$ in equation (A.1), we assume that:

- If the vectors $|A>$ and $|B>$ have unity norm, the probability of obtaining $a$ (or $b$) from the measurement of $X$ on $|C>$ is equal to $|\alpha|^2$ (or $|\beta|^2$).

## A.2 LINEAR OPERATORS

On the vector space $\mathcal{H}$ we can define linear operators, such that to every vector of $\mathcal{H}$ (contained in an appropriate region of $\mathcal{H}$ itself) we can associate another vector, which is related in a linear way to the first.[1]

$$|B> = X|A> \tag{A.7}$$

$$X(\alpha|A> + \beta|B>) = \alpha X|A> + \beta X|B> . \tag{A.8}$$

Given $X$, we can consider the complex number:

$$<A|X|B>^*$$

for every $<A|$ and $|B>$. This complex number depends antilinearly on $|B>$. Therefore we can write:

$$<A|X|B>^* = <B|V>$$

.

Moreover, because $|V>$ depends linearly on the ket $|A>$ which corresponds to bra $<A|$, we can also write:

$$<A|X|B>^* = <B|V> = <B|X^\dagger|A> \tag{A.9}$$

where $X^\dagger$ is a new linear operator, associated unambiguously to $X$ by (A.9), and known as the adjoint (or hermitian conjugate) operator of $X$. Clearly the relations:

$$(X^\dagger)^\dagger = X \tag{A.10}$$

$$(\alpha X)^\dagger = \alpha^* X^\dagger \tag{A.11}$$

[1] In what follows we will assume for simplicity that the operators are defined in all $\mathcal{H}$.

apply for every complex number $\alpha$.

An operator $H$ is hermitian, or self-adjoint, if $X^\dagger = X$. Hermitian operators have some properties with respect to their *eigenvalues* and *eigenvectors* which are crucial for the development of development of quantum mechanics. We recall that the eigenvalue of a linear operator X is a (real or complex) number $\lambda$ for which the equation:

$$X|v> = \lambda\,|v> \tag{A.12}$$

allows solutions $|v> \neq 0$. In this case we say that the eigenvector $|v>$ *belongs* to the eigenvalue $\lambda$ and we write $|v> = |\lambda, a, b, \ldots >$ ($a, b, \ldots$ are parameters which distinguish the eigenvectors which belong to the same eigenvalue).

The following properties apply:

- The *eigenvalues* of a hermitian operator are always *real numbers.*
- Eigenvectors, $|h'>$ and $|h''>$ which belong to two distinct eigenvalues, $h'$ and $h''$, are orthogonal to each other:

$$< h'|h'' > = 0, \quad \text{if}\, h' \neq h''. \tag{A.13}$$

- The eigenvectors of a hermitian operator form a *complete basis* in $H$.

Assuming, for simplicity, that $H$ has a discrete spectrum of eigenvalues, the final property implies that every vector can be expressed in the basis of the normalised eigenvectors of $H$:

$$|A> = \sum_n c_n |h_n> \tag{A.14}$$

with:

$$\begin{aligned} H|h_n> &= h_n|h_n> \\ < h_n|h_m> &= \delta_{n,m} \end{aligned}$$

from which it follows that:

$$\begin{aligned} c_n &= < h_n|A> \\ < A|A> &= \sum_n |c_n|^2. \end{aligned} \tag{A.15}$$

A useful concept, in connection to the base set of eigenvectors of $H$, is that of projection of one or more states. The projection operator of a given vector, for example of $|h_1>$, is one which has the property that:

$$P^2 = P \tag{A.16}$$

$$P|h_1> = |h_1> \tag{A.17}$$

$$P|V> = 0, \quad \text{if} \,< h_1|V> = 0 \tag{A.18}$$

i.e. if $|V>$ is orthogonal to $|h_1>$. It is easy to see that $P$ can be written formally as:

$$P = |h_1 >< h_1| \tag{A.19}$$

(assuming $< h_1|h_1 >= 1$). Thus:

$$\begin{aligned} P^2 &= |h_1 >< h_1|h_1 >< h_1| = |h_1 >< h_1| = P \\ P|h_1 > &= |h_1 >< h_1|h_1 >= |h_1 > \\ P|V > &= |h_1 >< h_1|V >= 0, \text{ if } < h_1|V >= 0. \end{aligned}$$

The projection operator in a multidimensional space, defined by a certain number of vectors orthogonal to each other, is simply given by the sum of the projections of single vectors:

$$P = \sum_n |h_n >< h_n| \tag{A.20}$$

and the *completeness condition* for the base set of eigenvectors of $H$ is expressed as:

$$\sum_n |h_n >< h_n| = 1 \tag{A.21}$$

(1 indicates the identity operator). Equation (A.15) is obtained formally from (A.21) in the following way:

$$|A >= 1|A >= \sum_n |h_n >< h_n|A >= \sum_n c_n|h_n >$$

.

Similarly,

$$< A|A >=< A|1|A >= \sum_n < A|h_n >< h_n|A >= \sum_n |c_n|^2.$$

## A.3 OBSERVABLES AND HERMITIAN OPERATORS

The relevance of the concepts just illustrated lies in the fact that in quantum mechanics every observable quantity, $\boldsymbol{O}$, is represented by a hermitian operator, $O$.

The eigenvectors of $O$ represent the physical states for which $\boldsymbol{O}$ assumes a definite value, equal to the eigenvalue which corresponds to the eigenvector in question. The spectrum of eigenvalues of $O$ therefore defines the set of possible results from a measurement of $\boldsymbol{O}$. The considerations of Section A.2 allow description of the results of a measurement of $\boldsymbol{O}$ on a general state $|A>$ in the following way.

- The measurement of $\boldsymbol{O}$ on $|A>$ gives one of the eigenvalues of $O$, for example $h_n$, as the result with a probability proportional to the squared modulus of the corresponding coefficient of the expansion, $c_n$.

- The sum of the probabilities for all possible cases must be equal to one. If $|A>$ is normalised to unity, equation (A.15) shows that:

$$P(h_n \text{ on } |A>) = |c_n|^2 = |<h_n|A>|^2. \tag{A.22}$$

  This result gives a physical meaning to the scalar product of two vectors. Let us suppose that $|A>$ and $|B>$ correspond to states in which two different quantities assume definite values, $x_a$ for the observable $X$ in $|A>$ and $y_b$ for the observable $Y$ in $|B>$. If $|A>$ and $|B>$ are normalised to unity, the probability that a measurement of $Y$ in $|A>$ has the result $y_b$ is given by the squared modulus of the corresponding scalar product:

$$P(y_b \text{ on } |A>) = |<B|A>|^2. \tag{A.23}$$

  For this reason, the scalar product is also known as the probability amplitude.

- The average result of many measurements of $O$ on $|A>$[2] is given by the formula:

$$<O>_A = \sum_n h_n P(h_n \text{ on } |A>) = \tag{A.24}$$

$$= \sum_n h_n |c_n|^2 = \sum_n h_n <A|h_n><h_n|A> =$$

$$= <A|O(\sum_n |h_n><h_n|)|A> = <A|O|A>$$

  by virtue of equation (A.21). For this reason, the matrix element of $O$ between $<A|$ and $|A>$ (diagonal matrix element) is also called the *expectation value* of $O$ on $A$.

## A.4 THE NON-RELATIVISTIC SPIN 0 PARTICLE

The spinless non-relativistic particle provides the simplest concrete example of the ideas explained above. The fundamental observables of this system are the coordinate, $x$, (for simplicity we consider just a single spatial dimension and we set $\hbar = 1$) and the conjugate momentum, $p$, with the commutation rules:

$$[x, p] = i. \tag{A.25}$$

[2]This means that each of these measurements is carried out on a new replica of the system, prepared in the state $|A>$ using appropriate experimental apparatus.

We can introduce the eigenstates of $x$ and $p$:

$$x|x>= x|x>$$
$$p|p>= p|p>$$

with the normalisation and completeness conditions:

$$<x|x'>= \delta(x - x') \tag{A.26}$$

$$\int dx|x><x| = 1 \tag{A.27}$$

$$<p|p'>= \delta(p - p') \tag{A.28}$$

$$\int dp|p><p| = 1. \tag{A.29}$$

The *wave function* of a given ket, $|A>$, is the component of $|A>$ in the base set of the eigenvectors $|x>$:

$$\psi_A(x) =<x|A> \tag{A.30}$$

with:

$$<B|A>= \int dx <B|x><x|A>= \int dx\psi_B(x)^*\psi_A(x) \tag{A.31}$$

from which:

$$1 =<A|A>= \int dx <A|x><x|A>= \int dx|\psi_A(x)|^2 \tag{A.32}$$

in agreement with the interpretation of $|\psi_A(x)|^2$ as the probability density to find the particle between $x$ and $x + dx$.

From equation (A.30) and from the commutation relation (A.25), one finds [111]:

$$(x\psi_A)(x) = x\psi_A(x)$$
$$(p\psi_A)(x) = -i\frac{d}{dx}\psi_A(x).$$

The wave function of the ket $|p>$ is obtained directly from this equation:

$$<x|p>= \frac{1}{\sqrt{2\pi}}e^{ipx} \tag{A.33}$$

with the normalisation factor determined from (A.28).

### A.4.1 Translations and Rotations

The results just illustrated permit us to characterise the operations of *spatial translation.*

We define the translation operators, $U(a)$, as those which, applied to a ket $|A>$, transform it into the ket which corresponds to the translated state, which is the state obtained by translating, through a length $a$, all the relevant apparatus necessary to produce the state $|A>$. The homogeneity of space requires that $U(a)$ should be a unitary operator (see Section 10.4)

$$U(a)^{\dagger} = U(a)^{-1} = U(-a). \tag{A.34}$$

From the definition of $U$, ignoring irrelevant phase factors, it follows that:

$$U(a)|x> = |x+a> \tag{A.35}$$

or:

$$<x|U(a)^{\dagger} = <x+a| \tag{A.36}$$

from which one obtains:

$$\begin{aligned} <x|U(a)^{\dagger}|p> = <x+a|p> = (2\pi)^{-1/2}e^{ip(x+a)} = \\ = e^{ipa} <x|p> = <x|e^{ipa}|p>. \end{aligned} \tag{A.37}$$

Because this relation must hold for every $|p>$ and for every $|x>$, we find:

$$U(a) = e^{-ipa}. \tag{A.38}$$

For infinitesimal transformations:

$$U(a) \equiv 1 - ipa. \tag{A.39}$$

*The momentum is the infinitesimal generator of spatial translation.* For a general state, $|A>$:

$$(U(a)\psi_A)(x) = <x|U(a)|A> = <x-a|A> = \psi_A(x-a). \tag{A.40}$$

The transition to the three-dimensional case allows discussion of spatial rotation symmetry. In analogy to the case of translation, we define the unitary operators, $U(R)$, according to the relation:

$$U(R)|\boldsymbol{x}> = |R\boldsymbol{x}> \tag{A.41}$$

where $R$ is the (orthogonal $3 \times 3$) rotation matrix, and moreover:

$$U(R)^{\dagger} = U(R)^{-1} = U(R^{-1}). \tag{A.42}$$

For rotations through an angle $\theta$ around the $z$ axis:

$$\begin{aligned}(R\boldsymbol{x})_x &= \cos\theta x - \sin\theta y\\ (R\boldsymbol{x})_x &= \sin\theta x + \cos\theta y\\ (R\boldsymbol{x})_z &= z.\end{aligned}$$

Proceeding as in (A.40) we find, for a general ket:

$$\begin{aligned}(U(R)\psi_A)(\boldsymbol{x}) &=< \boldsymbol{x}|U(R)|A>=< R^{-1}\boldsymbol{x}|A>=\\ &= \psi_A(R^{-1}\boldsymbol{x}).\end{aligned} \tag{A.43}$$

For infinitesimal rotations about the $z$ axis, we therefore find:

$$\begin{aligned}< \boldsymbol{x}|U(R)|A>&= \psi_A(\boldsymbol{x}) - \theta(x\frac{d}{dy} - y\frac{d}{dx})\psi_A(\boldsymbol{x}) =\\ &=< \boldsymbol{x}|[1 - i\theta(\boldsymbol{x}\times\mathbf{p})_z]|A>\end{aligned}$$

from which

$$U(R) = 1 - i\theta L_z \tag{A.44}$$

where $L_z$ is the angular momentum component along the $z$ axis:

$$\boldsymbol{L} = \boldsymbol{x}\times\boldsymbol{p}. \tag{A.45}$$

More generally, for an arbitrary rotation,

$$U(R) = e^{-i\mathbf{n}\cdot\boldsymbol{L}}$$

where $\mathbf{n}$ is a three-dimensional vector whose direction identifies the rotation axis while $|\boldsymbol{n}| = \theta$ is the rotation angle.

Therefore the generators of infinitesimal rotations are the angular momentum components. Making use of the canonical relations, (A.25), it can be verified that the operators $\boldsymbol{L}$ obey the commutation relations:

$$[L_i, L_j] = i\epsilon_{ijk}L_k. \tag{A.46}$$

It is important to be convinced that the commutation rules (A.46) are a direct consequence of the structure of the rotation group and the requirement that:

$$U(R_1)U(R_2) = U(R_1R_2) \tag{A.47}$$

for arbitrary infinitesimal rotations, $R_{1,2}$.

To obtain this result, we make use of the fact that every orthogonal $3\times 3$ matrix, $R$, can be written as:

$$R = e^{-i\mathbf{n}\cdot\boldsymbol{T}} \tag{A.48}$$

where $\boldsymbol{n}$ is the same vector which appears in equation (A.46) and the $3 \times 3$ matrices, $\boldsymbol{T}$, are given by:[3]

$$(T_k)_{ij} = i\epsilon_{ikj}. \tag{A.49}$$

An explicit calculation shows that the matrices $T_k$ satisfy commutation relations equivalent to (A.46):

$$[T_i, T_j] = i\epsilon_{ijk}T_k \tag{A.50}$$

For infinitesimal $\mathbf{n}_1$ and $\mathbf{n}_2$[4]:

$$\begin{aligned} R_1R_2 &= e^{-i\boldsymbol{n}_1\cdot\boldsymbol{T}}e^{-i\boldsymbol{n}_2\cdot\boldsymbol{T}} \\ &= e^{-i[(\boldsymbol{n}_1+\boldsymbol{n}_2)\cdot\boldsymbol{T}+(i/2)[\boldsymbol{n}_1\cdot\boldsymbol{T},\boldsymbol{n}_2\cdot\boldsymbol{T}]+\ldots]} \\ &= e^{-i[(\boldsymbol{n}_1+\boldsymbol{n}_2-(1/2)\boldsymbol{n}_1\times\boldsymbol{n}_2)\cdot\boldsymbol{T}+\ldots]} \end{aligned} \tag{A.51}$$

where the dots represent higher order terms in $\boldsymbol{n}_1$ or $\boldsymbol{n}_2$. On the other hand, we have:

$$\begin{aligned} U(R_1)U(R_2) &= e^{-i\boldsymbol{n}_1\cdot\boldsymbol{L}}e^{-i\boldsymbol{n}_2\cdot\boldsymbol{L}} = e^{-i[(\boldsymbol{n}_1+\boldsymbol{n}_2)\cdot\boldsymbol{L}+(i/2)[\boldsymbol{n}_1\cdot\boldsymbol{L},\boldsymbol{n}_2\cdot\boldsymbol{L}]+\ldots]} = \\ &= U(e^{-i[(\boldsymbol{n}_1+\boldsymbol{n}_2)\cdot\boldsymbol{T}\boldsymbol{T}+(1/2)[\boldsymbol{n}_1\cdot\boldsymbol{T},\boldsymbol{n}_2\cdot\boldsymbol{T}]+\ldots]}) \\ &= U(e^{-i[(\boldsymbol{n}_1+\boldsymbol{n}_2-(1/2)\boldsymbol{n}_1\times\boldsymbol{n}_2\cdot\boldsymbol{T}]+\ldots]}) \end{aligned} \tag{A.52}$$

where the last equalities follow from the composition rule (A.47). For comparison, we see that the operators $\boldsymbol{L}$ must obey the commutation relations (A.46). The fact that the canonical commutators lead to these relations shows that the operators $U$, equation (A.44), provide a *representation* of the rotation group, but there could be (and there are, as we will see) other independent representations.

The considerations we have just explained lead to a very general definition of angular momentum.

For any physical system, we can (operationally) identify operators $U(R)$ which describe the action of a rotation on that system and for which the equations (A.46) hold. For this system, *we define* the angular momentum from the equation:

$$U(R) = 1 - i\boldsymbol{n}\cdot\boldsymbol{J} \tag{A.53}$$

for infinitesimal transformations. From what was seen earlier, the components of $\mathbf{J}$ automatically satisfy commutation relations analogous to (A.46):

$$[J_i, J_j] = i\epsilon_{ijk}J_k. \tag{A.54}$$

---

[3]For example, it is easily shown that the matrix in equation (A.43) can be written as $e^{-i\theta T_3}$ with $(T_3)_{12} = -(T_3)_{21} = i\epsilon_{132}$, and all other elements equal to zero.

[4]This relation follows directly from expanding the exponentials in powers of $\boldsymbol{n}_1$ and $\boldsymbol{n}_2$.

Analogous arguments hold for the momentum components. For an arbitrary physical system, we can define the momentum by the relation:

$$U(\boldsymbol{a}) = 1 - i\boldsymbol{a} \cdot \boldsymbol{P} \tag{A.55}$$

valid for infinitesimal translations determined by the vector $\boldsymbol{a}$.

### A.4.2 Spin

The simplest example of the definition of angular momentum just given is that of a particle with spin. In this case, the kets which represent a particle localised in $\mathbf{x}$ are characterised by a further quantum number $\sigma$ such that the effect of a rotation, as well as turning $\mathbf{x}$ into $R\mathbf{x}$, is that of producing a linear combination of the states corresponding to various values of $\sigma$*[5]:

$$U(R)|\mathbf{x}, \sigma >= |R\mathbf{x}, \sigma' > S(R)_{\sigma'\sigma} \tag{A.56}$$

That $U$ is unitary implies that $S(R)$ should be a unitary matrix:

$$S(R)^{\dagger} S(R) = 1 \tag{A.57}$$

and furthermore the relation (A.56) implies that the matrices $S(R)$ must themselves provide a representation of the rotation group (for infinitesimal rotations):

$$S(R_1)S(R_2) = S(R_1 R_2). \tag{A.58}$$

Therefore, $S$ also must have the form:

$$S = 1 - i\boldsymbol{n} \cdot \boldsymbol{S} \tag{A.59}$$

with $\mathbf{S}$ being three suitable matrices in the space $\sigma$ which satisfy the angular momentum commutation rules:

$$[S_i, S_j] = i\epsilon_{ijk} S_k. \tag{A.60}$$

The possible realisations of $\boldsymbol{S}$ correspond, as is well known, to integer or half-integer values of angular momentum $s$. When $s$ has been fixed, $\sigma$ varies between $-s$ and $+s$ in unit steps. For example, for $s = \frac{1}{2}$, $\sigma = -\frac{1}{2}, +\frac{1}{2}$ and

$$S_i = \frac{1}{2}\sigma_i \tag{A.61}$$

where $\sigma_i$ are the three Pauli matrices.

The wave function of a general state $|A>$ is now a "spinor" with $2s + 1$ components:

$$\psi_\sigma(\mathbf{x}) = < \sigma, \mathbf{x}|A > \tag{A.62}$$

---

[5] In what follows the summation over repeated indices is understood.

and furthermore:

$$[U(R)\psi]_\sigma(\mathbf{x}) =< \sigma, \mathbf{x}|U(R)|A >= S(R^{-1})_{\sigma\sigma'}\psi_{\sigma'}(R^{-1}\mathbf{x})$$

from which, for infinitesimal transformations, it is found that:

$$< \sigma, \boldsymbol{x}|U(R)|A >=< \sigma, \boldsymbol{x}|1 - i\boldsymbol{n} \cdot (\boldsymbol{L} + \boldsymbol{S})|A > . \tag{A.63}$$

The operator associated with the generators of rotations, the total angular momentum, is the sum of two mutually commuting terms: the orbital angular momentum, $\boldsymbol{L}$, and the spin angular momentum, $\boldsymbol{S}$.

$$\boldsymbol{J} = \boldsymbol{L} + \boldsymbol{S}. \tag{A.64}$$

Before concluding this section, we note a characteristic of the representation of spin $\frac{1}{2}$. Neglecting the spatial variables, the action of a rotation through an angle $\theta$ around the $z$ axis on a spinor with, for example, $S_z = \frac{1}{2}$ is given by:

$$\begin{aligned} U(R)|\sigma = 1/2 >&= e^{-i\theta S_2}|\sigma = 1/2 >= \\ &= e^{-i\theta/2}|\sigma = 1/2 > . \end{aligned} \tag{A.65}$$

For $\theta = 2\pi$, the ket is multiplied by $-1$. This is completely consistent with the fact that the physical state should turn into itself after a complete rotation, since the kets $|A>$ and $-|A>$ represent the same physical state. However, this fact shows that the multiplication rule (A.58) cannot be satisfied for finite rotations in the case of representations corresponding to spin $\frac{1}{2}$ (more generally, for half-integer spins). In effect, as this example shows, quantum mechanics requires only that the representations of the rotation group should obey the law of group multiplications *at least to within a phase*:

$$U(R_1)U(R_2) = \omega(R_1, R_2)U(R_1, R_2). \tag{A.66}$$

The phase $\omega(R_1, R_2)$ can, without loss of generality, be chosen to be equal to $+1$ or $-1$. The operators $U(R)$ give, in this case, a *representation to within a phase* [113] of the group of rotations.

APPENDIX B

# The Non-Relativistic Hydrogen Atom

## B.1 FACTORISATION OF THE LAPLACIAN

We consider a particle of spin zero, with the Hamiltonian:

$$H = H_0 + V = \frac{p^2}{2m} + V \tag{B.1}$$

and with the canonical commutation rules:

$$[x_i, p_j] = i\delta_{ij}. \tag{B.2}$$

If $V$ has spherical symmetry, it is helpful to define the *radial momentum*, $p_r$, which allows factorisation of the free particle Hamiltonian in terms of constants of the motion. Classically, $p_r$ is simply the momentum component along the radial direction:

$$p_{r,cl} = \frac{1}{r}(\boldsymbol{x} \cdot \boldsymbol{p}). \tag{B.3}$$

However, if we substitute the quantum operators into (B.3) for the coordinates and momenta, we do not obtain a hermitian operator because the operators themselves do not commute. We obtain:

$$\begin{aligned}\left[\frac{1}{r}(\boldsymbol{x} \cdot \boldsymbol{p})\right]^\dagger &= (\boldsymbol{p} \cdot \boldsymbol{x})\frac{1}{r} = \frac{1}{r}(\boldsymbol{x} \cdot \boldsymbol{p}) + \left[p_i, x_i\frac{1}{r}\right] = \\ &= p_{r,cl} - i\frac{\partial}{\partial x_i}\left(\frac{x_i}{r}\right) = p_{r,cl} - \frac{2i}{r}.\end{aligned} \tag{B.4}$$

Therefore, if we define:

$$p_r = \frac{1}{r}(\boldsymbol{x} \cdot \mathbf{p}) - \frac{i}{r} = -i\left(\frac{\partial}{\partial r} + \frac{1}{r}\right) \tag{B.5}$$

DOI: 10.1201/9781003436263-B

we do obtain an hermitian operator, which coincides with $p_{rcl}$ in the limit $\hbar \to 0$. We note:

$$[r, p_r] = i. \tag{B.6}$$

The factorisation we seek can be obtained starting from the expression for the square of the orbital angular momentum[1]:

$$\boldsymbol{L} \cdot \boldsymbol{L} = (\epsilon_{ijk} x_i p_j)(\epsilon_{lsk} x_l p_s) = x_i p_j x_i p_j - x_i p_j x_j p_i = A - B. \tag{B.7}$$

Using the commutation rules (B.2) and the definition of $p_r$, we find:

$$\begin{aligned}
&A = r^2 p^2 - i(\boldsymbol{x} \cdot \mathbf{p}) = r^2 p^2 - i r p_r + 1 \\
&B = (\mathbf{p} \cdot \boldsymbol{x})(\boldsymbol{x} \cdot \mathbf{p}) + i(\boldsymbol{x} \cdot \mathbf{p}) = (\boldsymbol{x} \cdot \mathbf{p})(\boldsymbol{x} \cdot \mathbf{p}) - 2i(\boldsymbol{x} \cdot \mathbf{p})r = \\
&= r(p_r + \frac{i}{r})r(p_r + \frac{i}{r}) - 2irp_r + 2 = (rp_r + i)(rp_r + i) - 2irp_r + 2 = \\
&= r^2 p_r^2 - irp_r + 1
\end{aligned} \tag{B.8}$$

from which:

$$\boldsymbol{L} \cdot \boldsymbol{L} = r^2 (p^2 - p_r^2)$$

or:

$$p^2 = p_r^2 + \frac{\boldsymbol{L} \cdot \boldsymbol{L}}{r^2}. \tag{B.9}$$

(Being invariant under rotation, $r$ commutes with all components of $\boldsymbol{L}$). In terms of the Laplacian operator, we recover the well known formula:

$$\begin{aligned}
-\frac{\partial}{\partial x_i}\frac{\partial}{\partial x_i} &= -\left(\frac{\partial}{\partial r} + \frac{1}{r}\right)\left(\frac{\partial}{\partial r} + \frac{1}{r}\right) + \frac{\boldsymbol{L} \cdot \boldsymbol{L}}{r^2} = \\
&= -\frac{1}{r^2}\frac{\partial}{\partial r} r^2 \frac{\partial}{\partial r} + \frac{\boldsymbol{L} \cdot \boldsymbol{L}}{r^2}.
\end{aligned} \tag{B.10}$$

## B.2 SEPARATION OF VARIABLES

The Hamiltonian of the electron, in the centre of mass of the electron–proton system and ignoring the spin variables, is written:

$$H = \frac{p^2}{2m_r} - \frac{\alpha}{r} = \frac{p_r^2}{2m_r} + \frac{\boldsymbol{L} \cdot \boldsymbol{L}}{r^2} - \frac{\alpha}{r}. \tag{B.11}$$

$m_r = m_e M_p/(m_e + M_p)$ is the reduced mass of the electron and $\alpha = e^2/\hbar c \simeq 1/137$ is the fine structure constant.

Given that $r$ and $p_r$, and therefore $H$, commute with $\boldsymbol{L}$, we can choose a

[1] We recall that $\epsilon_{ijk}\epsilon_{lsk} = \delta_{il}\delta_{js} - \delta_{is}\delta_{jl}$.

basis in which $H$, $L^2 = \boldsymbol{L} \cdot \boldsymbol{L}$ and $L_z$ are diagonal and are represented by their eigenvalues, which we denote, respectively, as $E$, $l(l+1)$, $m$.

In this coordinate system, the wave function of the electron factorises into two terms which contain respectively the radial and angular dependence:

$$\psi(r, \theta, \phi) = R(r) Y_l^m(\theta, \phi) \tag{B.12}$$

with:

$$\begin{aligned} L^2 Y_l^m(\theta, \phi) &= l(l+1) Y_l^m(\theta, \phi); \\ L_z Y_l^m(\theta, \phi) &= m Y_l^m(\theta, \phi). \end{aligned} \tag{B.13}$$

The radial wave function satisfies the equation:

$$\left[ -\frac{1}{2m_r} \frac{1}{r^2} \frac{\partial}{\partial r} r^2 \frac{\partial}{\partial r} + \frac{l(l+1)}{2m_r r^2} - \frac{\alpha}{r} \right] R(r) = E R(r). \tag{B.14}$$

At this point, it is usual to define a new wave function, $\chi$, according to the relation:

$$R(r) = \frac{1}{r} \chi(r). \tag{B.15}$$

The equation for $\chi$ takes the form of a one-dimensional Schrödinger equation:

$$-\frac{1}{2m_r} \chi''(r) + \left[ \frac{l(l+1)}{2m_r r^2} - \frac{\alpha}{r} \right] \chi(r) = E \chi(r). \tag{B.16}$$

Boundary Conditions. The condition that $\chi(r)$ does not extend into non-physical negative values of the radial variable is imposed by assigning a constant potential, $V_0$, to the region $r < 0$ and making $V_0 \to +\infty$. In this situation it is well known that:

$$\chi(r) \to 0 \quad \text{for } r \to 0^+ \tag{B.17}$$

and (B.17) provides the boundary condition at $r = 0$. The second condition is that:

$$\chi(r) \to 0 \quad \text{for } r \to +\infty \tag{B.18}$$

such that the wave function is normalisable, as is necessary for a bound state:

$$\int d^3x |\psi(x)|^2 = \int d\Omega |Y_l^m|^2 \int r^2 dr |R(r)|^2 = \int_0^{+\infty} dr\, |\chi(r)|^2 < +\infty. \tag{B.19}$$

**Transformation to Dimensional Variables.** We define the natural scales of length and energy by means of the *Bohr radius* and the *Rydberg*:

$$R_B = \frac{\hbar c}{m_e \alpha} = 0.529\ 10^{-8}\ \text{cm};\quad r = R_B\ \rho$$
$$Ry = \frac{1}{2} m_e \alpha^2 = 13.6\ \text{eV};\quad E = Ry\ \epsilon. \tag{B.20}$$

In terms of the new variables (now the primes denote derivatives with respect to $\rho$), $\chi$ satisfies the equation:

$$-\chi''(\rho) + \left[\frac{l(l+1)}{\rho^2} - \frac{2}{\rho}\right]\chi(\rho) = \epsilon\chi(\rho). \tag{B.21}$$

## B.3 EIGENVALUES OF THE HAMILTONIAN

For $\rho \to \infty$ we can approximate the equation (B.21) with:

$$-\chi''(\rho) = \epsilon\chi(\rho)$$

which has the general solution:

$$\chi = Ae^{-\sqrt{-\epsilon}\rho} + Be^{+\sqrt{-\epsilon}\rho}. \tag{B.22}$$

To take account of the normalisability condition, (B.19), we set:

$$\chi(\rho) = e^{-\sqrt{-\epsilon}\rho}\left(a_0\rho^s + a_1\rho^{s+1} + \cdots a_\nu\rho^{s+\nu} + \cdots\right) \tag{B.23}$$

with $s > 0$ to satisfy the condition (B.17). Inserting this expression into (B.21), the term in $a_0\rho^s$ generates a term proportional to $\rho^{s-2}$ which should not exist, if we wish the series to start with a positive power. This term is:

$$a_{-2} = [-s(s-1) + l(l+1)]a_0 \tag{B.24}$$

and its absence requires that:

$$s(s-1) = l(l+1) \text{ or } s = l+1 \tag{B.25}$$

(we have discarded the root with $s < 0$). Now, the condition that the term in $\rho^{s-1}$ should also be absent determines $a_1$ starting from $a_0$, the cancellation of the term in $\rho^s$ determines $a_2$ in terms of $a_1$, etc. Proceeding in this way, one arrives at the solution which satisfies (B.25).

For general values of $\epsilon$, however, the solution found in this way tends asymptotically to the form (B.22), which is not normalisable unless $B = 0$. We now show that this can occur only if the series terminates for a finite value of $\nu$.

We write the various terms of equation (B.21) in the form (B.23) and we identify the coefficients of $\rho^{\nu-1}$. One finds:

$$
\begin{aligned}
-\chi(\rho)'' &\to -(s+\nu+1)(s+\nu)a_{\nu+1} + 2\sqrt{-\epsilon}(s+\nu)a_\nu + \epsilon a_{\nu-1} + \frac{l(l+1)}{\rho^2} \\
&\to l(l+1)a_{\nu+1} \\
-\frac{2}{\rho} &\to -2a_\nu \\
-\epsilon\chi(\rho) &\to -\epsilon a_{\nu-1} \qquad \text{(B.26)} \\
\text{total} &\to \\
0 &= [l(l+1) - (s+\nu+1)(s+\nu)]a_{\nu+1} + 2[\sqrt{-\epsilon}(s+\nu) - 1]a_\nu. \qquad \text{(B.27)}
\end{aligned}
$$

For general values of $\epsilon$ equation (B.27) allows $a_{\nu+1}$ to be obtained from $a_\nu$. In the limit of large $\nu$, we find:

$$a_{\nu+1} = \frac{2\sqrt{-\epsilon}}{\nu+1}a_\nu \qquad \text{(B.28)}$$

or:

$$a_\nu = \frac{(2\sqrt{-\epsilon})^\nu}{\nu!} \qquad \text{(B.29)}$$

which, summed, takes us to the positive exponential of $2\sqrt{-\epsilon}\rho$, or takes us back to the singular solution:

$$\chi \simeq e^{-\sqrt{-\epsilon}\rho}e^{+2\sqrt{-\epsilon}\rho} = e^{+\sqrt{-\epsilon}\rho}. \qquad \text{(B.30)}$$

To remain with a function which is normalisable the series must terminate, or there exists a value $\bar{\nu}$ for which $a_{\bar{\nu}+1} = 0$. This happens if:

$$\sqrt{-\epsilon}(s+\bar{\nu}) = 1 \qquad \text{(B.31)}$$

or:

$$\epsilon = -\frac{1}{(l+\bar{\nu}+1)^2}. \qquad \text{(B.32)}$$

The energy eigenvalues are therefore characterised by an integer number (the *principal quantum number*):

$$E_n = -Ry\,\frac{1}{n^2}, \quad n = 1, 2, \cdots \qquad \text{(B.33)}$$

For a given $n$, $E_n$ takes the same value in states with angular momentum $l$:

$$l = 0, \cdots, n-1 \qquad \text{(B.34)}$$

(since $\bar{\nu} \geq 0$) and a total number of states equal to:

$$N(E_n) = \sum_{l=0}^{l=n-1} (2l+1) = n^2. \tag{B.35}$$

In spectroscopic notation, the states with $l = 0, 1, 2, 3, \cdots$ are denoted by the letters $S$, $P$, $D$, $F$, etc. The order of the states is therefore:

$$\begin{aligned}
&E_1 = -Ry: \quad 1S, \\
&E_2 = -Ry/4: \ 2S, 2P, \\
&E_3 = -Ry/9: \ 3S, 3P, 3D \\
&\ldots
\end{aligned}$$

To the orbital angular momentum multiplicity it is, naturally, necessary to add contributions due to the spin of the electron and proton so that, in total, $N(E_n) = 4n^2$.

## B.4 EIGENFUNCTIONS

We start from the radial equation, Eq. (B.21), assuming the eigenvalue $\epsilon = -1/n^2$ found in Eq. (B.33). For $r \to +\infty$ the radial equation reads

$$\chi'' = \frac{1}{n^2}\chi \tag{B.36}$$

that is $\chi(\rho) = e^{\pm\rho/n}$. A normalizable wave function requires choosing the minus sign and we set

$$\chi(\rho) = \rho^{l+1} P(\rho) e^{-\frac{\rho}{n}} \tag{B.37}$$

where we have used the behaviour of $\chi$ for $r \to 0^+$ indicated by Eq. (B.25), $P(\rho)$ to be determined presently.

Setting temporarily

$$\bar{P}(\rho) = \rho^{l+1} P(\rho) \tag{B.38}$$

Eq. (B.21) becomes

$$\bar{P}'' - \frac{2}{n}\bar{P}' + \frac{1}{n^2}\bar{P} - \frac{l(l+1)}{\rho^2}\bar{P} + \frac{2}{\rho}\bar{P} = \frac{1}{n^2}\bar{P} \tag{B.39}$$

Simplifying the terms in $1/n^2$ and using (B.38), one finds

$$\rho P'' + P'(2l + 2 - \frac{2\rho}{n}) + 2P(-\frac{l+1}{n} + 1) = 0 \tag{B.40}$$

To connect to the notation of Landau-Lifshitz [86], we set

$$P(\rho) = w(\frac{2\rho}{n}) \tag{B.41}$$

so that

$$P' = \frac{2}{n}w', P'' = (\frac{2}{n})^2 w'', \text{ etc.} \tag{B.42}$$

and find the equation:

$$\xi w''(\xi) + w'(\xi)(2l + 2 - \xi) + w(\xi)(n - l - 1) = 0 \tag{B.43}$$

As discussed in Ref. [86] (p. 119 and Mathematical Appendix d) this equation admits solutions[2] which are exponentially bound at infinity for *n and l non-negative integers* and[3]

$$n \geq l + 1 \tag{B.44}$$

In these cases, the solutions are polinomials related to the Laguerre polynomials. Not surprisingly, we have re-obtained the condition previously found by a direct calculation based on the power expansion of the reduced radial wave function, Eqs. (B.31) and (B.32).

**The lowest radial wave functions.** In terms of the polynomials $P(\rho)$ defined in (B.37), one has

$$R_{nl} = \rho^l P(\rho) e^{-\rho/n} \tag{B.45}$$

We derived earlier the degree of the polinomial part of $\chi$ to be $\bar{\nu}+l+1 = n$, Eq. (B.32), so that the degree of the polynomial part of $R_{nl}$ is $n-1$. From this, using Eq. (B.40) we may construct any $R_{nl}$. We list below the normalised radial wave functions up to $n = 3$ (see also [86], p. 120).

$$\begin{aligned}
R_{10} &= 2e^{-\rho}; \\
R_{20} &= \frac{1}{\sqrt{2}}(1 - \frac{\rho}{2})e^{-\rho/2}; \\
R_{21} &= \frac{1}{2\sqrt{6}}\rho e^{-\rho/2}; \\
R_{30} &= \frac{2}{3\sqrt{3}}(1 - \frac{2\rho}{3} + \frac{2\rho^2}{27})e^{-\rho/3}; \\
R_{31} &= \frac{8}{27\sqrt{6}}\rho(1 - \frac{\rho}{6})e^{-\rho/3}; \\
R_{32} &= \frac{4}{81\sqrt{30}}\rho^2 e^{-\rho/3}.
\end{aligned}$$

[2] A class of functions known as Confluent Hypergeometrical functions.

[3] The inequality is specific of the Coulomb potential. For example, it is not satisfied by the Cornell potential, Sect. 18.2.1, where we may have levels with $l \geq n$.

# Bibliography


[1] F. J. Belinfante, *Physica* **6**, 887 (1939).
L. Rosenfeld, *Memoires de l'Acad. Roy. Belgique* **6**, 30 (1940).

[2] N. N. Bogoliubov and D. V. Shirkov, *Introduction to the Theory of Quantized Fields*, Interscience Publishers Inc., New York, 1959.

[3] F. Mandl and G. Shaw, *Quantum Field Theory*, John Wiley & Sons Ltd., Chichester, 1984.

[4] H. A. Lorentz, *The Theory of Electrons and Its Applications to the Phenomena of Light and Radiant Heat*, Dover Publications Inc., New York, 1952.

[5] P.A.M. Dirac, *Proc. Roy. Soc.* **112**, 661 (1926).

[6] E. Schrödinger, *Ann. Physik* **81**, 109 (1926).

[7] W. Pauli and V. Weisskopf, *Helv. Phys. Acta* **7**, 709 (1934).

[8] C. D. Anderson, *Phys. Rev.* **44** (1933), 406.

[9] P. A. M. Dirac, *The Principles of Quantum Mechanics*, Oxford University Press, Fourth Edition 1958, p. 254.

[10] L. Schiff, *Quantum Mechanics*, McGraw-Hill, New York, 1968.

[11] L. D. Landau and E. M. Lifshitz, *The Classical Theory of Fields* (Course of Theoretical Physics, Vol. 2). Butterworth–Heinemann (Oxford) 1980.
E. M. Lifshitz, V. B. Berestetski, and L. P. Pitaevskii, *Quantum Electrodynamics* (Course of Theoretical Physics, Vol. 4). Butterworth–Heinemann (Oxford) 1980.

[12] M. Tanabashi *et el.* (Particle Data Group Collaboration), *Phys. Rev. D* **98**, 030001 (2018).

[13] L. Maiani, *Electroweak Interactions*, to be published.

[14] O. Benhar, N. Cabibbo, and L. Maiani, *Gauge Theories*, to be published.

[15] The experiments are due to G. Gabrielse and collaborators, see, e.g., G. Gabrielse, *Extremely Cold Antiprotons*, *Scientific American*, December 1992 p. 78-89.

[16] K. Nakamura *et al.* [Particle Data Group Collaboration], *J. Phys. G* **37**, 075021 (2010).

[17] E. Amaldi, *Physics Reports* **111**, 1 (1984).

[18] E. Fermi, *La Ricerca Scientifica* **4**, 491 (1933).

[19] G. Luders, *Kongelige Danske Videnskabernes Selskab Matematisk-Fysiske Meddelelser*, **28** 1 (1954).

[20] W. Pauli in *N. Bohr and the Development of Physics*, Pergamon Press, Oxford, 1955.

[21] J. D. Bjorken and S.Drell, *Relativistic Quantum Fields*, McGraw-Hill, New York, 1965.

[22] E. Majorana, *Il Nuovo Cimento* **14**, 171 (1937).

[23] T. D. Lee and C. N. Yang, *Phys. Rev.* **104**, 254 (1956).

[24] C.S. Wu, E. Ambler, R.W. Hayward, D.D. Hoppes, and R.P. Hudson, *Phys.Rev.* **105**, 1413 (1957). Parity violation was subsequently observed in muon decays by R.L. Garwin, L.M. Lederman, and M., Weinrich, *Phys. Rev.* **105**, 1415 (1957). For a short history of the discovery of parity violation, see K. Myneni, http://ccreweb.org/documents/parity/parity.html.

[25] R. P. Feynman and M. Gell-Mann, *Phys. Rev.* **109**, 193 (1958); see also S. S. Gershtein, and J.B. Zeldovich, *Sov. Phys. JETP* **2**, 576 (1957).

[26] J.C. Hardy and I.S. Towner, *Phys.Rev. C* **7**, 055501 (2005); e-print arXiv nucl-th/0412056.

[27] N. Cabibbo, *Phys. Rev. Lett.* **10**, 531 (1963).

[28] S. L. Glashow, J. Iliopoulos, and L. Maiani, *Phys.Rev. D* **2**, 1285 (1970).

[29] M. Kobayashi and T. Maskawa, *Progr. Theor. Phys.* **49**, 652 (1973).

[30] J. Schwinger, *Ann. Phys.* **2**, 407 (1957).

[31] S. L. Glashow, *Nucl. Phys.* **22**, 579 (1961).

[32] F. Reines and C. L. Cowan, *Nature* **178**. 446 (1956).

[33] G. Danby, J. M. Gaillard, K. A. Goulianos, L. M. Lederman, N. B. Mistry, M. Schwartz, and J. Steinberger, *Phys. Rev. Lett.* **9**, 36 (1962).

[34] K. Kodama *et al.* [DONuT Collaboration], *Phys. Rev. D* **78**. 052002 (2008); e-print arXiv:0711.0728 [hep-ex].

[35] B. Pontecorvo, *Sov. Phys. JETP* **26**, 984 (1968) [*Zh. Eksp. Teor. Fiz.* **53**, 1717 (1967)].

[36] Z. Maki, M. Nakagawa, and S. Sakata, *Prog. Theor. Phys.* **28**, 870 (1962).

[37] https://handwiki.org/wiki/Astronomy:Proton%E2%80%93proton_chain_reaction

[38] J. N. Bahcall, A. M. Serenelli, and S. Basu, *ApJ* **621**, L85 (2005); e-print arXiv astro-ph/0412440.

[39] V. N. Gribov and B. Pontecorvo, *Phys. Lett. B* **28**, 493 (1969).

[40] L. Wolfenstein, *Phys. Rev. D* **17**, 2369 (1978).

[41] S. P. Mikheyev and A. Y. Smirnov, *Prog. Part. Nucl. Phys.* **23**, 41 (1989).

[42] S. Abe *et al.* [KamLAND Collaboration], *Phys. Rev. Lett.* **100**, 221803 (2008); e-print arXiv:0801.4589 [hep-ex].

[43] O. Q. R. Ahmad *et al.* [SNO Collaboration], *Phys. Rev. Lett.* **89**, 011301 (2002); e-print arXiv nucl-ex/0204008.

[44] G. L. Fogli, E. Lisi, A. Marrone, A. Palazzo, and A. M. Rotunno, *Phys. Rev. D* **84**, 053007 (2011); e-print arXiv:1106.6028 [hep-ph].

[45] T. Schwetz, M. Tortola, and J. W. F. Valle, *New J. Phys.* **13**, 109401 (2011); e-print arXiv:1108.1376 [hep-ph].

[46] W. C. Louis [LSND Collaboration], *Prog. Part. Nucl. Phys.* **40**, 151 (1998).

[47] W. C. Louis [LSND Collaboration], *Prog. Part. Nucl. Phys.* **40** (1998) 151.

[48] M. Goeppel-Mayer, *Phys. Rev.* **48**, 512 (1935).

[49] W.H. Furry, *Phys. Rev*, **56**, 1184 (1939).

[50] R. Saakyan, *Ann. Rev. Nucl. Part. Sci.* **63**, 503 (2013).

[51] N. Aghanim *et al.* (Plank Collaboration), Astronomy & Astrophysics **641**, A6 (2020).

[52] M. Aker *et al.* (The KATRIN Collaboration), *Nature Physics* **18**, 160 (2022).

[53] M. Agostini *et al.* (GERDA Collaboration), *Phys. Rev. Lett.* **125**, 252502 (2020).

[54] I.J. Arnquist *et al.* (Majorana Collaboration), *Phys. Rev. Lett.* **130**, 062501 (2023).

[55] S. Abe *et al.* (KamLAND-Zen Collaboration). *Phys. Rev. Lett.* **130**, 051801 (2023).

[56] G. Anton *et al.* (EXO-200 Collaboration), *Phys. Rev. Lett.* **123**, 161802 (2019).

[57] D.Q. Adams *et al.* (The CUORE Collaboration), *Nature* **604**, 53 (2022).

[58] C. Augier *et al.*, *Eur. Phys. J. C* **82**, 1033 (2022).

[59] O. Azzolini *et al. Phys. Rev. Lett.* **129**, 111801 (2022).

[60] S. Ajimura *et al.* (CANDLES Collaboration) *Phys. Rev. D* **103**, 092008 (2021)

[61] J. Engel and J. Menéndez, *Rep. Prog. Phys.* **80**, 046301 (2017).

[62] O. Benhar, R. Biondi, and E. Speranza, *Phys. Rev. C* **90**, 065504 (2014).

[63] I. Esteban *et al.*, *JHEP* **01**, 106 (2019).

[64] E. Fermi, C. N. Yang, *Phys. Rev.* **76** (1949) 1739.

[65] S. Sakata, *Progress of Theoretical Physics.* **16** (1956) 686.

[66] M. Gell-Mann, *Phys. Lett.* **8** (1964) 214-215.

[67] G. Zweig, *An SU(3) Model For Strong Interaction Symmetry And Its Breaking. 2*, CERN-TH-412.

[68] M. Gell-Mann, California Institute of Technology Synchrotron Laboratory Report No. CTSL-20, 1961 (unpublished), *Phys. Rev.* **125** (1962) 1067; Y.Ne'eman, Nuclear Phys. **26** (1961) 222.

[69] M.Y. Han and Y. Nambu, *Phys. Rev.* **139B** (1965) 1006.

[70] W. A. Bardeen, H. Fritzsch and M. Gell-Mann, [arXiv:hep-ph/0211388 [hep-ph]].

[71] C. N. Yang and R. L. Mills, *Phys. Rev.* **96** (1954), 191-195.

[72] D. J. Gross and F. Wilczek, *Phys. Rev. D* **8** (1973) 3633; Phys. Rev. D **9** (1974) 980.

[73] H. D. Politzer, *Phys. Rev. Lett.* **30** (1973) 1346.

[74] G. Altarelli, G. Parisi, *Nucl. Phys.* **B**126 (1977) 298; Y. L. Dokshitzer, *Sov. Phys. J.E.T.P.* **46** (1977) 691.

[75] W. E. Caswell, *Phys. Rev. Lett.* **33** (1974), 244

[76] S. Bethke, G. Dissertori and G.P. Salam, *Quantum Chromodynamics* in C. Patrignani *et al.* [Particle Data Group], *Chin. Phys. C* **40** (2016) 100001.

[77] T. Appelquist and H. D. Politzer, *Phys. Rev. Lett.* **34** (1975), 43 doi:10.1103/PhysRevLett.34.43

[78] A. De Rujula and S. L. Glashow, *Phys. Rev. Lett.* **34** (1975), 46-49 doi:10.1103/PhysRevLett.34.46

[79] C. A. Dominguez and M. Greco, *Lett. Nuovo Cim.* **12** (1975), 439 doi:10.1007/BF02815956

[80] E. Eichten, K. Gottfried, T. Kinoshita, J. B. Kogut, K. D. Lane and T. M. Yan, *Phys. Rev. Lett.* **34** (1975), 369-372 [erratum: *Phys. Rev. Lett.* **36** (1976), 1276] doi:10.1103/PhysRevLett.34.369

[81] E. Eichten, K. Gottfried, T. Kinoshita, K. D. Lane and T. M. Yan, *Phys. Rev. D* **17** (1978), 3090 [erratum: *Phys. Rev. D* **21** (1980), 313] doi:10.1103/PhysRevD.17.3090

[82] E. Eichten, K. Gottfried, T. Kinoshita, K. D. Lane and T. M. Yan, *Phys. Rev. D* **21** (1980), 203 doi:10.1103/PhysRevD.21.203

[83] N. Brambilla *et al.* [Quarkonium Working Group], [arXiv:hep-ph/0412158 [hep-ph]].

[84] N. R. Soni, B. R. Joshi, R. P. Shah, H. R. Chauhan and J. N. Pandya, *Eur. Phys. J. C* **78** (2018) 592, arXiv:1707.07144 [hep-ph].

[85] M. B. Voloshin, *Prog. Part. Nucl. Phys.* **61** (2008), 455, [arXiv:0711.4556 [hep-ph]].

[86] L. D. Landau and E. M. Lifshitz, *Quantum Mechanics (Nonrelativistic Theory)*, 3rd edition. (Pergamon Press, Oxford, 1977), p. 96.

[87] P. Falkensteiner, H. Grosse, Franz F. Schoeberl, P. Hertel, *Computer Physics Communication* **34** (1985) 287; W. Lucha and Franz F. Schoeberl, 1999. *Solving The Schroedinger Equation For Bound States With Mathematica 3.0*, International *Journal of Modern Physics C* (IJMPC), World Scientific Publishing Co. Pte. Ltd., vol. 10(04), pages 607-619. The corresponding program schroedinger.nb can be obtained from franz.schoeberl@univie.ac.at.

[88] H. Schopper, *Subnucl. Ser.* **15**, 203-355 (1979) DESY-77-79.

[89] W. Braunschweig *et al.* [DASP], *Phys. Lett. B* **57**, 407-412 (1975) doi:10.1016/0370-2693(75)90482-7

[90] J. S. Whitaker, W. M. Tanenbaum, G. S. Abrams, M. S. Alam, A. Boyarski, M. Breidenbach, W. Chinowsky, R. DeVoe, G. J. Feldman and C. E. Friedberg, *et al.* *Phys. Rev. Lett.* **37**, 1596 (1976) doi:10.1103/PhysRevLett.37.1596

[91] C. J. Biddick, T. H. Burnett, G. E. Masek, E. S. Miller, J. G. Smith, J. P. Stronski, M. K. Sullivan, W. Vernon, D. H. Badtke and B. A. Barnett, *et al.* *Phys. Rev. Lett.* **38**, 1324 (1977) doi:10.1103/PhysRevLett.38.1324

[92] W. Braunschweig *et al.* [DASP], *Phys. Lett. B* **67**, 243-248 (1977) doi:10.1016/0370-2693(77)90114-9

[93] N. Brambilla, doi:10.1007/978-981-15-8818-1_26-1 [arXiv:2204.11295 [hep-ph]].

[94] [LHCb], [arXiv:2204.12597 [hep-ex]].

[95] E. Braaten and M. Kusunoki, *Phys. Rev. D* **69** (2004), 074005 doi:10.1103/PhysRevD.69.074005 [arXiv:hep-ph/0311147 [hep-ph]].

[96] N. A. Tornqvist, Z. *Phys. C* **61** (1994), 525-537 doi:10.1007/BF01413192 [arXiv:hep-ph/9310247 [hep-ph]].

[97] L. Maiani, F. Piccinini, A. D. Polosa and V. Riquer, *Phys. Rev. D* **71** (2005) 014028.

[98] R. Aaij *et al.* [LHCb], *Phys. Rev. Lett.* **122**, no.22, 222001 (2019) doi:10.1103/PhysRevLett.122.222001 [arXiv:1904.03947 [hep-ex]].

[99] R. Aaij *et al.* [LHCb], *Sci. Bull.* **65** (2020) no.23, 1983-1993 doi:10.1016/j.scib.2020.08.032 [arXiv:2006.16957 [hep-ex]].

[100] A. Ali, L. Maiani and A. D. Polosa, Cambridge University Press, 2019, ISBN 978-1-316-76146-5, 978-1-107-17158-9, 978-1-316-77419-9 doi:10.1017/9781316761465

[101] *Lectures on Quantum Mechanics*, Cambridge University Press (2015)

[102] P. Bicudo, [arXiv:2212.07793 [hep-lat]].

[103] L. Pauling, *Chem. Rev.*, **5**, 173-213 (1928), DOI: 10.1021/cr60018a003, see also L. Pauling and E. B. Wilson Jr., *Introduction to Quantum Mechanics with Applications to Chemistry.* Dover Books on Physics (1985).

[104] E. Braaten, C. Langmack and D. H. Smith, *Phys. Rev. D* **90** (2014) 014044.

[105] N. Brambilla, G. Krein, J. Tarrús Castellà and A. Vairo, *Phys. Rev. D* **97** (2018) no.1, 016016 doi:10.1103/PhysRevD.97.016016 [arXiv:1707.09647 [hep-ph]].

[106] P. Bicudo, M. Cardoso, A. Peters, M. Pflaumer and M. Wagner, *Phys. Rev. D* **96** (2017) 054510.

[107] L. Maiani, A. D. Polosa and V. Riquer, *Phys. Rev. D* **100** (2019) no.7, 074002 doi:10.1103/PhysRevD.100.074002 [arXiv:1908.03244 [hep-ph]].

[108] M. J. Savage and M. B. Wise, *Phys. Lett. B* **248** (1990), 177-180 doi:10.1016/0370-2693(90)90035-5

[109] N. Brambilla, A. Vairo and T. Rosch, *Phys. Rev. D* **72** (2005), 034021 doi:10.1103/PhysRevD.72.034021 [arXiv:hep-ph/0506065 [hep-ph]].

[110] S. Fleming and T. Mehen, *Phys. Rev. D* **73** (2006), 034502 doi:10.1103/PhysRevD.73.034502 [arXiv:hep-ph/0509313 [hep-ph]].

[111] P. A. M. Dirac, *The Principles of Quantum Mechanics*, Oxford University Press, 1930.

[112] J.Bell, *Speakable and Unspeakable in Quantum Mechanics.* Cambridge University Press, 2004.

[113] For a full review, see H.Weyl, *The Theory of Groups and Quantum Mechanics.* Dover Publications Inc., New York, 1950.

# Index


$J/\Psi$ discovery, 296
$\beta$ decay, 186, 250
  CP invariance, 187
  inverse reaction, 257
  strange particle, 247
$\gamma\gamma$ annihilation, 224, 225
$\pi$ meson, 144, 188, 239
$\tau$ lepton, 246, 259
  lifetime, 246

action, 13, 18, 37
  invariance under Lorentz transformation, 28
  invariance under Lorentz transformations, 13, 20, 28
  principle of least a., 13, 18, 19, 53, 67, 141
adjoint spinor, 85, 196
Adone storage ring, 223
angular momentum, 27, 82, 99, 100, 320, 324
  under CPT, 187
angular momentum tensor
  based on $\theta^{\mu\nu}$, 35
  canonical, 34
anticommutation rules, 99, 115, 117, 118, 122, 128, 164, 196
  at space-like separation, 123
  canonical, 117, 195
  equal time, 117, 126
  Fermi oscillator, 114
  gamma matrices, 80, 82
  Pauli matrices, 81
antilinear operator, 174
antiparticle, 12, 83, 110, 119, 188
  neutrino, 142
  neutron, 119
  proton, 119
antiunitary operator, 174
ATLAS, 145

baryon, 124, 144
baryon number, 119, 285
Bevatron, 119
boost, 89, 94, 120
Born-Oppenheimer approximation, 300
  basic assumption, 301
  hydrogen molecule, 301
Born-Oppenheimer equation, 302
bra, ket notation, 311
  scalar product, 312, 313, 315, 316
branching ratio, 166

CERN, 145
charge
  under CPT, 187
charge conjugation-see discrete symmetries, 168
charged current, 254, 259, 265
charmonium, 291
  Cornell potential, 292
  spectrum, 296
Clebsch–Gordan coefficient, 100, 181, 182
CMS, 145
colour, 126
  Casimir operators, 303
  multiplets, 302
  symmetry generators, 302
commutation rules, 72, 82, 323
  angular momentum, 74, 320
  canonical, 48, 68
  equal time, 24, 47
  of $x$ and $p$, 316, 323
  orbital angular momentum, 319
  spin, 74, 321

Compton
cross section, 221
effect, 73, 221
inverse scattering, 221
scattering, 214, 257
wavelength, 131
conjugate momentum, 20, 47, 58, 63, 66, 117
conservation
angular momentum, 34, 155, 243
baryon number, 120, 196
electric charge, 120
energy, 21, 58, 149, 202, 236
energy and momentum, 14, 33, 59, 73, 161, 222
lepton number, 245
momentum, 155, 220
Noether charge, 30
of probability: see unitarity, 148
conservation laws, 136, 160
conserved current, 30, 47, 116, 196, 200
continuity equation, 57, 78, 86
Cornell potential, 302
Coulomb part, 302
correspondence principle, 138
Coulomb
divergence, 205
gauge, 68
interaction, 134
coupling constant, 140, 144, 204, 247, 248
fine structure, 99, 150, 212, 324
covariant derivative, 138
CPT theorem, 6, 183, 184, 187, 268
experimental status, 188
creation and destruction operators, 119
cross section, 161, 162
differential, 165, 202, 203, 211
Klein–Nishina, 221
Mott, 204, 206
relativistic invariance, 165
Rutherford, 204
Thomson, 221
crossed channel, 225
current
× current, 245
electromagnetic, 133, 141, 142, 196, 208, 228, 245, 247
in neutron decay, 233
lepton, 233
weak, 244, 247

decay lifetime, 165
degrees of freedom, 18, 22, 54, 56, 60, 63, 74, 144, 195, 197
density matrix, 234
Dirac
algebra, 86
bilinear covariants, 86, 143, 169, 185, 260
transformation under C, P, T, 172, 175, 183, 184
equation, 79, 98, 110
free particle, 91
general solution, 94
in electromagnetic field, 95
negative energy solutions, 91, 92, 95, 110
hole, 110, 193
mass, 198, 199
matrices, 80–82, 98, 169, 175, 260
Majorana representation, 195
Pauli representation, 175, 192
trace relations, 106
Dirac Lagrangian, 115, 137, 198
global symmetry, 116
in electromagnetic field, 139
under P, C, T transformations, 171
discrete symmetries
C, 168, 170, 172, 177, 182, 195, 226
CP, 187, 268
violation, 255, 267, 269, 270
P, 5, 168, 169
of $E$ and $B$, 26
particle-antiparticle, 177

violation, 143, 239, 245
T, 154, 168, 173, 175, 179
double-beta decay, 272
neutrinoless, 272, 274, 277, 281
experimental limits, 282
half-life, 280
with two neutrinos, 273, 277
doubly charmed baryon, 304
numerical results, 308
error estimate, 308
Dyson formula, 159, 248

electromagnetic interaction, 137
electron
Hamiltonian in Coulomb field, 99
scattering in Coulomb field, 202, 203
electron positron annihilation
in $\gamma\gamma$, 226
in $\mu^+\mu^-$, 227, 231
electrostatic potential, 62
energy-momentum tensor, 60
canonical, 31, 116
symmetric, 34, 35
Euler–Lagrange equations, 19, 20, 115
Euler–Lagrange equations, 39

Fermi constant, 142, 244, 247, 248
experimental value, 239
muon, 240, 242
tau lepton, 246
Fermi interaction, 142, 186
Fermilab, 271
Feynman
boundary condition, 46, 55, 130
gauge, 133, 135
matrix element, 164, 234, 241
propagator, 46, 130, 131, 249
field
CPT transformation, 183
Klein–Gordon, 42
scalar, 24, 39
spinor, 33
tensor, 24
Fierz transformation, 260
form factor
Dirac, 207
electric and magnetic, 208
electromagnetic, 206
nucleus, 205
Pauli, 207
proton measurement, 214
Sachs, 208, 214
functional derivative, 22
Furry's theorem, 180, 182

GALLEX experiment, 257, 264
gauge
invariance, 53, 138, 215
transformation, 53, 138
GIM mechanism, 286
gluon, 145
Gordon decomposition, 207, 212
Gran Sasso laboratory, 257, 259
Green's function, 42, 133, 180
electromagnetic field, 54
Feynman, 55
retarded and advanced, 45
vector potential, 55
group
little, 12, 121
Lorentz-see L. transformations, 3
non-compact, 87
O(2), 12
O(3), 12, 319, 322
Poincaré, 28, 136
representation, 87
gyromagnetic ratio
$\tau$ lepton, 246
electron, 97, 98, 106, 139, 140
muon, 140
nucleon, 140

hadron, 142, 144
exotic, 297
Hamilton's equations, 21, 40, 48, 65
Hamiltonian, 20, 39, 47, 64, 68, 72,

78, 82, 114, 117, 148, 155, 157, 324
$\nu - e$, 260
Dirac, 99
eigenvalues, 103
interaction, 159
neutrino, 253
Hamiltonian density, 20
Heisenberg equal-time relations, 24
Heisenberg equations, 117
helicity, 83, 188, 244
$\nu$, $\bar{\nu}$, 143
Higgs boson, 145
discovery, 145
Hilbert space, 312
Homestake mine, 257, 265
hydrogen atom
energy levels, 104, 327
non-relativistic, 323
relativistic, 98
stability, 120

IMB experiment, 258
inertial frame, 1
infinitesimal generators
spatial rotations, 319
spatial translations, 318
interaction
Lagrangian, 136
invariance
Lorentz group, 136
of Lagrangian, 136
Poincaré group, 136
spatial rotation, 34
time translation, 149
isotopic spin, 285

Jacobi identity, 23

K2K experiment, 259, 267
Kamiokande experiment, 258, 259
KAMLAND experiment, 259, 263
Klein–Gordon equation, 39, 46, 78, 129, 131, 133
solution, 40

Lagrange equations of motion, 13
Lagrangian, 19, 60, 63, 67, 185
current × current, 248
Fermi, 142, 233, 244
free classical particle, 14
interaction, 138, 142, 249
intermediate vector boson, 247, 249
Klein–Gordon, 42, 46
Maxwell, 53, 57, 60, 139
muon decay, 240
neutron decay, 143
relativistic invariance, 20, 29
Lamb shift, 106
Landau gauge, 135
Larmor frequency, 66, 98
lepton, 140, 145, 245
flavour, 240, 252
number, 200, 252
Levi–Civita tensor, 25, 26, 53, 106
little group, 12, 121
local observable, 6, 122, 155, 156, 185
Lorentz force, 58
Lorentz transformations, 3, 11, 83, 88, 94, 120, 196
boost, 89
orthochronous, 5
proper, 5
S-matrix, 159
Lorenz gauge, 54, 55, 133
LSND experiment, 271

magnetic moment, 141, 170, 188, 208
$\tau$ lepton, 246
electron, 67, 95, 139, 189
muon, 140, 189
nucleon, 140
Majorana mass, 196, 197, 199
Majorana neutrino mass, 270
experimental limits, 283
Mandelstam variables, 215, 224, 226
mass-energy relation, 15
Maxwell Lagrangian, 137
Maxwell tensor, 25, 52, 139
dual, 26

under CPT, 183
under parity, 169
Maxwell's equations, 52, 54
Maxwell–Lorentz equations, 57
meson
$\pi$ meson, 144, 284, 285
microcausality, 6, 122
Mikheyev–Smirnov–Wolfenstein effect
see MSW effect, 262
MiniBooNe experiment, 271
minimal substitution, 64, 95, 137, 138, 140, 160
MINOS experiment, 259, 267
momentum
under CPT, 187
momentum transfer, 206
MSW effect, 143, 262
muon
decay, 240
Fermi constant, 240
helicity in decay, 244
neutrino, 245
spin asymmetry, 242

neutral current, 145, 254, 259, 265
neutrino
$e, \mu, \tau$, 252
cosmic ray, 258
CP asymmetry, 267
flavour, 252
helicity, 143, 194
limit to mass, 240
long baseline beams, 259
Majorana, 195, 261, 270
mass hierarchy, 270
mass matrix, 199, 262
mixing, 252, 253, 255, 262, 263, 267
oscillations, 250, 252, 253, 264
solar, 255
solar deficit, 257, 258, 264
sterile, 254, 271
Weyl, 188, 191, 261
neutrinoless double $\beta$ decay, 270
neutron
$\beta$ decay, 142, 186
lifetime, 237
parity violation, 239
spin asymmetry, 238
Noether current, 138
proton, 141
Noether's theorem, 29, 47
normal product, 50, 122

occupation number, 49
OPERA experiment, 259
orthocharmonium, 291
orthopositronium, 181
lifetime, 226

paracharmonium, 291
parapositronium, 181
lifetime, 226
parity-see discrete symmetries, 168
Pauli
matrices, 81, 176, 262, 321
spinor, 189
term, 139, 141, 160, 170
perturbation theory, 152, 157, 209, 233, 248
photon
spin, 13, 73, 74
pion decays, 285
Poisson bracket, 23, 48
polarisation
asymmetry, 245
circular, 74
neutron, 234
photon, 73, 215
positron, 111, 188
positronium, 180
charge conjugation, 182
lifetime, 183
parity, 182
Poynting vector, 72
propagator, 128
Dirac field, 131
fermion, 128
intermediate boson, 248, 249

photon, 133, 210, 228
scalar field, 129
proper time, 10

QCD, 287
asymptotic freedom, 289
infrared confinement, 288
Lagrangian, 288
renormalization group equation, 290
running coupling constant, 290
scale $\Lambda_{QCD}$, 290
string, 289
strong coupling, 289, 291
weak coupling, 288, 291
QED, 59, 106, 137, 145, 150, 168, 170
C invariance, 172
non minimal, 160, 170
P invariance, 172
spinor, 139, 140, 159
quantum
expectation values, 147, 316
states-under time inversion, 173
quantum electrodynamics-see QED, 137
quark, 124, 145, 247, 285
beauty, 287
charm, 247, 286
colour, 287
confinement, 288
flavour, 286
heavy quark, 286
mass, 286
quantum numbers, 286

relativistic invariance, 83
renormalisation, 160
representation
Dirac, 147, 150
Heisenberg, 147, 149
interaction, 147, 150, 157
Lorentz group, 27
unitary, 87, 88, 120, 121
O(3), 320, 322
Schrödinger, 147

Rosenbluth formula, 208, 213

S-matrix, 158, 180, 201, 227, 240, 248
relativistic invariance, 159
SAGE experiment, 257, 265
scalar potential, 53
Schrödinger equation, 78, 97, 148, 262, 325
second quantisation, 112, 113
see-saw mechanism, 200
SNO experiment, 257
solar fusion reactions, 16
space-time translation, 28
spherical harmonics, 100, 325
spin
asymmetry, 238, 242
spin-statistics theorem, 124
spinor, 94, 100, 234, 321
equation of motion, 139
massless, 192
orthonormality, 93
parity, 101
Pauli, 92
radial, 101
representation, 87
under parity, 169
statistics
Bose–Einstein, 49, 124
Fermi–Dirac, 113, 124
strangeness, 285
strong interaction, 137, 143
SuperKamiokande experiment, 257, 258, 267
superposition principle, 311
symmetry
antiunitary operator, 174
continuous, 28
global, 36, 116, 138, 245
Hamiltonian, 153
local see gauge, 53
positronium, 180

tensor, 25, 26
rank, 25
time reversal-see discrete symmetries, 168

time-ordered product, 128, 153, 159, 209, 228, 248
  scalar field, 131
time-ordering operator, 201
total inversion
  see CPT, 184
transition
  Fermi, 234
  Gamow–Teller, 234

unitarity, 158
  evolution operator, 148
universality, 140, 247
  lepton, 244
unstable particle
  decay width, 166

V–A interaction, 143, 240, 243, 246, 247, 260
vacuum, 110
  energy, 49
  expectation value, 128
  fluctuations, 106
  state, 49, 72, 118
vector boson, 145, 247
  mass, 249
vector mesons, 188
vector potential, 25, 53, 67, 97
  under CPT, 183
  under time inversion, 175
vector space, 311

W boson, 145, 259
weak interaction, 137, 142, 245
Weyl equation, 192

Yukawa meson, 144

Z boson, 145, 249, 259
Zeeman effect, 65, 98
zero point energy, 114

Printed in the United States
by Baker & Taylor Publisher Services